300 ASTRONOMICAL OBJECTS

A VISUAL REFERENCE TO THE UNIVERSE

300 ASTRONOMICAL OBJECTS

A VISUAL REFERENCE TO THE UNIVERSE

JAMIE WILKINS

ROBERT DUNN

FIREFLY BOOKS

A FIREFLY BOOK

Published by Firefly Books Ltd. 2011

First printing

Publisher Cataloging-in-Publication Data (U.S.)
Wilkins, Jamie, 1981-
   300 astronomical objects : a visual reference to the universe / Jamie Wilkins and Robert Dunn.
Originally published 2006.
[264] p. : col. photos. ;   cm.
Includes index.
Summary: A reference guide to astronomical objects and tour guide through the galaxy; data sidebars with each entry provide facts and figures on every object including: mass, magnitude, density, radius, rotation period, and surface and core temperatures.
ISBN-13: 978-1-55407-812-7 (pbk.)
1. Astronomy--Observers' manuals. I. Three hundred astronomical objects : a visual reference to the universe / Jamie Wilkins and Robert Dunn. II. Dunn, Robert. III. Title.
522 22   QB64.W555 2011

Library and Archives Canada Cataloguing in Publication
Wilkins, Jamie, 1981-
   300 astronomical objects : a visual reference
to the universe / Jamie Wilkins and Robert Dunn.
Includes index.
Previous ed.: Buffalo, N.Y., 2006.
ISBN 978-1-55407-812-7
1. Astronomy--Observers' manuals. I. Dunn, Robert, 1981-
II. Title. III. Title: Three hundred astronomical objects.
QB64.W54 2011   522   C2011-901352-5

Published in the United States by
Firefly Books (U.S.) Inc.
P.O. Box 1338, Ellicott Station
Buffalo, New York 14205

Published in Canada by
Firefly Books Ltd.
66 Leek Crescent
Richmond Hill, Ontario L4B 1H1

Printed in China

Brown Bear Books Ltd
First Floor, 9-17 St Albans Place,
London N1 0NX

For Brown Bear Books Ltd:
Editorial Director: Lindsey Lowe
Project Editor: Graham Bateman
Editor: Virginia Carter
Design: Steve McCurdy

Advisory Editor
Carolyn Crawford,
Royal Society University Research Fellow and Senior Outreach Officer at the Institute of Astronomy, Cambridge, United Kingdom

Front cover: Pleiades, Photo © Terence Dickinson.
Spine: Nik Szymanek
Back cover: (from left to right) NASA-JPL; Nik Szymanek; NASA-JPL; NASA-HQ-GRIN

Page 1: False-color composite image of the Moon take by the Galileo spacecraft
Page 2-3: The Bubble Nebula (NGC 7635) imaged by the Hubble Space Telescope.

Contents

GALAXIES AND BEYOND
172–253

INTRODUCTION

One of the biggest challenges to humanity in understanding the universe is coming to terms with the immense scales involved. Our minds have developed to deal with local geography, such as the rivers and hills around a settlement. As we made the transition from a tribal to a global species, we explored Earth. About 3,000 years ago, at the time of the Greek poet Homer, maps of the world consisted only of the Mediterranean and its neighboring land. Earth itself was thought to be a flat body of land surrounded by unknown seas.

Scientific advancement pushed back the geographical boundaries. While investigating shadows cast by sticks at noon on a summer solstice at two places 500 miles (800 km) apart, Greek philosopher Eratosthenes (c.276–c.194 B.C.) determined that Earth's surface must be curved. On a flat Earth, the sticks' shadows would have been equal, but they were not. Furthermore, he was able to estimate Earth's circumference to be about 25,000 miles (40,000 km). This new understanding of the world prompted sailors and explorers to imagine circumnavigating the globe.

Of course humanity looked not only north, south, east, and west, but also up into the sky. The regularity of the passage of the Sun, the Moon, and the stars was understood well before people began keeping records. This led to an understanding of time and the passage of the seasons.

Ancient monuments such as Stonehenge in England are aligned in a way that suggests they were used to calculate the annual calendar.

The motion of the skies was crucial to the survival of early societies and they attached great significance to them. Changes in the skies were thought to be omens. The practice of interpreting omens from the heavens is known as astrology. Throughout most of their recorded history, rulers in China paid great attention to their court astrologers. Huge effort was put into observing the heavens and recording the various phenomena, and China produced the earliest-known records of events such as solar eclipses and comets. Texts describing every return of Halley's Comet for 2,000 years have helped modern astronomers refine their understanding of its orbit. References to sunspots provide us with information on potential long-term variations in the output of power from the Sun. The famous Crab Nebula is the remnant of a supernova explosion that was documented by Chinese court astronomers in 1054.

Earth-Centered Universe

Astrology was by no means unique to the Chinese. Scientific study of the universe was typically driven by a desire to understand the future rather than the universe itself. The Ancient Greek philosophers used the stars

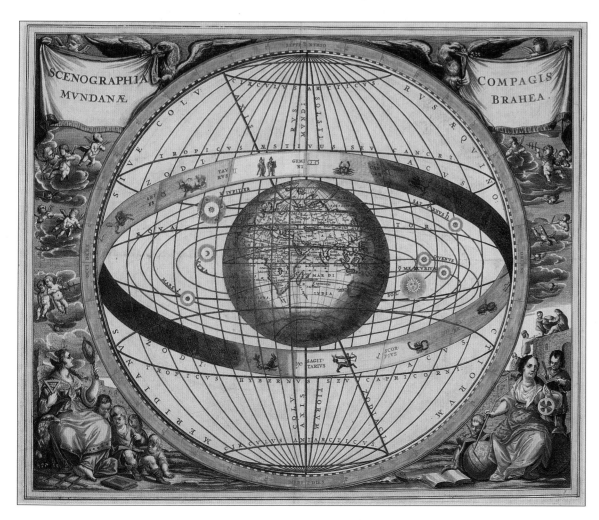

An illustration from 1708 depicting the geocentric (Earth-centered) system of the universe. Proposed by Ptolemy in the 2nd century A.D., it remained the accepted model for almost 1,700 years.

to predict events. One of the greatest was Aristotle (384–322 B.C), who studied almost every subject known at the time and produced many books in which he put forward his views. His concept of the universe was geocentric: Earth was stationary at the center, and the planets, the Moon, and the Sun were in orbit around it. He stated that the heavens (the sky) were perfect and incorruptible, unlike Earth. The path of the planets must therefore take the form of a circle, which was the most perfect shape known. Not all of Aristotle's pronouncements relied on real evidence, and in this case the observed motion of the planets was incompatible with his simple model. Viewed over several weeks, the path of a planet in Aristotle's system would simply trace a curved line across the sky. In reality, Mars, Jupiter, and Saturn all appear to slow down at times in their prescribed routes, and even go backward for a while (retrograde motion). To reconcile this apparent discontinuity, the concept of epicycles was proposed, whereby each planet moved in a circle (an epicycle), the center of which (known as the deferent) itself described a larger circle around Earth. This theory of the universe was codified by Ptolemy (c.100–c.170) in his treatise on astrology.

Arabic Astrology

The line of Greek philosophers died out at this point, and Europe underwent a period known as the Dark Ages. The great library at Alexandria, the storehouse of ancient knowledge, was lost. Ptolemy's treatise survived, however, as an Arabic translation entitled the *Almagest*. Arabic civilizations drew on Ptolemy's work to refine their own brand of astrology and to achieve accurate timekeeping that was essential for their prayer times. Elaborate and accurate astrolabes followed, enabling followers of Islam to know the direction of Mecca.

Arabic astronomers soon improved on the *Almagest*, especially in the area of star catalogs. Most of the names we use for stars today come from the Arabic language, although often corrupted by the passage of time. (For example, the name Acubens, the Alpha star of Cancer, comes from the Arabic *al zubanah*, meaning "the claws.") While the precision of astronomical knowledge was greatly enhanced during this time, the geocentric cornerstone of the Ptolemaic system was not successfully challenged. After the chaos resulting from the fall of the Roman Empire had cleared, astronomical knowledge flooded back into Europe through texts translated from Arabic. The survival of Greek geometry alongside Arabic numerals and algebra was essential for what followed.

Sun-Centered Universe

Although Polish astronomer Nicolaus Copernicus (1473–1543) was not the first person to conceive of a heliocentric (Sun-centered) universe, he is credited with beginning the paradigm shift away from the beliefs of Ptolemy and Aristotle. In his controversial book *De Revolutionibus* he argued that the Earth moved, rotating on its own axis as well as orbiting the Sun. The other

Early astrologers used astrolabes to calculate the position of the Sun. As seen in this page from a 14th-century Arabic manuscript, astrolabes were important in the Islamic world because they could be used to determine the time of day and, therefore, prayer times.

Above: *Astronomer Nicolaus Copernicus, who challenged the established view of Earth as the center of the universe. He proposed a Sun-centered system, in which Earth is merely one of the planets revolving around the Sun and rotating on its axis.*

planets also orbited the Sun, and Copernicus correctly calculated their Order, placing Earth between Venus and Mars. This simple alteration solved several of the issues surrounding the Ptolemaic system, which had remained the accepted model for 1,700 years. Unsurprisingly, it was not well received, and Copernicus was criticized in print by many who believed his theories to be an attack on religion itself.

Elliptical Orbits of the Planets

Despite the advances made after Ptolemy's time, the charts and tables that were used to predict planetary motions were still inaccurate, partly because they used circular orbits and partly because they were based on imprecise observations. A Danish astronomer, Tycho Brahe (1546–1601), improved on previous attempts, in particular with measurements of the movement of Mars. His accuracy was all the more amazing since the telescope had not yet been invented. He was also a witness to the supernova of 1572, known thereafter as Tycho's Star. It was another nail in the Ptolemaic view of the world—the Ancient Greeks had taught that the heavens were constant and unchanging (comets were seen as simply atmospheric phenomena).

Shortly before he died, Tycho was joined by an assistant, the German mathematician Johannes Kepler (1571–1630). Armed with Tycho's observational data, Kepler was able to construct a new model of the universe. Kepler's breakthrough was the realization that the orbits of the planets are not perfect circles but are slightly elliptical. He formulated his ideas as a set of mathematical laws that also related each planet's orbital period to its distance from the Sun.

Above: *Johannes Kepler, who discovered that the orbits of planets were elliptical.*

Left: *Copernicus's model is demonstrated in this 18th-century illustration.*

Galileo

Overlapping the careers of Tycho and Kepler was the life of Italian scientist Galileo Galilei (1564–1642). His use of experimentation and observation was in direct contrast to the Aristotelian methods of arguing the nature of reality from pure reason. Sometime in the first decade of the 17th century the telescope was invented in Holland. Galileo heard about this new invention and proceeded to construct telescopes of his own, improving the quality and power in each one he built.

Galileo Galilei, whose firm belief in the Sun-centered model of the universe angered the Catholic church and led to him being placed under house arrest.

The increased vision from the telescopes showed Galileo that there were many more stars in the sky than any unaided observer would ever have guessed. The Milky Way was resolved into a dense region of stars rather than the cloudlike object it was previously believed to be. Galileo saw changing spots on the surface of the Sun and craters on the Moon. He could see the closer planets as disks, suggesting that they were much closer to Earth than the stars that remained point sources of light. He also saw the rings of Saturn, although his telescopes were not powerful enough to fully reveal their nature. Most importantly for the events that followed, he saw four tiny satellites flanking the planet Jupiter. Through repeated observations Galileo became convinced that these objects were in orbit about Jupiter, proving beyond any doubt that the Earth was not the only center of motion in the universe.

Living in Italy, home of the influential Catholic church, Galileo was bound to come into conflict with the prevailing Ptolemaic philosophy. He made no secret of his theories, which were technically heretical. He was, however, friends with many important members of the Church. At first, these friendships gave him protection, but pressure to reassert the dominance of the geocentric model prevailed. Galileo was ordered by his friend Cardinal Bellarmine to cease teaching the theory of a moving Earth. He kept quiet for a while, but the promotion to the papacy of another of his friends (Cardinal Barberini, who became Pope Urban VIII) gave Galileo fresh encouragement to push the boundaries.

He wrote a dialog pitting a proponent of the Ptolemaic system against a believer in Copernicanism. It was a popularization of Galileo's theories and beliefs and was construed as an effort to present both sides of the argument. This enabled the dialog to be printed and distributed, but the Papacy turned against it and forced Galileo to renounce his ideas under threat of the Inquisition. But with the book in circulation, the damage to the geocentric model of the universe had been done.

Isaac Newton and Edmund Halley

In his masterwork *Principia Mathematica,* English mathematician and physicist Isaac Newton (1642–1727) completed the shift away from an Earth-centered to a Sun-centered universe. His universal theory of gravity gave the heliocentric model mathematical foundations, surpassing the laws of Kepler. Newton's theory of the force of gravity explained phenomena such as the fall of an apple, the changing of the tides, and the paths of the planets. Newton's contemporary Edmund Halley (1656–1742) used the theory to predict correctly the periodic return of a particular comet that both he and Kepler had observed—now known as Halley's comet. Newton's other major advances in the theory of light would lead to spectrographic analysis, an invaluable tool for astrophysics right up to the present day.

It was not until about the 18th century that our knowledge of the universe extended beyond Saturn's orbit. German-born English astronomer William Herschel (1738–1822) discovered the next farthest planet—Uranus—in 1781. His study of the motion of stars in the sky independent of Earth's movement also led him to realize that the solar system itself was moving through space. His discovery of binary stars was the first evidence of "centers of motion" outside the solar system. With each increase in knowledge, Earth's place in the universe became less and less "special." Spectroscopy revealed that the light from the distant stars was very similar to that of the Sun, suggesting that it was not a unique object as the ancients had assumed.

In the same way that improvements in the engineering discipline of boat building enabled the early explorers to expand our geographical knowledge, the advances in telescope manufacture were revealing more and more clues about the nature of the universe. French astronomer Charles Messier (1730–1817) discovered a number of fuzzy "nebulae" with his telescope; later observers with better equipment began to determine individual stars in some of these clouds and spiral shapes in others. Eventually, in 1924, these spiral shapes were shown to be vastly more distant from Earth than ordinary stars—they were galaxies. The size of humanity's known universe took another enormous leap forward. The man who determined the nature of the "spiral nebulae" was American astronomer Edwin Hubble (1889–1953). He was able to show that not only were the galaxies located at a tremendous distance from our own, but they were also moving away from us. Hubble's observations, combined with the newly developed general theory of relativity devised by German-born American physicist Albert Einstein (1879–1955), produced the concept of an expanding universe. Extrapolating these theories into the past resulted in the big bang theory, which is now well established as the dominant scientific theory of the origin of the universe.

In just 6,000 years or so our view of the universe has developed beyond recognition. Our forebears perceived a world that was small and flat with lights in the night sky above. After numerous revisions and theoretical blind alleys we have come to acknowledge the existence of a mind-bogglingly huge universe that is 13.7 billion years old, filled with galaxies and stars and planets. In spite of all our advances, there is still a vast amount about this universe that is unknown. We have not yet reached the end of our revelations.

Two 30-foot (10-m) telescopes at the summit of Mauna Kea, Hawaii, have been linked to create the world's most powerful optical telescope system, known as the Keck Interferometer.

THE SOLAR SYSTEM

View over Earth at sunrise with Mars and Venus rising.

Family Portrait

Born out of a collapsing dust cloud over four and a half billion years ago, the solar system is our home, beyond which neither we nor our probes have gone. The "solar system" refers to all the objects that are gravitationally bound to our Sun. The Sun generates energy, including heat and light, from nuclear reactions in its core. The planets shine by reflecting the sunlight—they do not generate light by nuclear reactions.

Parts of the Solar System

Earth is one of the inner four "rocky planets," together with Mercury, Venus, and Mars. Farther out are four much larger planets (the so-called gas giants), two of which (Jupiter and Saturn) are composed mainly of gas, and two predominantly of gas and liquid (Uranus and Neptune). Pluto—historically regarded as the ninth planet—now classified as a Dwarf planet, the second largest in Kuiper Belt.

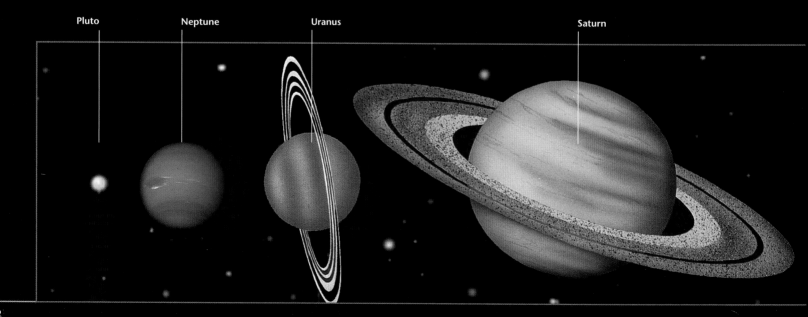

Pluto   Neptune   Uranus   Saturn

Small rocky asteroids are concentrated in the asteroid belt that lies between Mars and Jupiter. Many such objects crashed into the solid surfaces of the planets during early years of the solar system, forming impact craters and large basins. The Sun's family is completed by icy comets that originated in the far reaches of the solar system—in the Kuiper Belt outside the orbit of Neptune or the even more distant Oort Cloud. Some come close to Earth and occasionally appear as spectacular objects in the night sky.

Meteors are tiny dust or ice fragments that impact on Earth's atmosphere. Many of the planets have rock and ice moons or orbiting systems of rings. Our moon is one of the largest natural satellites, and in 1969 was the first extraterrestrial world to be visited by astronauts.

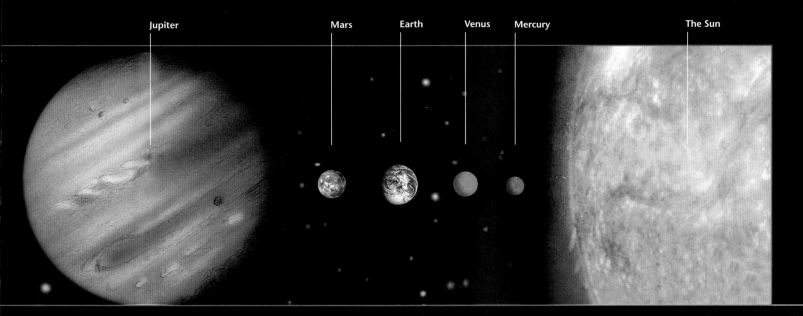

Jupiter   Mars   Earth   Venus   Mercury   The Sun

Orbits, Shapes, and Rotation

Each planet orbits the Sun in an ellipse rather than in a circle. When closest to the Sun, a planet is said to be at perihelion; at its most distant, at aphelion. A simple equation is used to calculate the ellipticity of the orbit: $(r_a-r_p)/(r_a+r_p)$ (where ra is the distance to the center of mass at aphelion and rp is the distance at perihelion).

The mean radius of Earth's orbit is 93 million miles (149.6 million km). This distance is used frequently by astronomers as a convenient unit of measurement, known as the astronomical unit (AU).

With the exception of Mercury and Pluto and its moon Charon, the planetary orbits all lie in the same plane, owing in part to the gravitational attraction exerted by the Sun on the solar nebula. Because the Sun and the planets share this orbital plane, each moves against the star background along the same path, known as the ecliptic. Material that did not get swept up into major bodies (such as comets) does not follow this pattern. Instead, their orbits are inclined to the plane.

The shapes of the inner rocky planets depart little from the spherical. In contrast, the rapidly rotating outer gaseous worlds are highly oblate, or squashed, their polar diameters being significantly less than the equatorial ones. The larger moons are also spherical, but the smaller ones often have irregular shapes, like many asteroids. Comet nuclei are also believed to be irregular in shape.

The relatively long rotation periods of the inner group (measurable in days) contrast with those of the outer group, which are generally measurable in hours (Pluto's being the exception).

Individual planets rotate on axes that have varying inclinations with respect to the plane of the ecliptic—a property known as obliquity. Thus Earth's rotational axis is inclined at 23.5 degrees and that of Jupiter is 3.2 degrees. Most extreme is Uranus, which is tipped on its side and has an inclination of 97.86 degrees. The varying obliquities may be the result of ancient collisions, which may also have affected the ultimate distances of the planets from the Sun. Venus is unusual in that it rotates in the opposite direction to Earth.

The strength of gravity on the surface of an object depends on its mass and—to a lesser extent—on the distance from the surface to the center of mass. Surface gravity is usually given compared to Earth (which has a value of 1g). The Moon, being lighter than Earth, has a surface gravity of 0.17g.

Magnetic Fields

Modern research shows that Earth and several other planets act as giant electromagnets. Earth's magnetic field originates in motions within the fluid outer core that are partly maintained by the planet's rotation. Earth behaves as if a giant bar magnet lies within, its length aligned nearly parallel to the spin axis. At different latitudes its magnetic inclination differs: at the equator it lies horizontal, but at the poles it dips vertically. The magnetic field is vital to life. It shields Earth's surface from harmful radiation, such as cosmic rays, as well as from the steady stream of plasma or ionized particles that emanates from the Sun, by trapping it outside Earth's atmosphere.

An aurora is a diffuse but striking glow, characterized by streamers, that is sometimes seen in the sky at high latitudes near both the north and south poles. It is caused by charged particles entering the upper atmosphere along magnetic lines of force, interacting with it, and fluorescing. Very intense auroras are associated with high solar activity.

Mercury and Venus have weak magnetic fields, while Mars and the Moon effectively have none. Saturn, Uranus, and Neptune have magnetic fields comparable with Earth's, but Jupiter's field is more than 19,000 times more powerful than Earth's. Its magnetosphere is so large that it extends well beyond the orbit of Saturn.

Right: *A single meteor is captured (indicated by an arrow) above Hawaii in this long-exposure photo. The slow motion of the stars produces the numerous curved trails seen in the background.*

View of Aurora Borealis, the aurora seen around the north pole. It is also known as the northern lights.

THE SUN

DATA

Classification: G

Mass: 1.99×10^{30} kg

Magnitude: −26.8

Equatorial radius:
433,000 mi
(695,000 km)

Density: 1.4 g/cm³

Rotation period:
25–36 days

Average surface temperature:
6000 K

Central temperature:
15,000,000 K

The Sun is a star, similar to the others in the night sky. Powered by nuclear-fusion reactions at its heart, transforming hydrogen into helium, it generates 3.9×10^{26} watts.

The Sun is not a particularly unusual star. Its mass is greater than the mass of the average star in the near vicinity and makes up 99 percent of the mass of the solar system. Most stars are paired up in binary systems, but the Sun stands alone. It is currently in the middle of its life span, having formed 4.6 billion years ago out of a swirling mass of dust and gas. Its components are hydrogen (74 percent), helium (24 percent), and other elements (2 percent). The fusion reactions that power the Sun turn hydrogen into helium but cannot form heavier elements, which must have been created by another, larger star whose remnants have helped form the Sun. This makes the Sun at least a second-generation star.

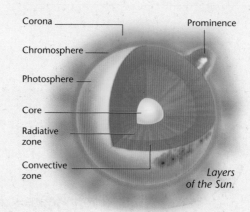

Corona — Prominence
Chromosphere —
Photosphere —
Core —
Radiative zone —
Convective zone —

Layers of the Sun.

Sunburst over Earth seen from the space shuttle Endeavor, with its robot arm visible in the foreground.

Solar Power

For fusion to occur, there must be extremely high temperature and pressure. The gravitational effect of so much mass in the same place creates the necessary conditions in the Sun's core. Every second the Sun turns 700 million tons of hydrogen into 695 million tons of helium. The missing five million tons of mass are converted directly into energy in the form of gamma rays. These rays make their way out of the Sun's core very slowly, and are constantly absorbed and retransmitted by the intervening particles. Near the surface great columns of plasma rise up, carrying the internal heat to the surface. Known as convection currents, they give the surface of the Sun its ever-changing patterns. They also generate a strong magnetic field, the influence of which extends past the orbit of Pluto.

The Sun will last for another five billion years, using up its hydrogen fuel. At present, the energy generated by fusion in the core creates an outward force that balances against gravity, which is crushing inward. As hydrogen runs out in the core, fewer fusion reactions will take place. The core will contract, and the gravitational energy released by contraction will heats things up further. When the temperature reaches 300 million K, the core will be hot enough to cause the fusion of helium nuclei. The outer layers of the Sun will expand, possibly as far out as the orbit of Earth. It will change color from yellow to red as the outer gases cool. Stars in this phase of their lives are known as red giants. The helium fusion process is unstable, and the Sun will begin pulsing, throwing off huge layers of plasma into a planetary nebula. The core will contract even more and will eventually run out of helium fuel. It will remain as a small, very dense sphere of carbon—the "ashes" of helium fusion. In that phase it will be a white dwarf, and it will eventually become a black cinder when all the heat has been radiated into space.

A huge, handle-shaped prominence of plasma erupts from the Sun's surface.

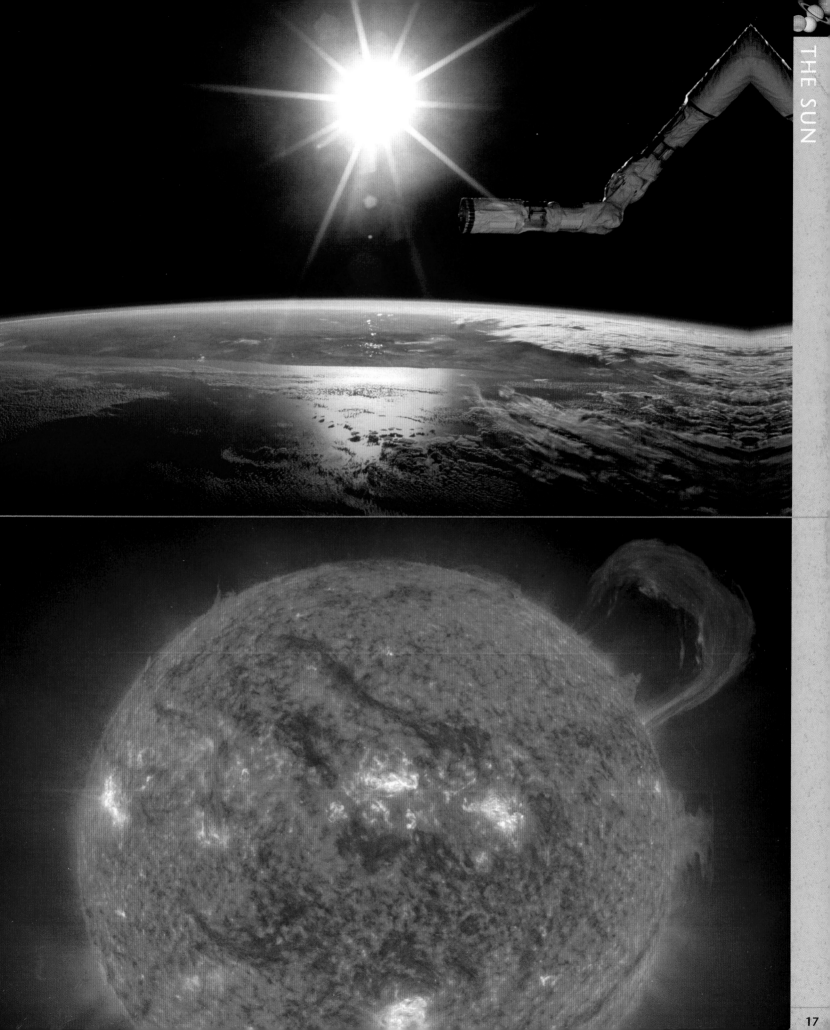

Sunspots

Made up of gas and plasma, the Sun does not rotate as a solid body. A point on the Sun's equator will travel around the Sun in 25 days. Near the poles it takes longer—up to 35 days. This differential rotation affects the magnetic field lines, twisting and distorting them. Regions of dense magnetic activity inhibit the rising of heat from the interior, and are therefore cooler. These cool spots appear darker than the rest of the surface of the Sun and are known as sunspots.

The appearance of sunspots is tied to an 11-year cycle. Periods of low sunspot activity are known as "solar minimums," and large numbers of sunspots appear at the "solar maximums." Each cycle begins with the reversal of the magnetic field of the Sun. From 1645 to 1715—a period known as the Maunder Minimum—almost no sunspots were recorded. The period coincided with a spell of particularly cold weather, a fact that has led some people to suggest that sunspots and the heat received from the Sun are linked.

Photographed in March 2001, a very large area of sunspots extends across a region more than 13 times the diameter of Earth, shown as a circled dot on the inset image. The inset image also indicates how spots change even during a single day.

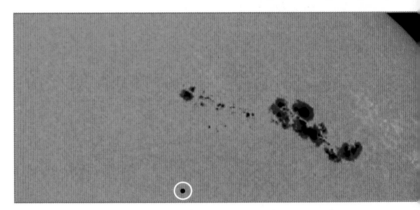

Solar Flares

The magnetic field lines that create sunspots are also responsible for solar flares. These are effectively huge explosions in the corona that surrounds the Sun like an atmosphere. Huge amounts of plasma (charged particles) are blasted from the corona. These streams of particles would be very damaging to life on Earth were it not for the planet's own magnetic field. The magnetic field acts as a protective shield, deflecting hazardous particles away from the surface. Astronauts who are caught outside this protection need shielding of their own—the speed at which a flare travels means that they may get only 15 minutes' warning of its arrival.

When a solar flare is very strong, it is sometimes accompanied by a Coronal Mass Ejection. This occurs when a chunk of the Sun's corona is ripped away and projected into space. When it hits Earth's magnetic field, the interactions between the corona and the atmosphere generate light, producing the famous auroras. More destructive events can occur on Earth's surface—distortion of the magnetic field lines around Earth can cause electrical surges in power lines and orbiting satellites. In the aftermath of large Coronal Mass Ejection events, communications disruptions on Earth are common.

A less intense but constant stream of plasma emanates from the Sun, known as the solar wind. Common to all stars (and therefore also known as stellar wind), it is driven by radiation from the Sun pushing particles from the corona out into space. This wind affects the direction of the tails of comets.

A large solar flare photographed by the Solar and Heliospheric Observatory (SOHO) satellite. An opaque circle is used to block the direct light from the Sun (the size of which is given by the white circle), so that the fainter corona can be seen.

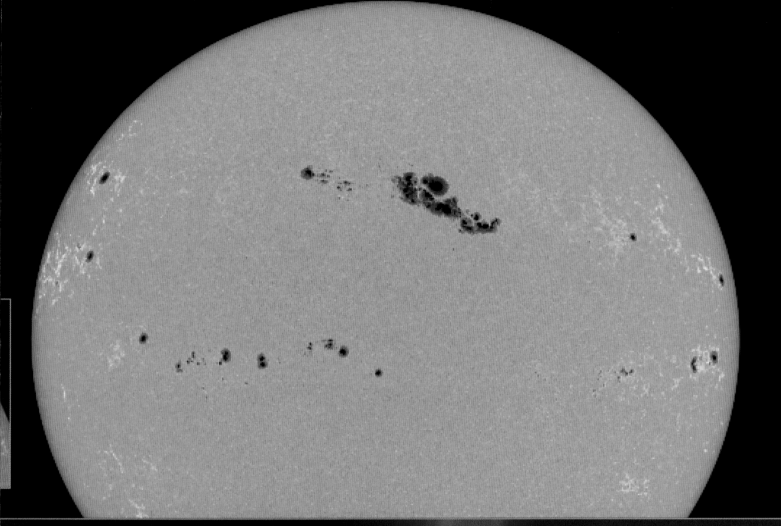

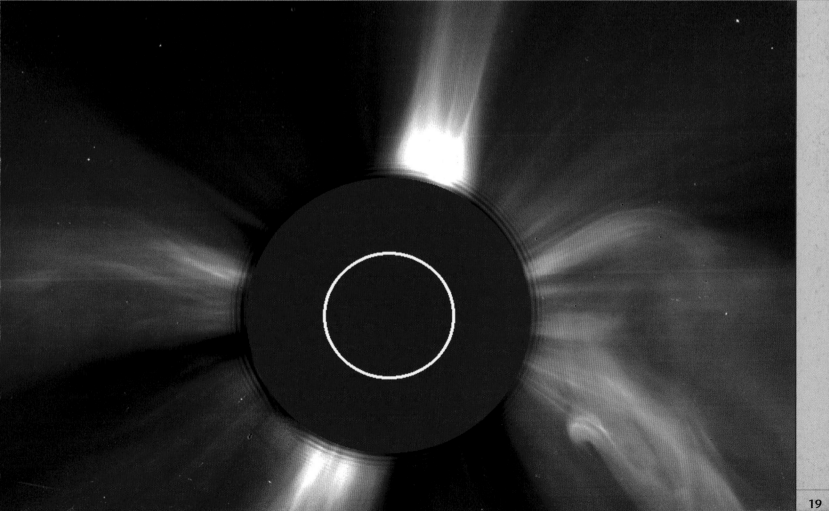

MERCURY

Type: Rocky planet

Mass: 3.3 x 10²³ kg

Equatorial diameter:
3,032 mi (4,880 km)

Density: 5.42 g/cm³

Surface gravity: 0.39

**Mean surface tempera-
ture:** 452 K
(354°F/179°C)

Magnitude: –1.9

**Mean distance from the
Sun:** 36,000,000 mi
(58,000,000 km)

Orbital eccentricity: 0.21

Mercury year: 87.97
Earth days

Mercury day: 58.65
Earth days

Mercury is the closest planet to the Sun and is smaller than all the other planets apart from Pluto. It was named for the Roman god of travel because it seems to move so quickly across the sky.

Mercury is a planet in an extreme environment. It orbits at an average of 40 percent of the Earth–Sun distance, and the heat received from being so close to a star raises its surface temperature to as much as 700 K (800°F/427°C).

The solar wind, consisting of charged particles blown from the Sun, is constantly striking the planet, and the atomic debris creates a thin atmosphere. The atmosphere does not have time to build up because the combination of intense heat and low gravity effectively allows the gases to evaporate into space. To an observer on the surface of Mercury, the Sun would appear 2.5 times larger then it does on Earth—a blinding region of light in an otherwise black sky.

First close-up image of Mercury taken by the Mariner 10 space probe in March 1974

Mercury's Orbit

Apart from observations by the space probe *Mariner 10* in 1974, we have little detailed information about the planet Mercury. Viewed from Earth, its maximum angular separation from the Sun is 28 degrees. The Hubble Space Telescope cannot point at Mercury for fear of destroying its delicate optics. A probe named *Messenger* arrived at Mercury March 17, 2011, and is due to provide much more detailed views during its mission.

Mercury's orbit is highly eccentric. It varies from 43 million miles (70 million km) at the farthest point (aphelion) to 29 million miles (46 million km) at the closest point (perihelion). The gravitational effects of the other planets cause the perihelion to precess slowly around the Sun at a rate of 2 degrees every century. These measurements of Mercury's orbit did not match up to a previously stated theory—Newton's universal theory of gravitation (1687)—that predicted a lower rate of precession. To make theory fit reality, an undiscovered planet was predicted to be lying between Mercury and the Sun. The hypothetical planet, named Vulcan, was never found. The mystery of the precession remained until Einstein's general theory of relativity (1915) provided a subtle alteration to the calculations, which matched observations exactly. This was one of the most important tests of general relativity.

Initially it was thought that, like Earth's Moon and some moons of other planets, Mercury rotated at the same speed as it orbited. Known as being "tidally locked," this orbit would keep one side of the planet facing the intense heat of the Sun.

The dark side, on the other hand, would be exceptionally cold because Mercury did not have an atmosphere capable of redistributing heat. Measurements taken in 1962 showed that, although cold at 90 K (–298°F/–183°C), the dark side of the planet was too warm never to have seen the Sun. Further examination showed that there was a 3:2 ratio between the rotational period and the orbital period. For every two full orbits, Mercury rotates three times.

Transit of Mercury (seen toward top right) as it passes in front of the Sun. The larger dark spot near the center is a sunspot.

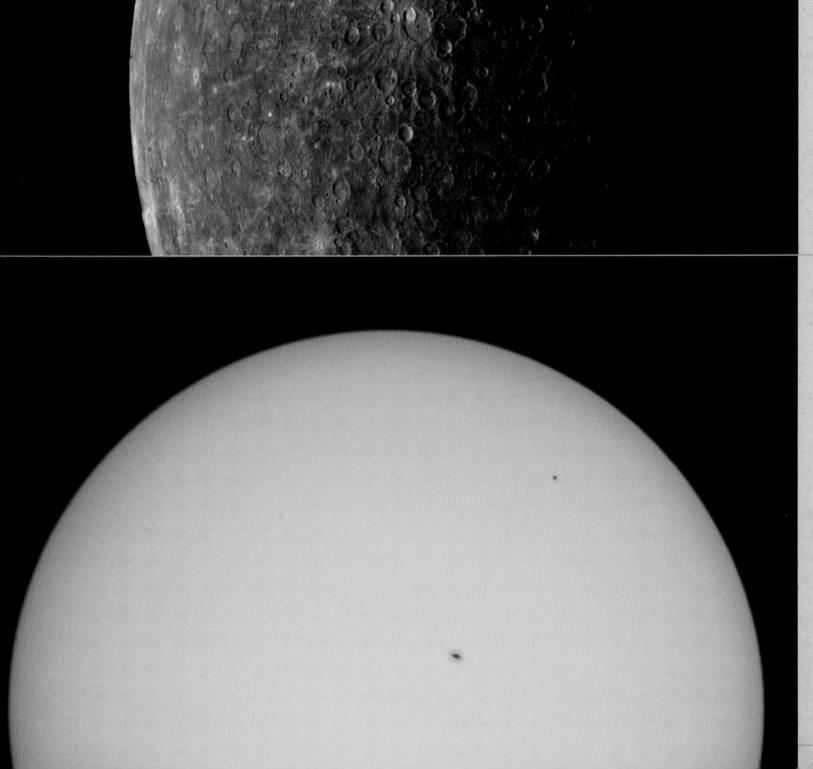

Mercury's Surface

The surface of Mercury is covered in craters and fault lines. There is little evidence of large-scale resurfacing resulting from lava flows, suggesting a planet whose geological structures have experienced no significant movement for most of its lifetime. In the past—probably shortly after it formed—it appears that the planet contracted slightly, causing a wrinkling of the crust. The contraction produced the fault lines, some over 0.6 miles (1 km) high and hundreds of miles long. In detailed images some smooth plains can be seen, formed from lava that was brought to the surface by large meteorite strikes.

Mercury has a high density and the largest proportional metal core of any planet in the solar system. Metal makes up between 60 and 70 percent of its mass, with rock accounting for the rest. Water was not expected to exist at all on such a hot planet, but radar observations made in 1991 revealed a bright reflection from the north pole, which is characteristic of water ice. Since Mercury has very little tilt to its axis, the north pole would see the Sun continually circling on the horizon. Craters could therefore contain areas of permanent shadow, maintaining a temperature below the freezing point of water. Ice from cometary impacts would be able to collect in these areas. Anywhere else on the planet the water would boil and evaporate into space.

Photomosaic of the Caloris Basin region of Mercury, which is some 800 miles (1,300 km) across, photographed by Mariner 10.

VENUS

DATA

Type: Rocky planet

Mass: 4.9×10^{24} kg

Equatorial diameter: 7,521 mi (12,103 km)

Density: 5.20 g/cm³

Surface gravity: 0.91

Mean surface temperature: 735 K (900°F/480°C)

Magnitude: −4.4

Mean distance from the Sun: 67,240,000 mi (108,200,000 km)

Orbital eccentricity: 0.0068

Venus year: 224 Earth days

Venus day: 243 Earth days

Named for the Roman goddess of love and beauty, Venus is the brightest planet in the sky. It has been known both as the morning star and the evening star, because in ancient times few sky watchers knew that these two "stars" were, in fact, the same planetary object.

Venus is an inferior planet, meaning that it is closer to the Sun than Earth is. For this reason, Venus can show phases in the way that the Moon does. Italian scientist Galileo Galilei observed these phases and understood that they provided yet more evidence for a heliocentric (Sun-centered) theory of the universe, as opposed to the prevailing Ptolemaic (Earth-centered) theory.

Venus is unusual in that its day (the rotation about its center) takes longer than its year (a full orbit around the Sun). Furthermore, its rotational direction is retrograde, meaning that it is reversed from the common direction for the other inner planets. This suggests that some large event, perhaps a collision, occurred in the past and changed the way Venus rotates.

View of Venus's surface generated from radar data produced by the Magellan space probe. The image is tilted, with the poles at 2 o'clock and 8 o'clock.

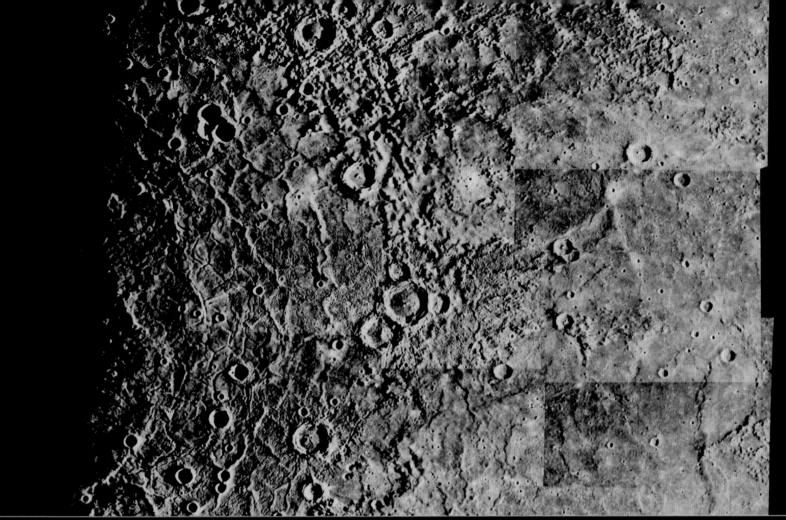

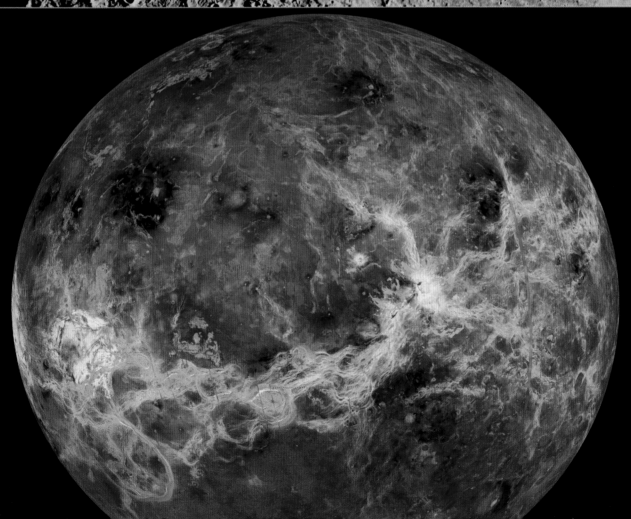

Venus's Clouds

Venus is often called Earth's sister planet. It has a similar size, mass, and distance from the Sun. However, it is one of the most hostile places in the solar system for life as we know it. There are no oceans on Venus. In fact, there is no water on the surface at all. The atmosphere is chiefly carbon dioxide, and the thick clouds that permanently cover the surface consist of sulfuric acid. The atmospheric pressure at ground level is a crushing 92 times that of our own atmosphere, which is equivalent to the pressure exerted at a depth of 0.6 miles (1 km) in Earth's oceans.

Venus's surface is heated by rays of sunlight that penetrate the clouds. The carbon dioxide keeps the heat from escaping into space. This is the greenhouse effect at work on a planetary scale. The dense atmosphere raises the surface temperature to an average of 755 K (900°F/480°C), hotter than the Sun-facing side of Mercury. A person standing unprotected on the surface of Venus would be burned alive, crushed by the atmosphere, suffocated by the carbon dioxide, and finally poisoned and corroded by the clouds of sulfuric acid.

A thick veil of clouds cloaks the surface of Venus, as photographed by Mariner 10. *The image is tilted with the north-pole top right.*

Venus's Surface

The first probe ever to return data from the surface of another planet was *Venera 7*, a Soviet craft. It was heavily damaged as it landed, but was able to send back information about the temperature and pressure before succumbing to those very factors. Later probes were able to take photos of their landing grounds. Since optical observation of the planet was impossible owing to the thick clouds, much more information was to come from radar mapping of the surface.

The U.S. space probe *Magellan* spent four years in orbit around Venus and sent back a comprehensive and detailed map. It was able to view features as small as 1000 feet (300 m) in size. The images revealed a relatively young surface, no older than 800 million years. The terrain is divided into lowland plains that are covered with the results of lava flows. There are highland regions with hard rock that forms steep mountains even taller than Earth's Mount Everest. Active volcanic regions still exist, with calderas as large as 62 miles (100 km) across.

Impact craters are less common than volcanoes, because old craters have been obliterated from the surface by lava flows. Individual craters of fewer than 1.2 miles (2 km) in width do not occur—a meteorite as small as that would be burned up in the thick atmosphere before it hit the ground. Larger meteorites are also affected by the atmosphere, breaking up into smaller chunks just before impact. Strikes like these cause crater clusters.

Three-dimensional computer-generated image of western Eistla Regio on the surface of Venus, produced from Magellan *probe data. Seen on the right is the volcano Gala Mons some 435 miles (700 km) away.*

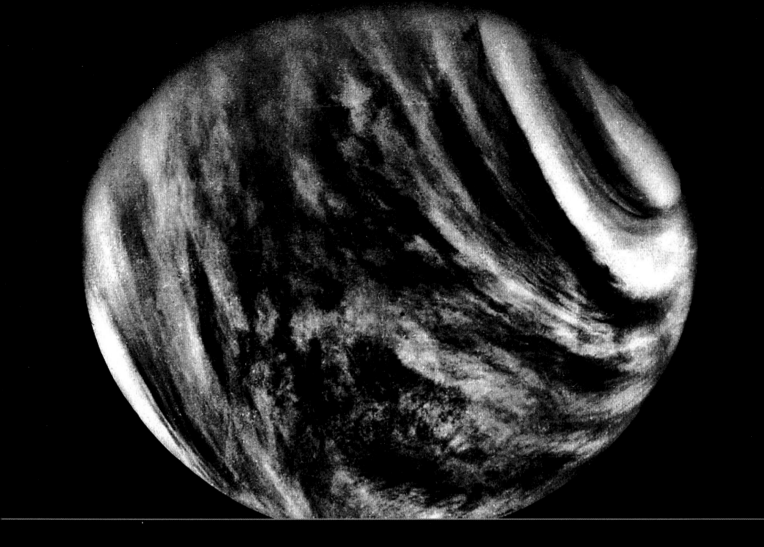

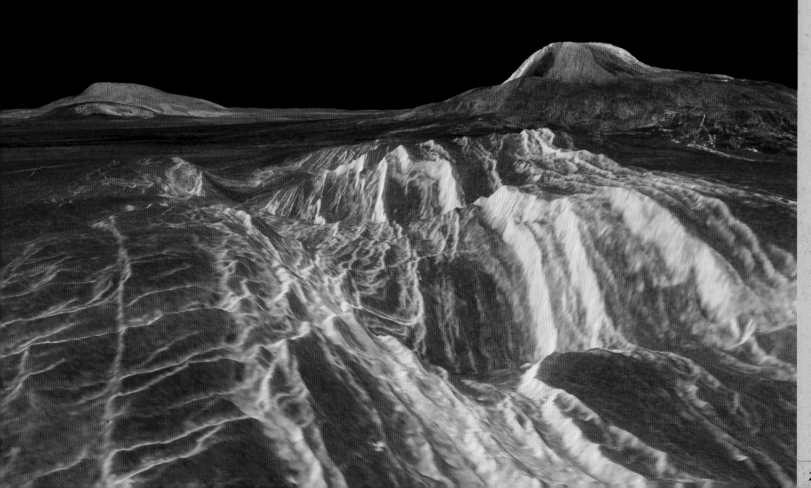

EARTH

Earth is the planet about which we know the most and the only planet upon which humankind has set foot. Although it is no longer thought of as the center of the universe, it is set apart: Our planet is to date the only place proven to harbor life.

The single most important condition for life on Earth to exist is water. The combination of Earth's atmospheric composition and its distance from the Sun keeps the surface at a reasonable temperature, which allows water to flow on Earth.

Earth photographed by Apollo 17 crew in 1972.

Earth's Interior

Earth is the densest planet known. It has a solid inner core made of iron and nickel. At temperatures of up to 7500 K, the core should be molten, but the gravitational pressure at 3.5 million atmospheres compresses the metal back into a solid. As the pressure drops farther from the planet's center in the outer core, the metal mixture becomes liquid. The liquid is capable of flowing and moving through convection and the effect of the rotation of the whole planet. The movement of such a large amount of conductive material generates Earth's magnetic field.

Surrounding the core is the mantle, a region composed of rock rather than pure metal. It has a higher melting point than the core, and even with the high pressure it is closer to a solid than a liquid state. Nearer to the surface, the mantle flows slowly. Convection in this region drives the movement of the tectonic plates that make up the crust. These plates are made of huge slabs of rock ranging from 3 to 44 miles (5–70 km) in thickness. There are eight major plates and 20 smaller ones.

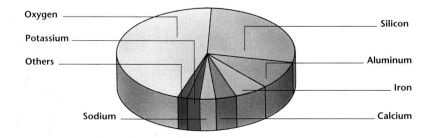

Relative abundance of elements in Earth's crust.

Oxygen

Potassium

Others

Sodium

Silicon

Aluminum

Iron

Calcium

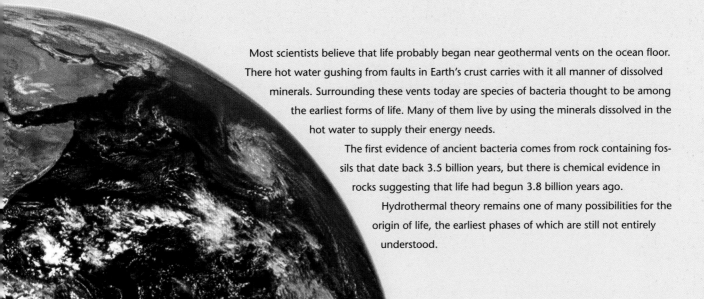

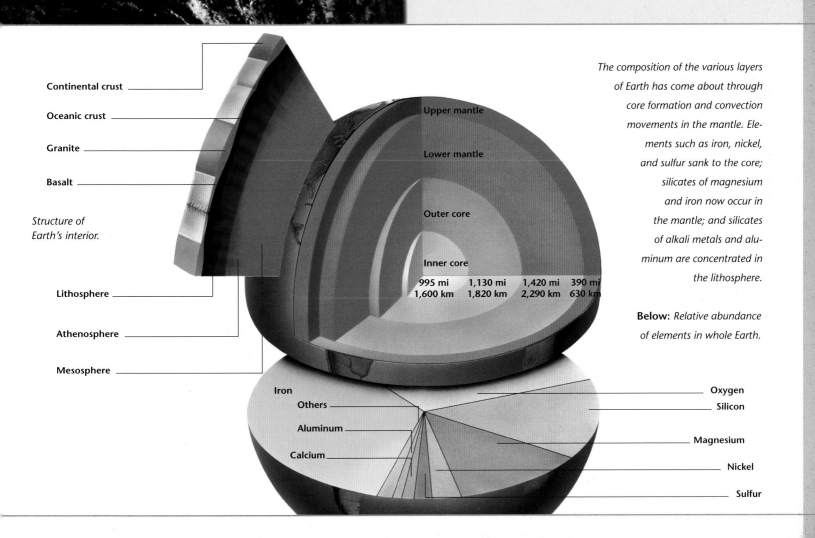

Most scientists believe that life probably began near geothermal vents on the ocean floor. There hot water gushing from faults in Earth's crust carries with it all manner of dissolved minerals. Surrounding these vents today are species of bacteria thought to be among the earliest forms of life. Many of them live by using the minerals dissolved in the hot water to supply their energy needs.

The first evidence of ancient bacteria comes from rock containing fossils that date back 3.5 billion years, but there is chemical evidence in rocks suggesting that life had begun 3.8 billion years ago.

Hydrothermal theory remains one of many possibilities for the origin of life, the earliest phases of which are still not entirely understood.

Continental crust

Oceanic crust

Granite

Basalt

Structure of
Earth's interior.

Lithosphere

Athenosphere

Mesosphere

Upper mantle

Lower mantle

Outer core

Inner core

| 995 mi | 1,130 mi | 1,420 mi | 390 mi |
| 1,600 km | 1,820 km | 2,290 km | 630 km |

The composition of the various layers of Earth has come about through core formation and convection movements in the mantle. Elements such as iron, nickel, and sulfur sank to the core; silicates of magnesium and iron now occur in the mantle; and silicates of alkali metals and aluminum are concentrated in the lithosphere.

Below: *Relative abundance of elements in whole Earth.*

Iron

Others

Aluminum

Calcium

Oxygen

Silicon

Magnesium

Nickel

Sulfur

Earth's Surface

Continued volcanic eruptions and earthquakes caused by movement of the tectonic plates have shaped the continents that we see today. Another major influence on the surface of the planet is erosion. Water and wind, over geological timescales, can level mountains and carve deep gullies. There are very few visible records of such impacts on Earth simply because the resulting craters and patterns are eroded. Add to that the effect of volcanism and the movement of the plates, and the result is one of the youngest surfaces in the solar system. Radar and gravitational field measurements have been able to pick out evidence of the sites of ancient impacts from the crust, now buried under miles of mud and sedimentary rock.

Left: *Radar image from space of an area of the Sahara desert in northern Chad, revealing a 10.5-mile (17-km) crater left by an asteroid or comet.*

Earth's Oceans and Atmosphere

Earth's oceans are a feature not found anywhere else in the solar system. Titan may have liquid methane on its surface, and Jupiter's moon Europa may have a layer of water forming an underground ocean, but only Earth has surface water in its liquid state. More than 70 percent of the surface is water. The oceans help regulate the surface temperature by absorbing and redistributing heat from the Sun. Water is so essential for life that without the oceans we would not exist.

The oceans themselves would not exist in liquid form were it not for the greenhouse effect. For a planet nearly 93 million miles (150 million km) from the Sun the expected surface temperature is 252 K (–6°F/–21°C)—well below freezing. Water vapor and the small amount of carbon dioxide present in the atmosphere provide enough trapped heat to raise the average surface temperature to 288 K (59°F/15°C). The ozone layer plays a part in maintaining the planet's water by blocking high-energy ultraviolet radiation. Further protection of the planet comes from its magnetic field. It diverts the solar wind and channels it into regions of dangerous radiation known as the Van Allen belts. They occur between 1600 miles (1000 km) and 104,000 miles (65,000 km) from the Earth's surface. The northern and southern lights are auroras, the atmosphere glowing through interaction with the streams of charged particles that are focused onto the magnetic poles.

Image from space showing vast cloud formations in the atmosphere. In particular, the swirl of hurricane Andrew (upper right) can be seen about to hit the Louisiana coast in 1992.

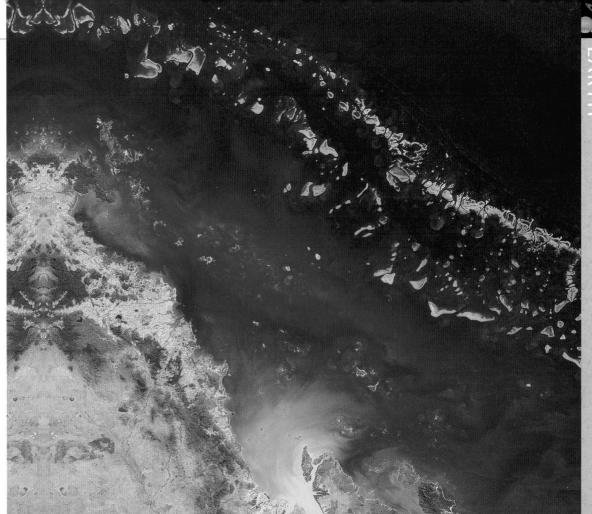

Right: *True color image from space of the Great Barrier Reef off the coast of Queensland in eastern Australia.*

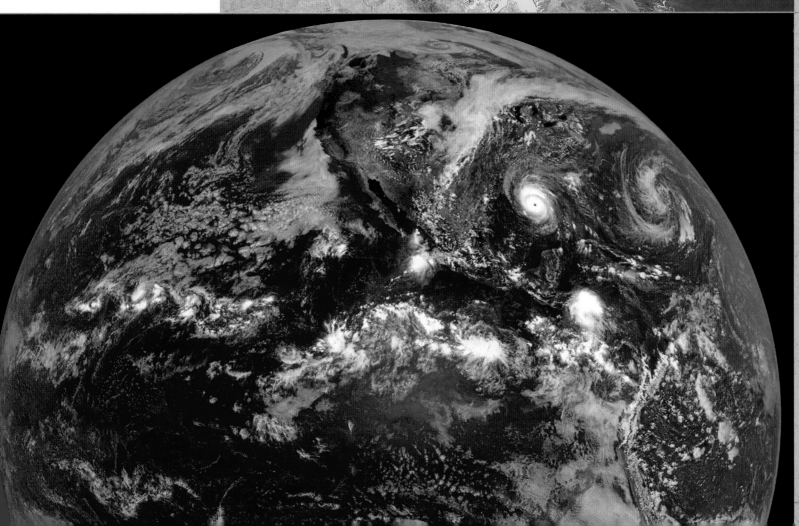

THE MOON

Mass: 7.35 x 10²² kg

Magnitude: –12.74

Equatorial diameter:
2,159 mi (3,474.8 km)

Density: 3.34 g/cm³

Surface gravity: 0.17

Rotation period:
27.32 days

Orbital period: 27.32 days

Radius of orbit:
238,900 mi (384,400 km)

Maximum surface temperature: 396 K
(253°F/123°C)

Minimum surface temperature: 40 K (–388°F/–233°C)

There are many satellites of Earth, but only one has existed for more than 50 years—the Moon. The Moon is the only natural satellite of Earth. Most natural satellites in the solar system are believed to have formed through accretion at the same time as the parent planet (for example, the Galilean moons of Jupiter). Others were originally asteroids captured by the planet's gravity (for example, the moons of Mars). In contrast, our Moon was formed by the collision of two planetary bodies early on in the formation of the solar system. One of them was Earth, which scientists think was hit hard by a mass of rock roughly the size of Mars. A huge amount of debris would have been flung into orbit, and the debris would over time coalesce into the Moon.

As the Moon formed and cooled, Earth's gravity pulled it into a slightly distorted shape. The center of mass is 1.2 miles (2 km) closer to Earth than the geometric center, and the crust on the Earth-facing side is thinner on average. The Moon orbits Earth at a mean distance of 238,900 miles (384,400 km). It takes one month to complete each revolution and the same period to rotate once on its axis. As a result, we always see the same side of the Moon.

The Moon's surface shows the results of considerable cratering. The side we see has a pattern of dark and light areas. The dark areas are known as maria (Latin for seas). They are ancient lava flows that rose from inside the Moon and flooded large impact craters. They appear only on the near side of the Moon, which protects them a little from further bombardment. Most impacts occur on the far side of the Moon, which is covered in overlapping craters. Maria may have existed here once but have since been obliterated by the erosion of the impact sites.

Far side of the moon showing its heavily cratered surface.

Moon's Surface

The surface of the Moon is coated with a layer of rock fragments and dust known as regolith, ranging in thickness from 10 to 66 feet (3–20 m). Below that the crust is 44 miles (70 km) thick on average, and the rest is rocky mantle all the way to a very small core, which makes up less than 2 percent of the Moon's total mass. There is no real atmosphere to speak of, just very small amounts of gas escaping from the rock and briefly captured elements from the solar wind. Comets and meteorites hitting the surface of the Moon will deposit small amounts of water ice. Most of it will boil off into space, split into hydrogen and oxygen by the cosmic rays, but it is possible that some remains in the form of ice. Because of the orientation of the Moon, there are craters around the poles that are so deep that their floors never see the light of the Sun. In those areas of permanent shadow, ice could remain undisturbed indefinitely.

Far right: *Close-up of the Aristarchus crater, which is about 26 miles (42 km) in diameter and 2 miles (3.2 km) deep.*
Right: *Astronaut's footprint in the dusty Moon's surface, photographed in 1969 by an Apollo 11 crew member.*

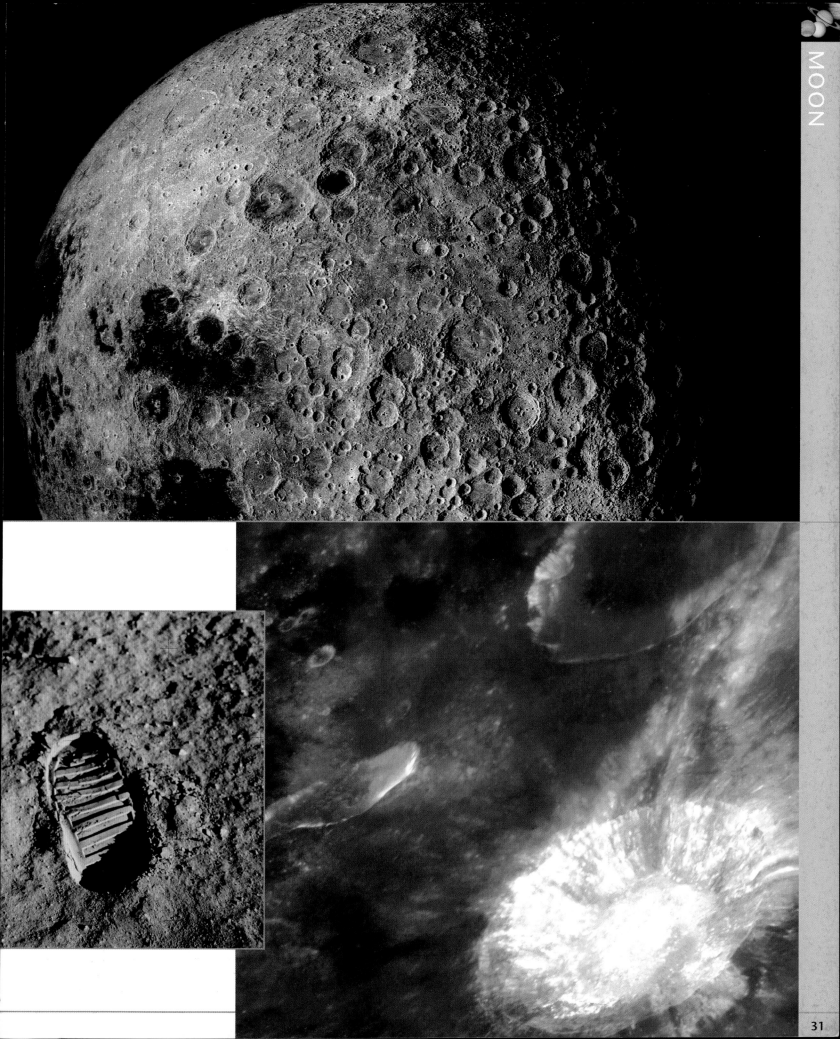

Moon and Earth

Compared with its parent planet, the Moon is one of the largest in the solar system, and it orbits only 60 Earth radii away. It is therefore not surprising that it exerts gravitational effects on Earth in the form of tides. Liquid water in the oceans is free to move around the planet and is pulled in the direction of the Moon. Earth spins much faster than the Moon orbits, so the high tides in the oceans are rotated away from the Moon as it is attempting to pull back on them. This leads to a slight braking effect on Earth's rotation. Currently, the rate is such that the day gets longer by 1.5 milliseconds every century. The loss of energy to braking also causes the Moon to spiral slightly out, moving 1.5 inches (38 mm) away from Earth each year.

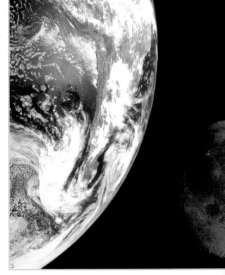

Right: *Earth (foreground) and the Moon, seen by the* Galileo *spacecraft in 1992 as it started its journey to explore the Jupiter system.*
Far right: *Earth appears over lunar horizon, seen from the* Apollo 11 *command module as it nears the surface of the Moon in 1969.*

Man on the Moon

The Moon is the only extraterrestrial body to have been visited by humans. Twelve men walked on its surface during the years 1969 to 1972, but none since. All of them were Americans working on the Apollo program. It began with tests of the technology in Earth orbit, progressed to sending crewed ships for the first time to lunar orbit, and then to *Apollo 11*'s historic landing in 1969. Following that success, another six Apollo missions went to the Moon. Only the ill-fated *Apollo 13* mission failed to land on the lunar surface.

Once on the Moon, astronauts were able to undertake scientific exploration. They left behind seismographs. No moonquakes were recorded, but meteorite and spacecraft impacts provided data that could reveal the structure of the Moon's interior. Other equipment left on the Moon is still in use today. The Lunar Laser Ranging Experiment consists of reflectors built to send light back exactly the way it came. Lasers are shone from Earth-bound observatories, and the time for the light to be reflected back gives a measurement of the distance to the Moon that is accurate to within 0.8 inches (20 mm). Over the course of the Apollo program 842 pounds (382 kg) of moon rock was removed and taken back to Earth for analysis. A rock from *Apollo 11* contained a mineral not previously found on Earth. It was named armalcolite after the crew: Armstrong, Aldrin, and Collins.

Since the Apollo missions space agencies have not sent humans back to the Moon. Crewed missions were phased out in favor of safer, cheaper robotic probes. *SMART-1* is a small lunar orbiter launched by the European Space Agency (ESA) in 2003, costing only $126 million. Its mission is to map the Moon's surface using X-ray and infrared frequencies in order to gather data on its composition.

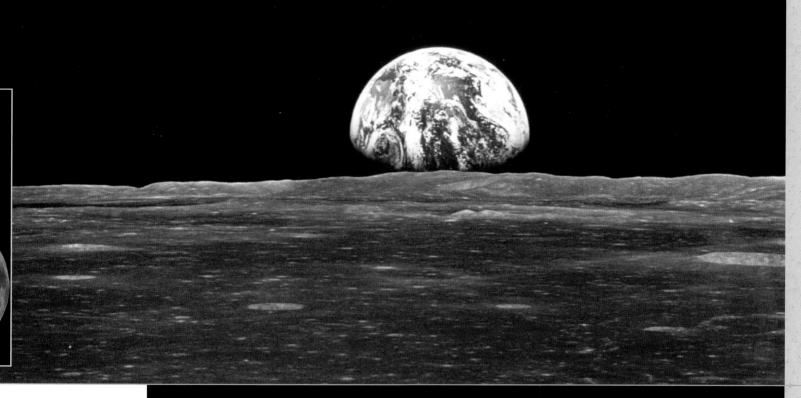

Apollo 12 *astronaut Alan L.
Bean during extravehicular
activity on the Moon's
surface, November 1969.*

MARS

DATA

Type:	Rocky planet
Mass:	6.42 x 10²³ kg
Equatorial diameter:	4,222 mi (6,794 km)
Density:	3.94 g/cm³
Surface gravity:	0.38
Maximum surface temperature:	293 K (68°F/20°C)
Mean surface temperature:	210 K (−82°F/−63°C)
Minimum surface temperature:	133 K (−220°F/−140°C)
Magnitude:	−2.01
Mean distance from the Sun:	142,000,000 mi (227,940,000 km)
Orbital eccentricity:	0.0934
Mars year:	687 Earth days
Mars day:	24.6 hrs
Moons:	Deimos, Phobos

Named for the Roman god of war, Mars is visible as a bright reddishorange planet in the night sky. The red color comes from iron oxides within the surface soil and gives it the alternative name "The Red Planet." Mars has long been the subject of stories of life on planets other than Earth. Some stories were rooted in (erroneous) scientific discovery.

In the late 1800s astronomers examined Mars through telescopes and thought they saw evidence of life. American astronomer Percival Lowell made detailed studies of so-called canals—huge linear features on Mars that had been described previously by Italian astronomer Giovanni Schiaparelli. Lowell believed them to be artificial. Patchy color changes across the surface were interpreted as seasonal variation in vegetation growth—perhaps the canals were part of a vast irrigation system? By the start of the 20th century further observations by other astronomers revealed the "canals" to be optical illusions. The changes in color were caused by the vast dust storms that cross the planet. The space probe *Mariner 4*, which landed on Mars in 1965, removed the final hopes of finding extensive life on Mars. It returned images of a rocky, dusty crater-covered landscape that resembles the Moon in many respects.

Mars's Atmosphere

The conditions on the surface of Mars indicate that life as we know it would not be possible there at the present time. The surface is very cold, almost permanently below the freezing point of water. There are no significant amounts of liquid water on the surface. The sun's damaging ultraviolet rays are not blocked by the atmosphere, and they bombard the surface. The atmosphere is 95 percent carbon dioxide and very thin. The average atmospheric pressure is more than 100 times lower than on Earth.

Clouds and fog have been seen on Mars, the result of the tiny amount of water in the atmosphere (0.03 percent) condensing out. The polar caps are made mainly of solid carbon dioxide that freezes directly from the gas phase at 195 K (−109°F/−78°C). Mars's distance from the Sun varies considerably, and when the planet is at its farthest, more of the carbon dioxide is trapped in the polar caps, reducing the low atmospheric pressure even more. As on Earth, the carbon dioxide creates a greenhouse effect, but it is very weak. The wind on Mars can be very strong and can cover the surface of the planet in dust storms. Features normally visible from Earth disappear, and the entire planet can look almost as featureless as Venus.

Right: *Polar storm located just to the west of the north polar ice cap.* **Far right:** *Frosty-white water ice clouds and swirling orange dust storms in the atmosphere of Mars.*

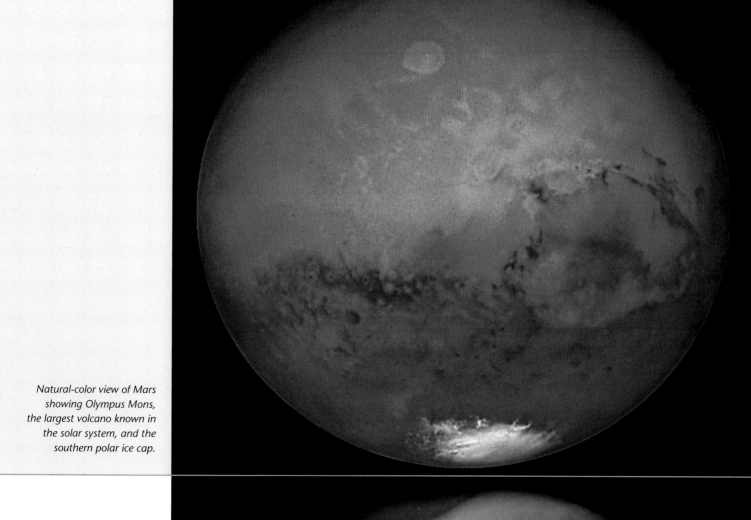

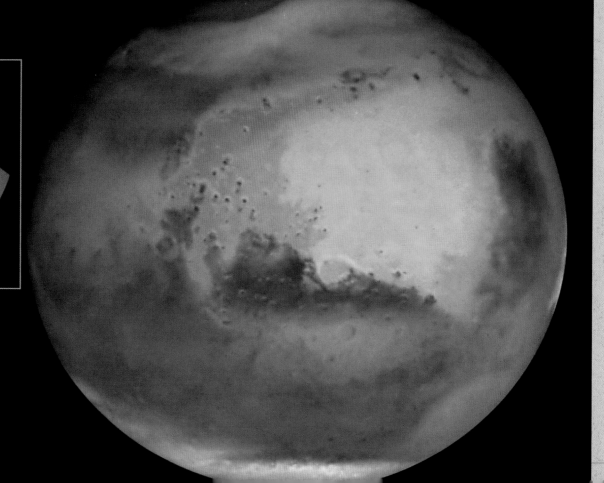

Natural-color view of Mars showing Olympus Mons, the largest volcano known in the solar system, and the southern polar ice cap.

Mars's Surface

The interior structure of Mars is typical of the inner rocky planets. It has a metallic core, a rocky mantle, and a thin crust. It is less dense than the other rocky planets, indicating that it is less metal-rich. Unlike Earth, it appears to be tectonically inactive. There is no evidence of recent movement of tectonic plates. Mars boasts the largest mountain in the solar system, the volcano Olympus Mons, but it appears long extinct. Like the Moon, Mars has two different types of terrain: low-lying plains indicating old lava flows contrasting with the older crater-covered highlands. Strangely, the northern hemisphere consists almost entirely of lowlands and the southern hemisphere of highlands. On average, the altitude difference is about 3.7 miles (6 km). Recent images taken by the *Mars Global Surveyor* show features on the surface that look like dried-up gullies, implying that Mars was warmer and wetter in the past.

Right: *Frost-cloaked Lowell Crater.*

Deimos and Phobos

Phobos and Deimos are two tiny moons of Mars. They were first discovered in 1877 by American astronomer Asaph Hall and have since been viewed closely by probes sent to study the parent planet. The moons are named for two attendants of the Roman god of war. *Phobos* translates as "fear," *Deimos* as "terror." They are small, irregular satellites made up of a mixture of rock and ice. They closely resemble carbonaceous asteroids, which has led to the theory that they originally came from the asteroid belt and were captured by the gravitational pull of Mars. However, they both have very circular orbits aligned to Mars's equatorial plane, much like conventional moons. Captured asteroids would be more likely to have highly eccentric and tilted orbits.

To an observer on Phobos Mars would cover one-quarter of the sky. Phobos is so close to Mars that it orbits faster than the planet rotates. An observer on the surface of Mars would see Phobos rise in the west and set in the east twice a day.

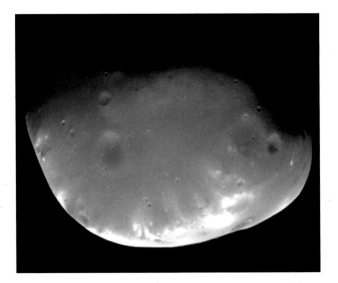

View of Martian moon Deimos.

DEIMOS

Mass: 1.48 x 10^{15} kg

Magnitude: 12.4

Dimensions (diameter):
9.3 x 7.6 x 6.8 mi
(15 x 12.2 x 11 km)

Density: 1.47 g/cm³

Surface gravity:
0.0006

Rotation period:
1.26 days

Orbital period:
1.26 days

Radius of orbit:
14,580 mi
(23,460 km)

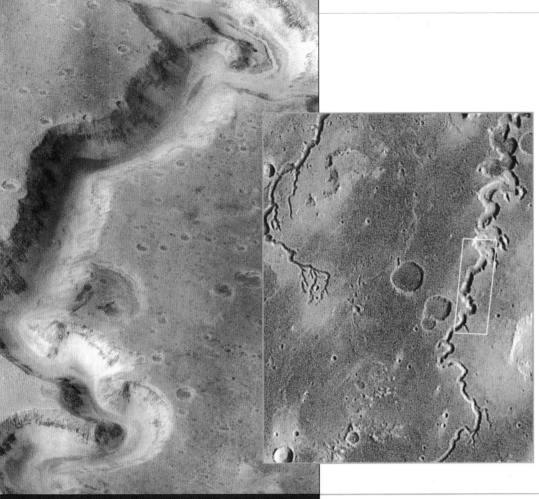

Left: *Canyon of Nanedi Vallis, which is about 1.6 miles (2.5 km) wide.*
Left (Detail): *Detail of the canyon cutting through the cratered plains in the Xanthe Terra region of Mars.*

Phobos is orbiting so closely to Mars that gravitational tidal forces exert a braking effect on the satellite, reducing the altitude of the orbit by 6 feet (1.8 m) per century. In a few tens of millions of years (a short time in the history of the solar system) Phobos will either crash onto the surface or break up into a ring structure. It has many small craters and one huge crater 6.2 miles (10 km) in diameter. Deimos is the smaller outer moon, which is visible only as a bright point of light from the surface of Mars. Neither Phobos nor Deimos can be seen from high latitudes on Mars, because their close orbit places them below the horizon.

Close-up of Phobos, showing the large crater Stickney toward the upper right.

PHOBOS

Mass: 1.08 x 10¹⁶ kg

Magnitude: 11.3

Dimensions (diameter):
16.8 x 13.4 x 11.7 mi
(26.8 x 22.4 x 18.4 km)

Density: 1.9 g/cm³

Surface gravity: 0.001

Rotation period:
0.32 days

Orbital period:
0.32 days

Radius of orbit:
5,830 mi (9,380 km)

THE ASTEROID BELT

Between the orbits of Mars and Jupiter lies the asteroid belt. Asteroids are made from material left over from the formation of the solar system. The rock and ice that was never swept up into a planet could provide important clues as to what conditions were like when the Sun was young.

Millions of small bodies, as well as the eight planets of the solar system, are in orbit around the Sun. Among them are the asteroids, or "minor planets." Most orbit the Sun between Jupiter and Mars in the asteroid belt, but some have paths that cross the orbit of Earth. The Trojan groups move along the same orbit as Jupiter, one group ahead of Jupiter, the other behind.

The largest asteroid, Ceres, was discovered by the Italian astronomer Giuseppe Piazzi in 1801and is now considered the smallest Dwarf planet. It measures 580 miles (930 km) across. This planet is round and believed to have an interior similar to that of the terrestrial planets. It is believed that there are at least one million asteroids larger than 0.6 miles (1 km) across. Many have very irregular shapes, such as Gaspra. This asteroid measures about 12 miles (19 km) long by 7 miles (11 km) wide and orbits the Sun at the inner edge of the main asteroid belt.

*Right: Location of the asteroid belt and Trojan groups in relation to the planets. **Above top:** The spherical form of asteroid Ceres is clear from this sequence, which also indicates that it is rotating. **Above bottom:** The irregular form of the asteroid Gaspra is typical of many asteroids.*

Eros

DATA

Mass: 6.7×10^{15} kg

Density: 2.4 g/cm³

Dimensions (diameter):
8 x 8 x 20.5 mi
(11.2 x 11.2 x 34.4 km)

Surface Temperature:
337 K (147°F/64°C) to
123 K (–238°F/–150°C)

Magnitude: 11.16

Rotational period:
5 hrs 16 min

Orbital period:
643.2 days

Mean distance from the Sun:
107,400,000 mi
(172,800,000 km)

Eccentricity: 0.223

Eros (asteroid 433) orbits mainly between Earth and Mars, but its path crosses those of both planets. Asteroids that cross Earth's orbital path are known as near-Earth objects, and Eros is the second-largest one known. Near-Earth objects are relatively easy to reach with spacecraft. For that reason Eros was chosen to be examined by the Near-Earth Rendezvous Probe (known as *NEAR Shoemaker* for the planetary scientist Eugene Shoemaker).

Eros, whose namesake was the Greek god of love, is composed of stone. It is highly irregular, shaped like a fat banana with sections gouged out. One particular gouge looks as if it almost broke the asteroid in two, and close-up pictures from the *NEAR Shoemaker* spacecraft reveal a layered structure. This could indicate that the asteroid formed as part of a planetoid (small planet) and was broken off as a result of some huge impact. Unlike some asteroids, Eros is a solid body rather than a loose pile of rubble. It has a density similar to that of Earth's crust.

Surface of Eros covered in regolith from past impacts.

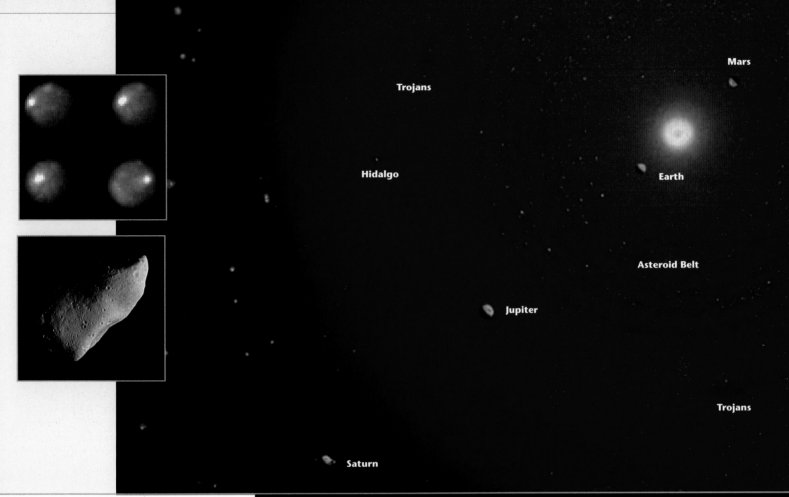

Mars

Trojans

Hidalgo

Earth

Asteroid Belt

Jupiter

Trojans

Saturn

As expected, Eros has many craters, but there is one area where the craters are small and sparse. This area is centered on a large crater. The theory is that when this impact occurred, it caused all the existing craters in the area to be filled in. Debris resulting from the impact, as well as the seismic shock, could have been responsible for filling in some of the craters.

After staying in orbit around the asteroid for a year, *NEAR Shoemaker* landed successfully on Eros—the first time that a soft landing (as opposed to a crash landing) had occurred on an asteroid. The last few pictures showed a dusty, pebble- and boulder-strewn surface.

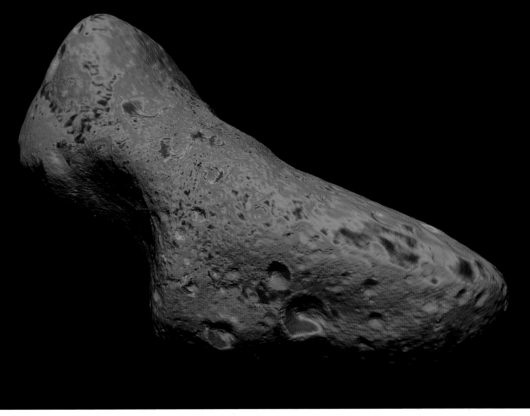

False-color-enhanced image of Eros.

Ida and Dactyl

Ida (asteroid 243) orbits in the main asteroid belt between Mars and Jupiter. It is named for a nymph who looked after the infant Zeus on a mountain in Crete that bears her name. Discovered in 1884, Ida was looked at closely during a flyby by the space probe *Galileo*. It is made principally from silicates and is a member of the Koronis family of asteroids. They are all believed to have come from the breakup of the same planetoid, which possibly measured more than 62 miles (100 km) in diameter. Ida has many craters, enough to suggest that Ida is older than the breakup of the original Koronis object.

While looking over the images of Ida returned by *Galileo*, scientists made a surprising discovery. Located a mere 56 miles (90 km) above Ida's surface was another body, essentially a satellite of the asteroid. It was roughly spherical, with a diameter of only 0.9 miles (1.4 km). It was given the name Dactyl for a mythological race of beings that were said to have lived on

Mount Ida. Scientists are not certain how this binary asteroid system came to be. Dactyl may have been just another chunk of rock in the Koronis breakup. Another theory suggests that a heavy impact on Ida could have thrown up enough debris to coalesce in orbit, in a similar way to Earth's Moon. Dactyl was the first object found orbiting something as small as an asteroid. Since then many other asteroid moons have been discovered. Sylvia (asteroid 87) has two satellites.

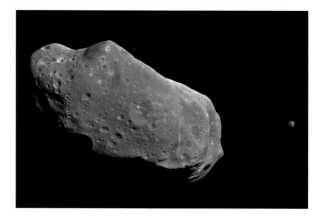

Ida and its satellite, Dactyl.

DATA	
Mass: 4.2 x 10¹⁶ kg	
Density: 2.6 g/cm³	
Dimensions (diameter): 34 x 15 x 9 miles (54 x 24 x 15 km)	
Surface Temperature: 158 K (–175°F/–115°C)	
Magnitude: 9.94	
Rotational period: 4 hrs 37 mins	
Orbital period: 1767.7 days	
Mean distance from the Sun: 266,000,000 mi (428,000,000 km)	
Eccentricity: 0.046	

JUPITER

Jupiter is the largest planet in the solar system, heavier than all the other planets combined, although still only 0.1 percent of the mass of the Sun. It has over 60 moons ranging from the huge satellites first observed by Italian astronomer Galileo Galilei to the numerous smaller bodies, some of which have only recently been discovered.

Jupiter is the first of the gas giant planets in the solar system. Named for the Roman version of the "king of the gods" from Greek mythology, it is the largest object in our

region of space (apart from the Sun) and is very different from the rocky inner planets.

Despite its huge size, Jupiter has the shortest rotation period of all planets, spinning once every 9 hours 55 minutes 30 seconds, a phenomenon that causes its equatorial regions to bulge. Parallel bands of light and dark clouds can be seen, as well as one semipermanent atmospheric feature, the Great Red Spot (GRS).

DATA	
Type: Gas giant	
Mass: 1.9 x 10²⁷ kg	
Equatorial diameter: 88,733 mi (142,796 km)	
Density: 1.31 g/cm³	
Surface gravity: 2.64	
Mean surface temperature: 129 K (–227°F/–144°C)	
Magnitude: –2.70	
Mean distance from Sun: 483,654,000 mi (778,330,000 km)	
Orbital eccentricity: 0.048	
Jupiter year: 11 years 314.84 days	
Jupiter day: 9 hrs 55 mins 30 secs	

True color image of Jupiter, showing cloud features and the Great Red Spot.

Mathilde

On its way to Eros the *NEAR Shoemaker* probe flew past Mathilde (asteroid 253) at a distance of 745 miles (1200 km). Despite passing it at a speed of 6 miles (10 km) per second, the probe was able to gather useful information. Mathilde lies in the main asteroid belt between Mars and Jupiter and has an eccentric orbit. It orbits between 1.9 and 3.3 times farther than Earth does from the Sun. Discovered in 1885, Mathilde was named for the wife of the astronomer Moritz Loewy. It is larger than any other asteroid visited by a spacecraft and has several unusual characteristics.

Mathilde has one of the longest rotational periods for an asteroid. Only two other asteroids are known to take longer than Mathilde's 17 days and 9 hours to complete a full revolution. For that reason, *NEAR Shoemaker* was only able to examine 60 percent of the asteroid's surface. As a carbonaceous asteroid, Mathilde is very black, reflecting less light than either coal or charcoal. It is about half as dense as the major-

Asteroid Mathilde.

ity of other carbonaceous asteroids, however. It is believed that Mathilde's interior is particularly porous. The entire asteroid is more of a rubble pile held together by its own gravity than a solid body. It has huge craters, some more than half as wide as the object itself. It is not fully understood how these craters came to be without the impact shattering the asteroid. Mathilde appears to be homogeneous (made from the same material throughout). If it had a different composition below the surface, the craters would reveal a color change.

DATA

Mass: 1.0×10^{17} kg

Density: 1.3 g/cm³

Dimensions (diameter):
41 x 30 x 29 mi
(66 x 48 x 46 km)

Surface Temperature:
174 K (−146°F/−99°C)

Magnitude: 10.2

Rotational period:
17 days 9 hrs 30 mins

Orbital period:
1,572.7 days

Mean distance from the Sun: 245,000,000 mi
(394,000,000 km)

Eccentricity: 0.266

Jupiter's Atmosphere

Much of our understanding of Jupiter's atmosphere comes from the Galileo mission. A probe was sent down into the atmosphere and reached a depth of 100 miles (160 km) before losing contact. Jupiter's atmosphere is 600 miles (965 km) deep. It is composed of hydrogen (80 percent), helium, and small amounts of heavier elements that give the clouds their characteristic colors. Methane is another important component. Ammonia condenses in the upper atmosphere to form white patches. The brown and red areas may be caused by trace sulfur or phosphorus compounds. The colors are arranged in bands across the planet parallel to its equator. The dark bands are areas of descending gases known as belts, and the light bands, known as zones, are rising gases.

Over time, belts and zones can vary in width and appearance, even disappearing completely for short periods. Variable rates of rotation among the bands make the atmosphere very turbulent, with winds well in excess of 400 miles per hour (640 km/h). There are also frequent oval-shaped storms, the largest of which is the Great Red Spot (GRS).

Close-up views of Jupiter. **Right:** *The atmosphere in natural colors.* **Far right:** *Artificial color image showing the height of the clouds—red are deep water clouds, bright blue are high haze, as over the Great Red Spot. The small yellowish white spots are lightning storms.*

Great Red Spot

The Great Red Spot (GRS) is the largest-known weather phenomenon on any planet. 25,000 miles (40,000 km) wide by 12,000 miles (19,300 km) long from top to bottom, it protrudes 5 miles (8 km) above the surrounding atmosphere. It is a huge anticyclonic storm, with gas rising in a spiral and cooling, then sliding to the edge of the oval. Its deep red color is variable. The spot has been known to vanish entirely for several months, but it has been observed for as long as Jupiter itself, making the storm over 300 years old. It is not clear what gives it its longevity, but it is possible that its size gives it stability.

Other interesting features include the White Oval Spots, which are found south of the latitude of the GRS. They are a group of smaller anticyclonic disturbances that have been seen to interact in recent years, two storms combining to form a larger one. During the first half of the 20th century a shaded area flanked by white spots existed at roughly the same latitude as the GRS, but circumnavigated the planet at a different speed. It was called the South Tropical Disturbance, and it may have been the precursor to the White Oval Spots. It faded away in about 1940 and has not returned. The same could happen to the GRS—it has shrunk slightly over time and may eventually disappear.

This mosaic of the Great Red Spot shows a white cloud around the northern boundary, which extends east of the region. Another white oval cloud is seen south of the Great Red Spot.

Close-up view of the Great Red Spot.

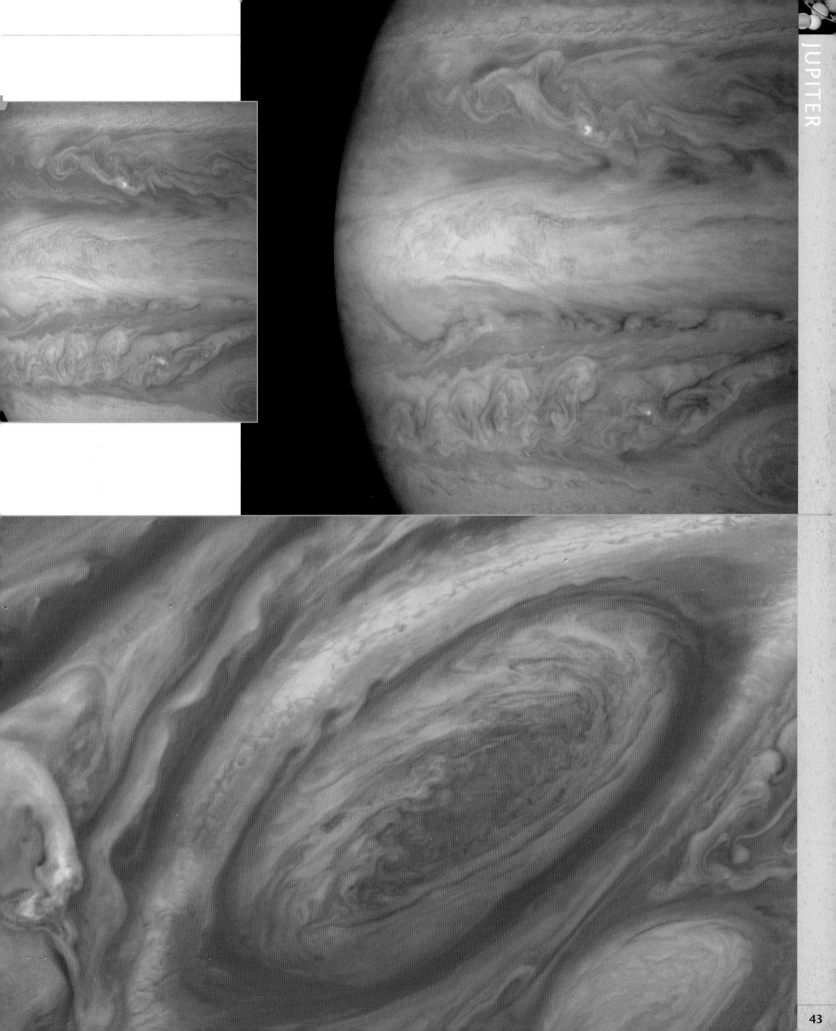

Jupiter's Center and Auroras

It is difficult to imagine the conditions at the center of a gas giant, but current theories suggest there is a core of rock and ice five to ten times more massive than Earth. The temperature of the core is approximately 30,000 K, far short of the temperature that is needed to initiate fusion and turn Jupiter into a small star.

The gradual settling of the core into a more compact center releases enough gravitational energy to keep the center hot. In fact, Jupiter radiates almost twice as much energy into space as it receives from the Sun. This internal heat drives the complex weather systems that produce the patterns we see from Earth.

Outside the rocky core is a layer of liquid metallic hydrogen. Although hydrogen is normally a gas, the tremendous pressure and temperature has changed it into a liquid sea of protons and electrons, which is very conductive. This layer possibly accounts for as much as 75 percent of Jupiter's total mass, and is the cause of the planet's strong magnetic field. Jupiter's magnetic field is roughly 10 times as strong as Earth's, (measured at the planets' surfaces). It creates a magnetosphere that extends a few million miles toward the Sun and over 400 million miles (650 million km) away from the Sun—far enough to reach Saturn's orbit. This field interacts strongly with Io, the nearest large moon and first of the Galilean satellites. Io is a highly volcanic body and throws off large quantities of sulfur-based compounds that form a plasma torus centered on Io's orbit. Some of this material is channeled by the magnetic field, causing a current of over one million amps to connect Jupiter and Io, and forming auroras on Jupiter's poles.

Comet Shoemaker-Levy 9 Impact

Like Saturn, Jupiter has rings, but they are thin and almost invisible. They are dark and rocky and probably consist of debris from meteors hitting the moons. Jupiter, being the heaviest object in the solar system apart from the Sun, is a regular target for impacts. Many scientists believe that without the "gravitational shield" provided by Jupiter, life on Earth would not have evolved far before being wiped out by an impact from a large meteor.

A comet hit Jupiter in 1994. Shoemaker-Levy 9 had been orbiting the planet for some time and was getting closer and closer to hitting it. Its last pass had been two years earlier, and it had approached so close by that the gravitational forces had ripped it apart, forming a line of debris and chunks of rock. There were 20 large chunks, each no more than a few miles across. They hit the atmosphere of Jupiter on the far side. Our telescopes did not see the initial impact, but the rapid rotation of the planet soon brought the impact sites into view. The comet had created large dark-brown bruises on the light-brown cloudscape— the largest roughly the size of Earth. The bruises were the result of the comet chunks exploding in the upper atmosphere and ejecting material out over the tops of the clouds. The largest impact struck with the energy of six million megatons of TNT, and the material ejected was visible over the limb of the planet, even though the impact site was not. As the foreign material has dispersed through Jupiter's atmosphere, the bruises have faded from the southern hemisphere of the planet.

Sequence of images showing the progression of the impact of Shoemaker-Levy 9 chunk G on Jupiter. From bottom: Impact plume five minutes after impact; after 1.5 hours; after three days; after five days. The top two images also show the impact sites of chunk L.

Above: *Detailed ultraviolet image of the aurora centered on the magnetic north pole.*
Right: *Ultraviolet images that show both polar auroras.*

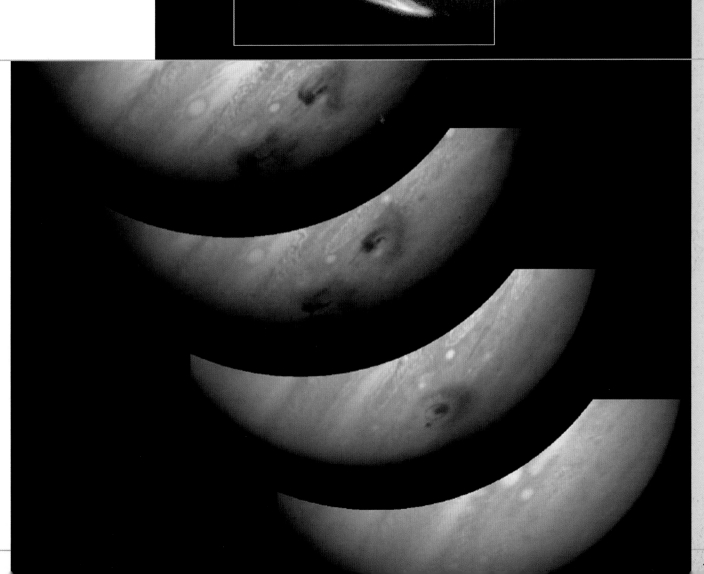

JUPITER'S MOONS

Jupiter has at least 63 moons—more than any other planet in the solar system. They range from the tiny outer satellites, some less than 1.2 miles (2 km) in diameter, to the massive inner Galilean moons. They are named after mythological figures who featured in the life of Jupiter (or Zeus).

Right: *Three of Jupiter's moons in orbit—Io over Jupiter's disk, Europa to the right, and Callisto bottom left.*
Far right: *False-color image showing shadows of three of Jupiter's moons on its surface (seen here as black circles)—Io left of center, Ganymede on left edge (partly obscured), and Callisto at near right edge.*

Jupiter's Rings and Inner Moons

The closest objects to Jupiter are its small inner regular satellites: the four rocky moons Metis, Adrastea, Amalthea, and Thebe. The group is known as regular because the orbits are all circular, prograde (meaning that they rotate counterclockwise), and roughly flat along Jupiter's equatorial plane. They orbit in the same area as Jupiter's faint rings. The rings are almost invisible structures consisting of microscopic dust particles that are probably debris caused by impacts from small meteorites on the inner satellites. There are four rings in total, with gaps swept clean by the passage of the moons.

In 1979 *Voyager I* caught sight of the inner satellites. Except for the larger Amalthea, they appeared as points of light or shadows cast on Jupiter's clouds. The *Galileo* probe showed much more detail, revealing nonspherical, heavily cratered bodies in the case of Metis and Thebe.

Metis and Adrastea have similar orbits, separated by only 620 miles (1,000 km) at the point of closest approach. They orbit Jupiter faster than the speed at which Jupiter rotates. Being so close to the planet, only their internal tensile strength keeps them from being ripped apart by tidal forces. The moons will lose energy slowly, their orbits will decay, and they will crash into Jupiter. It is thought that they were rogue asteroids ejected from the asteroid belt and captured by Jupiter's great gravitational field.

Thebe and Amalthea

Thebe is more spherical than the similar-sized moons in this group. It has three or four very large craters covering a large proportion of the surface. Like the other inner regular

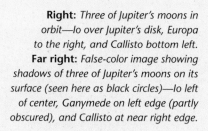

DATA			
Moon	**Radius(km)**	**Mass (kg)**	**Distance from Jupiter (km)**
Metis	30	9.56×10^{16}	127,969
Adrastea	12.5 x 10 x 7.5	1.91×10^{16}	128,970
Amalthea	131 x 73 x 67	7.17×10^{18}	181,300
Thebe	55 x 45	7.77×10^{17}	221,895

satellites, it has one face pointed at Jupiter at all times, and it takes two-thirds of a day to orbit. It was initially discovered as a shadow on Jupiter's surface in a picture from *Voyager 1*, and then found to be in earlier images as a tiny dark dot.

After the four massive Galilean satellites, Amalthea was the next of Jupiter's satellites to be discovered and was recorded in 1892 by the astronomer Edward Barnard. It is highly irregular—its length is nearly twice its height. It orbits synchronously with its longest axis always pointing at Jupiter. The surface is heavily cratered. It is covered with a red dust, believed to be sulfur compounds from Io's volcanoes that have been ripped away by the strong magnetic field of the planet. This covering makes Amalthea the reddest heavenly body in the solar system. The largest crater, Pan, is 62 miles (100 km) long, making it almost the size of the moon itself. Like Jupiter, Amalthea gives out more heat than it receives from the Sun, suggesting an internal heat source. It is likely to be from tidal forces stretching and squeezing the rock, although the strong magnetic field of Jupiter may create currents (and therefore heat) within the core of the moon.

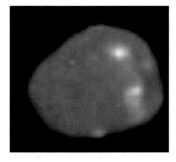

Left: *Jupiter's moon Amalthea.*

Above: *Jupiter's inner moons and rings. The innermost moons Adrastea and Metis orbit at the boundary of the narrow main ring (red), while the outer moons Thebe and Amalthea orbit at the edges of the thicker "gossamer" rings (green and yellow). All moons contribute dust to the rings.*

Io

Mass: 8.9 x 10²² kg

Magnitude: 5.0

Equatorial radius:
1,131 mi (1,821.6 km)

Density: 3.55 g/cm³

Surface gravity:
0.183

Rotation period:
1.769 days

Orbital period: 1.769 days

Radius of orbit:
262,000 mi (421,600 km)

Average surface tempera-ture: 130 K (−226°F/−143°C)

Jupiter's four largest moons are known as the Galilean satellites, named for Galileo Galilei, the Italian astronomer who based his Sun-centered model of the solar system on the fact that he could see four heavenly bodies orbiting Jupiter. The one closest to Jupiter is Io.

Io is the most volcanic body known in the solar system. Its surface is covered by debris from the eruptions of hundreds of volcanoes, the most powerful of which send plumes of sulfur-based compounds more than 300 miles (500 km) high. The compounds vary in color and create the mottled appearance of Io's face. The picture shows mountain ranges, volcanoes with fallout rings, lava flows, calderas, and what appear to be lakes of molten sulfur. The hottest points reach 2000 K, but the average temperature is much lower than that.

The constant volcanic activity has to be powered by a significant amount of energy. Io is in a resonant orbit with the moons Europa and Ganymede. They all orbit in the same plane, and for each orbit made by Ganymede, Europa orbits exactly twice and Io orbits exactly four times. The gravitational pull of Europa and Ganymede on Io perturbs it, causing the whole moon to flex. The flexing changes Io's shape by up to 330 feet (100 m) in the equatorial plane, and constant friction within the moon generates huge amounts of energy. Io's interaction with Jupiter's strong magnetic field also creates some heat.

Europa

Europa is a moon unlike any other. Almost devoid of volcanoes and impact craters, its surface is mostly highly reflective ice, with darker patches and streaks of rocky material. The ice forms a crust covering the entire moon to a depth of 62 miles (100 km). Evidence from the *Galileo* probe suggests that the solid ice may give way to liquid water beneath it, forming potentially the largest ocean in the solar system.

Europa is subject to tidal stretching forces, causing its interior to heat up enough to raise the temperature of the ice close to the rocky interior above freezing. Patterns in Europa's surface ice appear similar to tectonic plates on Earth—large slabs of solid material moving slowly under the influence of the tidal forces of a liquid beneath. Geysers

of liquid water reaching the surface and then freezing would provide explain the near absence of impact craters: the surface is continually being renewed with fresh ice. More evidence comes from measurements of Europa's weak magnetic field. The *Galileo* probe showed that it varied in a way that is consistent with the existence of a shell of conductive fluid (such as water with dissolved salts) across the whole moon.

The presence of the only known ocean outside Earth, together with heat from the interior, gives Europa the best conditions of any other planet in the solar system for supporting extraterrestrial life. Hydrothermal vents in Earth harbor life forms known as extremophiles that live in very similar environments.

Mass: 4.8 x 10²² kg

Magnitude: 5.29

Equatorial radius: 980 mi
(1,569 km)

Density: 3.01 g/cm³

Surface gravity: 0.135

Rotation period: 3.55 days

Orbital period: 3.55 days

Radius of orbit:
418,710 mi (670,900 km)

Average surface tempera-ture: 111 K (−260°F/−162°C)

Right: *Natural color image of Europa.* **Far right:** *Image of Europa enhanced to show dark brown rocky areas overlying fine-grained ice (pale blue) and coarse-grained ice (dark blue) in polar areas. A recent 31-mile- (50-km-) diameter crater named Pwyll is visible.*

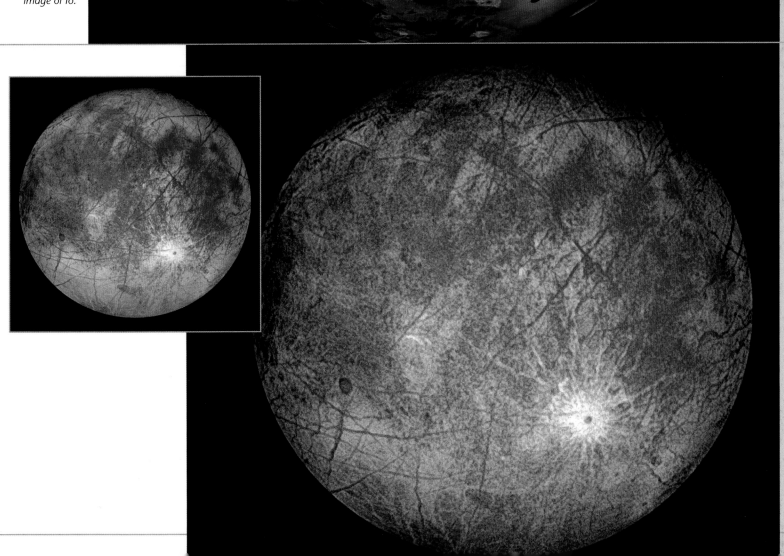

False-color image of Io.

Callisto

DATA

DATA

Mass: 1.1 x 10²³ kg

Magnitude: 5.65

Equatorial radius:
1,490 mi (2,400 km)

Density: 1.86 g/cm³

Surface gravity: 0.127

Rotation period:
16.69 days

Orbital period:
16.69 days

Radius of orbit:
1,170,100 mi
(1,883,000 km)

**Average surface tem-
perature:** 105 K (–
271°F/–168°C)

Callisto is the outermost of the Galilean satellites and the most cratered body in the solar system. Typically, the more craters there are on a surface, the older it is. Estimates of the cratering rate throughout the Callisto's history give an age of roughly four billion years. This indicates that no major tectonic movement or renewal of the surface has occurred since the moon first formed. It is said to be tectonically inactive, unlike bodies such as Europa and Earth that have young, tectonically active surfaces.

Callisto has a low density, which has led to the conclusion that it is made up of 40 percent ice and 60 percent rock and metal. Heavier substances settle to the center of the moon, so the percentage of rock and metal increases deeper within the moon, while the surface is dirty ice. White patches show where impacts have thrown up cleaner ice from under the surface. The largest impact craters, such as

Valhalla and Asgard, show interesting concentric circle (or "bull's-eye") patterns radiating out for a huge distance from the center. They look like closely spaced ridges and cracks, and in the case of Valhalla the circles cover a region 1,900 miles (3,000 km) in diameter. Several impact chains have also been found—similarly sized craters organized in straight lines, like the results of a "bombing run." They are caused by comets being pulled apart into strings of rocks by close interactions with the strong gravity of Jupiter, and then their orbit intersecting with Callisto's. The most famous event of this type occurred when the comet Shoemaker-Levy 9 split into more than 20 fragments and hit Jupiter.

Right: *Surface of Callisto, showing numerous impact craters with landslides at top right.* **Far right:** *Full view of Callisto.*

Ganymede

Ganymede is the largest moon in the solar system. It is larger than both Mercury and Pluto. Ganymede has a low density of 1.94 g/cm³, indicating that a lot of its mass is made up of water in the form of ice (density 0.9) rather than rock (commonly basalt, density 3.3) and metal (mostly iron, density 7.8). It also has a magnetic field, suggesting that, like most planets, it has a dense metallic core. The current accepted model of the internal structure of Ganymede is one of a rocky layer around a metal core, and then a deep layer of ice, with a thin crust of colder ice and rock on the surface. The moon is not thought to be tidally heated in the same way as Europa and Io. A thin oxygen-based atmosphere is present, probably caused by charged particles striking the ice and splitting it into hydrogen, which is lost to space, and oxygen, which remains mostly in the form of ozone.

The surface of Ganymede is split into light and dark regions with different characteristics. The dark areas are heavily cratered, indicating that the surface is very old, dating

back perhaps 3.5 billion years. The lighter areas are newer, with fewer impact craters and more ice. Parallel grooves and ridges similar to those on Europa's surface occur in these areas, suggesting tectonic renewal of the surface as the underlying deep ice layer shifts over long periods of time. Because ice flows more readily than rock, the oldest impact craters fade as the ring of cliffs subsides into the central depression. Circular marks that are otherwise flat on the surface are such ancient craters, known as palimpsests.

DATA

Mass: 1.5 x 10²³ kg

Magnitude: 4.61

Equatorial radius:
1,642 mi (2,631 km)

Density: 1.94 g/cm³

Surface gravity:
0.145

Rotation period:
7.15 days

Orbital period:
7.15 days

Radius of orbit:
668,000 mi (1,070,000 km)

**Average surface tem-
perature:** 117 K (–
249°F/–156°C)

Above: *Natural color image of Ganymede, showing darker, older, and heavily cratered areas, and the brighter spots of more recent impacts.*

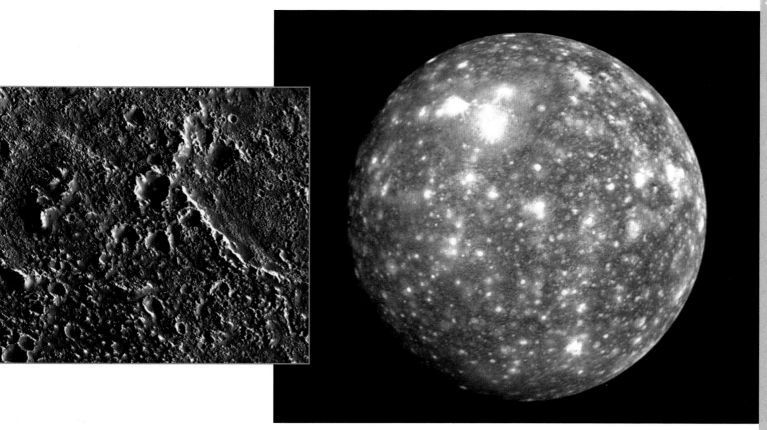

Jupiter's Minor Moons

The remaining satellites of Jupiter are tiny in comparison with the Galilean moons. The table lists those that were discovered and named prior to 1975. Many more were discovered in 1999 and 2000. The largest outer satellite by far is Himalia (shown right), with a radius of 58 miles (93 km). The picture was taken by the *Cassini* probe, and is the only view of any member of this group that shows it as more than a simple dot. The insets show the position of sunlight when the picture was taken and a zoomed-in portion.

The outer moons are divided into two groups. Leda, Himalia, Lysithea, and Elara orbit between 6.8 and 7.5 million miles (11 and 12 million km) from Jupiter in a plane that is tilted at 28 degrees to Jupiter's equatorial plane (where the inner and Galilean moons lie). Ananke, Carme, Pasiphae, and Sinope are much farther out at 13 to 14.9 million miles (21 to 24 million km), and their plane of orbit is tilted by 150 degrees. Unusually, their orbits are retrograde—they rotate around Jupiter in the opposite direction to the other moons and to Jupiter's own spin.

The grouping of bodies and the similarity of orbits has led to a theory that the outer moons are the remnants of asteroids that were captured by Jupiter's gravity and broken up into the fragments that exist today.

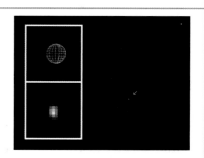

Himalia, the only one of Jupiter's minor moons that has been photographed in any detail.

DATA			
Moon	**Radius (km)**	**Mass (kg)**	**Distance from Jupiter (km)**
Leda	8	5.68×10^{15}	11,094,000
Himalia	85	9.56×10^{18}	11,480,000
Lysithea	18	7.77×10^{16}	11,720,000
Elara	43	7.77×10^{17}	11,737,000
Ananke	14	3.82×10^{16}	21,200,000
Carme	23	9.56×10^{16}	22,600,000
Pasiphae	30	1.91×10^{17}	23,500,000
Sinope	19	7.77×10^{16}	23,700,000

SATURN

DATA

Type: Gas giant

Mass: 5.7 x 10²⁶ kg

Equatorial diameter: 74,900 mi (116,000 km)

Density: 0.69 g/cm³

Surface gravity: 1.1

Mean cloud temperature: 148 K (−193°F/−125°C)

Magnitude: 0.67

Mean distance from the Sun: 892,000,000 mi (1,429,400,000 km)

Orbital eccentricity: 0.0560

Saturn year: 29.5 Earth years

Saturn day: 10.7 hrs

Saturn is possibly the most recognizable planet in the solar system after Earth. No other planet has such a bright ring system. Saturn is named after the Roman god, who was the father of Jupiter.

Saturn is the second largest planet in the solar system after Jupiter. It is similar to Jupiter in composition, consisting of approximately three-quarters hydrogen and one-quarter helium, plus small amounts of other elements. Both planets have a rocky core surrounded by a layer of liquid metallic hydrogen. Like Jupiter, Saturn produces more heat than it receives from the Sun. It is perhaps generated by gravitational settling in the core. Above the liquid hydrogen the planet is entirely gas, with molecular hydrogen (H_2) dominant. Saturn has the lowest overall density of all the planets. At 0.7 g/cm³, it is less dense than water.

Natural color view of Saturn.

Saturn's Rotation

Even a small telescope on Earth shows an observer the rings of Saturn. It also reveals that Saturn is not perfectly spherical. Because of its fast rotation, Saturn is wider around the equator and flattened at the poles. Since Saturn is a gas giant, it has more than one rotational speed. The equatorial regions rotate once every 10 hours 14 minutes, but regions farther north or south take 10 hours 39 minutes to rotate. One way of assigning a single rotation rate to a gas giant is to base it on the radio wavelength emissions. They are tied to the magnetic field of the planet, generated beneath the surface. When measured by *Voyager* probes at the start of the 1980s, Saturn's radio emission pattern repeated at the same rate as the rotation of the nonequatorial regions.

Measurements taken by the *Cassini* probe 20 years later show that the radio emission rotational period had lengthened by six minutes. It is not known why, but it is not expected to mean that the rotation of the entire planet is slowing down. A more likely explanation is the variation in the way in which Saturn's magnetic field is generated.

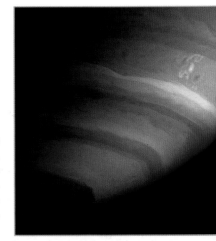

Right: *The Dragon Storm, a giant thunderstorm that appeared in Saturn's southern hemisphere in 2004.* **Far right:** *Ultraviolet view of Saturn's southern aurora. Auroras occur around each pole and are caused by charged particles entering the upper atmosphere.*

Saturn's Rings

In 1610 Galileo Galilei was the first person to observe Saturn with a telescope, but he could not interpret what he saw. He saw shapes on either side of Saturn's disk, but he thought that they were other bodies attached to the planet. He was even more confused when in another observation the shapes seemed to disappear—the relative position of Earth and Saturn had changed so that the rings appeared edge-on and invisible. Forty-five years later the Dutch scientist Christiaan Huygens used a better telescope and was able to identify the shapes as a ring. After a further 20 years Giovanni Cassini saw multiple rings with gaps between them. The rings are now known as A, B, and C, and the gap between A and B is the Cassini Division. More rings that are not visible from Earth have been discovered by the *Voyager* probes and labeled D to G. The B ring is the brightest of the rings, indicating that it contains a number of large icy particles. The outermost E ring is the widest and darkest, consisting of dust.

The rings themselves are made of lumps of rock, iron oxide, and ice, ranging in size from minute particles to several meters across. Each piece of the rings is a tiny satellite, orbiting Saturn. Although the rings are visually very bright, there is not much material in them. Gathered together into a moon, they would form a sphere only 62 miles (100 km) across. The total diameter of the rings is about 175,000 miles (282,000 km), but they are less than 0.6 miles (1 km) thick.

The structure of the rings is very complex. There are hundreds of tiny gaps within each ring, as well as wavelike variations in density, all created by gravitational interactions with Saturn's moons. Prometheus and Pandora are known as shepherd satellites, orbiting just outside the A ring. They maintain the sharp edge of the ring by attracting any particle that strays too close to their orbit. The rogue particle is then either flung out to space or collides with the moon. The cause for most of the gaps in the rings is gravitational resonance that

Image of Saturn's C ring, with B ring at top left.

SATURN'S MOONS

Saturn has 34 named moons and at least another 15 that are yet to be named. They range from the giant Titan to the tiny shepherd moons that orbit within the ring structure. The relationship between the rings and the moons is very complex.

Right: *Moons Mimas (at right) and Pandora in orbit around Saturn.*

occurs when the orbit of a moon and that of a particular section of ring are in a simple ratio of time. The Cassini Division is linked to the moon Mimas in this way. Any particle in the Cassini Division orbits twice for each orbit of Mimas. Once every two orbits, Mimas provides a gravitational pull, tugging the particle outward. Over a long period of time, this action removes the particle into a different orbit, creating the gap. This particular resonance has a 2:1 ratio—Mimas also has a 5:3 ratio with the Encke Division. The complexity of the rings is caused by the large number of moons and the large number of possible ratios for gravitational resonance.

The rings are unstable over very long periods of time, so it is thought that they were not created as long ago as the planets themselves. It is possible that a moon of Saturn was broken up into the pieces that now form the rings, either by gravitational interaction or by an impact with another body.

Right: *Saturn's rings with moons Tethys (above) and Dione. Shadows of the rings and Tethys can be seen on the planet's surface.*

The Shepherd Moons

Orbiting within the Encke Division embedded in the A ring, Pan is the innermost of Saturn's known satellites. Its passage through the rings has cleaned a corridor 200 miles (325 km) wide. As it moves through, it creates small disturbances in the division boundaries that look like waves. Pan was discovered only after these waves had been examined and used to predict its existence. A search through the photos from space probe *Voyager 2* revealed the tiny moon, only a few pixels across in any image. A similar method was used to discover the moon Daphnis in May 2005: the *Cassini* probe spotted the characteristic wavelike disturbances in the ring material surrounding the Keeler gap, which is just 22 miles (35 km) wide.

Atlas orbits just outside the A ring. It is named for the mythological titan who holds the heavens above the Earth—Atlas the moon holds up the edge of the A ring. Photos from *Cassini* reveal a very flattened shape, like two saucers placed together. It is a shepherd moon, creating the sharp cut-off edge of the A ring. Prometheus does a similar job for the inner edge of the F ring, a much thinner, fainter ring beyond the A ring. It is the largest of the inner shepherd moons and resembles an elongated potato with some cratering. Kinks and knots appear in the F ring alongside the moon as its gravity pulls material away from the rings. Pandora is the outer shepherd of the F ring and has more craters than Prometheus.

All these inner moons have very low densities of roughly 0.6 to 0.7 g/cm³, indicating that they are loosely packed and consist mainly of ice.

DATA			
Moon	**Diameter (km)**	**Mass (kg)**	**Distance from Saturn (km)**
Pan	35 x 32 x 21	5×10^{15}	133,583
Daphnis	9 x 9 x 6	8.4×10^{13}	136,505
Atlas	46 x 38 x 19	1.1×10^{16}	137,670
Prometheus	148 x 100 x 68	3.3×10^{17}	139,353
Pandora	110 x 88 x 62	1.94×10^{17}	141,520

Epimetheus and Janus

Outside the F ring there are two moons with an orbital relationship believed to be unique in the solar system. Epimetheus and Janus were observed in 1966. At first they were thought to be a single object, because two separate observations saw a satellite at the same distance from Saturn. Further investigation simply confused astronomers, since observation and predicted orbital paths failed to match up. In 1978 it was realized that the orbit was inhabited by two satellites. The *Voyager 1* probe was able to confirm this in 1980.

Epimetheus and Janus are coorbital moons. Their orbital radius differs by only 31 miles (50 km), a small enough difference for them to collide if one tried to pass the other. The moon in the orbit closer to Saturn is orbiting slightly faster than the moon farther away and will therefore catch it up.

As it does so, the gravitational interaction between the moons drags the inner one outward, raising its orbit and slowing it down. The outer moon is dragged down, lowering its orbit and speeding it up. The two moons therefore effectively swap orbits. Each moon orbits in less than 17 hours, but the orbit swap occurs only once every four years. It is not known whether this situation is stable.

Both moons are similar in composition. They are low density, probably porous, and made mostly of ice. They are heavily cratered, indicating that they are old. Several craters are over 18 miles (30 km) in diameter. It is possible that both moons came from the same parent body, split by a large impact. This would have happened early in the history of the Saturnian system since they are both relatively old moons.

JANUS
Diameter: 113 mi (181 km)
Mass: 1.98×10^{18} kg
Density: 0.65 g/cm³
Orbital radius: (94,125 mi (151,472 km)
Orbital period: 0.7 Earth days
Rotational period: 0.7 Earth days
Eccentricity: 0.007

Above: *Prometheus and Pandora orbiting on either side of the F ring, with Epimetheus farther out.*

Above right: *Atlas orbiting between Saturn's A and F rings.*
Bottom right: *Daphnis orbiting in the Keeler gap of Saturn's moons.*

EPIMETHEUS

Diameter: 72 mi (116 km)

Mass: 5.35×10^{17} kg

Density: 0.61 g/cm³

Orbital radius: 94,094 mi (151,422 km)

Orbital period: 0.7 Earth days

Rotational period: 0.7 Earth days

Eccentricity: 0.009

Below: *Epimetheus and Janus in their orbits around Saturn.*

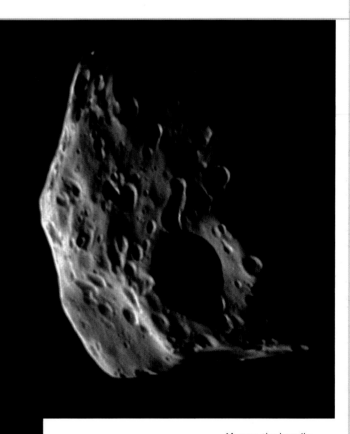

Above: *the heavily cratered and irregularly shaped Epimetheus.*

Mimas

William Herschel discovered the first of Saturn's large moons—Mimas—in 1789. Moons are often termed "large" when they contain enough mass to force the object to collapse into a spheroid, rather than being an irregular shape. It was named after a titan in Greek mythology—the son of Gaia, who was slain by Hercules.

Mimas has a very low density of only 1.17 g/cm³, indicating that it is made almost entirely from water ice, with perhaps some rock at the core. The surface is exceptionally cold—64 K (−344°F/−209°C)—which helps prevent ancient features such as crater walls from slowly slipping and flattening out. Because of this, Mimas is covered in impact craters from the early era of the solar system when meteorite bombardment occurred with greater frequency.

The face of Mimas is dominated by one huge crater: Herschel. It is 80 miles (130 km) across—nearly one-third of the whole moon's diameter. Its crater floor is 6 miles (10 km) deep, and it has a central peak 3.7 miles (6 km) high. Fracture marks on the opposite surface are thought to have been caused by the shockwaves propagating around the moon and meeting on the far side. Herschel is one of the biggest craters relative to its parent body. A larger impact could have smashed Mimas to pieces, perhaps forming another ring around Saturn.

Mimas has a major gravitational impact on Saturn's rings. A 1:2 resonance with its orbit creates the Cassini Division, the most visible gap. Other exact orbital ratios produce smaller gaps and wrinkles in the rings. The moon Tethys is also in a resonance with Mimas, orbiting once for every two orbits made by Mimas.

DATA

Diameter: 247 mi (397 km)

Mass: 3.84 x 10¹⁹ kg

Density: 1.17 g/cm³

Orbital radius: 115,210 mi (185,404 km)

Orbital period: 0.94 Earth days

Rotational period: 0.94 Earth days

Eccentricity: 0.02

Enceladus

Outside the F ring there are two moons with an orbital relationship believed to be unique in the solar system. Epimetheus and Janus were observed in 1966. At first they were thought to be a single object, because two separate observations saw a satellite at the same distance from Saturn. Further investigation simply confused astronomers, since observation and predicted orbital paths failed to match up. In 1978 it was realized that the orbit was inhabited by two satellites. The *Voyager 1* probe was able to confirm this in 1980.

Epimetheus and Janus are coorbital moons. Their orbital radius differs by only 31 miles (50 km), a small enough difference for them to collide if one tried to pass the other. The moon in the orbit closer to Saturn is orbiting slightly faster than the moon farther away and will therefore catch it up. As it does so, the gravitational interaction between the moons drags the inner one outward, raising its orbit and slowing it down. The outer moon is dragged down, lowering its orbit and speeding it up. The two moons therefore effectively swap orbits. Each moon orbits in less than 17 hours,

DATA

Diameter: 310 mi (499 km)

Mass: 1.08x10²⁰ kg

Density: 1.61 g/cm³

Orbital radius: 147,861 mi (237,948 km)

Orbital period: 1.37 Earth days

Rotational period: 1.37 Earth days

Eccentricity: 0.0045

The fractured surface of Enceladus.

Right: *False-color image in which colors indicate different surface materials.*
Below: *Mimas showing the crater Herschel.*

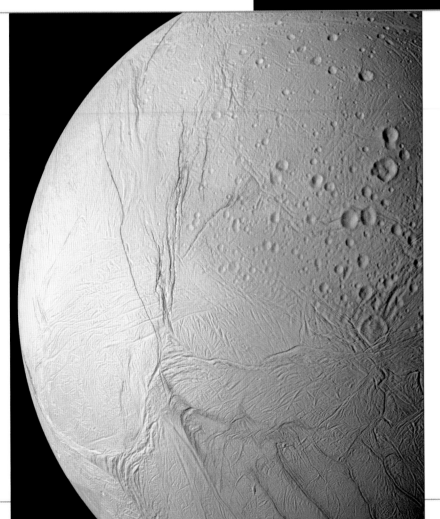

but the orbit swap occurs only once every four years. It is not known whether this situation is stable.

Both moons are similar in composition. They are low density, probably porous, and made mostly of ice. They are heavily cratered, indicating that they are old. Several craters are over 18 miles (30 km) in diameter. It is possible that both moons came from the same parent body, split by a large impact. This would have happened early in the history of the Saturnian system since they are both relatively old moons.

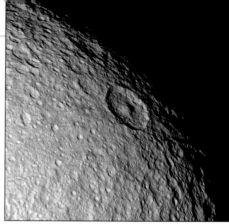

Tethys

Tethys is a heavily cratered and cracked ball of ice. Most areas are so saturated with impact craters that any new ones will obliterate older ones. There are some belts around the moon with lower cratering rates, indicating that some resurfacing has occurred. A major feature of Tethys is the 250-mile- (400-km-) wide impact basin called Odysseus. It is believed to have been created by an impact very early in the moon's history. Tethys would have been warmer shortly after its formation. That would have enabled it to absorb the impact, whereas a colder brittle body might have broken apart. In the past the effect would have been very similar to the Herschel crater on Mimas, but over time the cliffs and central peak of Odysseus have slowly collapsed.

Another massive feature of Tethys is Ithaca Chasma, a canyon that wraps around three-quarters of the moon's circumference. It reaches 62 miles (100 km) wide and is between 1.9 and 3.2 miles (3–5 km) deep. It is thought to have formed as Tethys froze. The heat generated by Tethys's formation meant that a large part of the moon was liquid. The surface froze, followed by the interior. As the interior froze, it expanded, cracking the crust and forming the canyon.

Two other moons of Saturn are coorbital with Tethys. Telesto and Calypso share the same orbit but remain 60 degrees ahead and behind of Tethys respectively at all times. These locations are known as Lagrangian points, and the satellites that inhabit them are called Trojan moons. They represent gravitationally stable locations for a small body in a system where a large body is orbiting an even bigger body.

DATA

Diameter: 660 mi (1,060 km)

Mass: 6.18×10^{20} kg

Density: 1.21 g/cm³

Orbital radius: 183,076 mi (294,619 km)

Orbital period: 1.88 Earth days

Rotational period: 1.88 Earth days

Eccentricity: 0.0000

Tethys's northern polar region showing a peaked crater named Telemachus.

Titan

Titan is the largest of Saturn's moons and the second largest in the solar system after Ganymede. It was discovered in 1655 by Dutch astronomer Christiaan Huygens—the first new moon to be found since the observations of Jupiter's moons by Italian astronomer Galileo Galilei in 1610. It can be seen with a small telescope from Earth. Its name comes from the mythological Greek race of titans, godlike beings who were overthrown by Zeus (Jupiter), the god of Olympus.

Titan is the densest of Saturn's moons, but not because it has a different composition. The ice and rock are being compressed by the stronger gravitational field of the large moon. It is also the only known moon to have a dense atmosphere. At the surface of Titan the atmospheric pressure is 50 percent greater than the pressure at sea level on Earth. It consists of 95 percent nitrogen, some methane and argon, and much smaller amounts of various hydrocarbon compounds, including ethane, acetylene, and hydrogen cyanide. The Sun's solar wind also interacts with the atmosphere of Titan because it has no magnetic field to shield itself and it is out of reach of Saturn's magnetosphere. Charged particles in the solar wind would certainly have an effect on the chemistry of the atmosphere and may also erode some of it.

It is not known why Titan has an atmosphere when similarly sized moons such as Ganymede and Callisto do not. It has made it very hard for us to understand the rest of Titan, since the gases in the atmosphere form something similar to smog. This prevents even the Sun's light from making much impression on the surface. Mapping the surface is therefore difficult and must be done by radar or light of a frequency that is not absorbed by the atmosphere. There is a very slight greenhouse effect, but it is not very efficient, so the surface temperature is kept low, at 94 K (–290°F/–179°C). Despite the cold, liquid exists on the surface of Titan, and it appears to have shaped the features of the planet. The liquid is not water, but methane.

One of the missions for the *Cassini* probe was to release a lander named *Huygens* onto the surface of Titan. The mission

DATA

Diameter: 3,200 mi (5,150 km)

Mass: 1.35×10^{23} kg

Density: 1.88 g/cm³

Orbital radius: 759 273 mi (1,221,931 km)

Orbital period: 15.9 Earth days

Rotational period: 15.9 Earth days

Eccentricity: 0.029

Dione

DATA

Diameter: 695 mi (1,118 km)

Mass: 1.1 x 10²¹ kg

Density: 1.50 g/cm³

Orbital radius: 234,514 mi (377,396 km)

Orbital period: 2.73 Earth days

Rotational period: 2.73 Earth days

Eccentricity: 0.0022

Dione is the second densest of Saturn's moons (after Titan). Its density stems from its rock core, which makes up roughly one-third of the total mass of the satellite. Like Saturn's other moons, the remainder outside the core consists of water ice. Cracks and many craters cover the surface, as well as younger areas with less cratering.

As with all Saturn's inner moons, Dione's rotational period is the same as the time it takes to orbit the planet, so one face always points at Saturn. The leading hemisphere of these moons usually has more craters than the trailing side, but in Dione's case this is reversed: the trailing hemisphere contains many more craters. This suggests that during the early stages of the solar system, when the bombardment of planets and moons was heavier, Dione faced the other way. A large impact billions of years ago may have spun it around. Since then, a gentle coating of the leading hemisphere by microscopic dust in the E ring has reduced its reflectivity.

Images from *Voyager* showed wispy features present on the trailing hemisphere. Snow or ash thrown up by ice volcanoes was suggested as the cause, but closer inspection by *Cassini* found numerous ice fractures. The cliffs created by the tectonic movements shone brightly because they consisted of cleaner ice than on the plains. Like Tethys, Dione has two small coorbital moons at its Lagrangian points, Helene and Polydeuces.

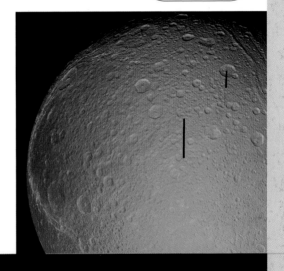

False-color image of Dione's cratered surface. The vertical black lines are image artifacts.

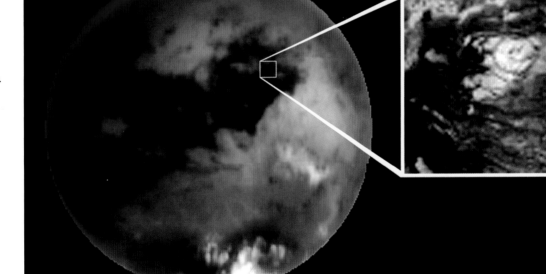

False-color image of Titan. The colors correspond to atmospheric (red) and surface (green and blue) features not visible to the naked eye.
Inset: *detailed view of what is probably an icy volcano.*

planners were unsure about what *Huygens* would land on. Large oceans filled with methane were considered as well as soft methane-based swamps. As it drifted down on a parachute through the thick atmosphere, *Huygens* saw landscapes that looked very much like shorelines and river deltas. However, there was no methane rain at the time, and no large bodies of liquid were detected. The ground, consisting mostly of rock-hard water ice, was wet with methane. *Huygens* may have landed on Titan during a dry season, but there was adequate evidence that methane rain and methane oceans had existed in the recent past.

Hyperion

Hyperion was first discovered by the American astronomer William Bond in 1848, a year after Saturn's previously discovered moons were named. The names referenced titans from Greek mythology, and Hyperion was named in the same vein.

Hyperion is the largest irregularly shaped body in the solar system. Any object larger than this would be naturally compressed into a spherical shape by its own gravity. Hyperion is believed to be a fragment of a larger object that was broken up by impact. Its very low density shows it to be a loosely bound pile of rubble—debris that may have resulted from a large impact.

Although Hyperion is chiefly composed of ice, it is not very reflective. It has a covering of dark reddish dust that may be the same material that coats parts of Iapetus. Dark regolith is particularly apparent on the floor of the many deep sharp craters. The spongelike appearance seen in the image is caused by steep craters. Perhaps the looseness of the structure helps create such steep craters and pits, but the low gravitational field also helps preserve them. The dark areas are not holes but dust on the crater floors.

Hyperion is the only moon in the solar system that has a chaotic rotation. Its rotational axis moves in all directions, making it impossible to predict its orientation from orbit to orbit. It is not known why this should occur, but the moon's irregular shape and noncircular orbit may have an effect. Hyperion is also close to the huge moon Titan, and is in a 3:4 orbital resonance with it.

False-color image of Hyperion.

DATA

Diameter: 224 x 174 x 140 mi (360 x 280 x 225 km)

Mass: 5.69×10^{18} kg

Density: 0.6 g/cm³

Orbital radius: 920,299 mi (1,481,009 km)

Orbital period: 21.3 Earth days

Rotational period: Chaotic

Eccentricity: 0.123

URANUS

DATA

Type: Gas giant

Mass: 8.7×10^{25} kg

Equatorial diameter: 31765 mi (51,118 km)

Density: 1.29 g/cm³

Surface gravity: 0.88

Mean cloud temperature: 80 K (–316°F/–193°C)

Magnitude: 5.52

Mean distance from the Sun: 1,784,000,000 mi (2,871,000,000 km)

Orbital eccentricity: 0.0461

Uranus year: 84 Earth yrs

Uranus day: 17.9 hrs

The third largest planet in the solar system after Jupiter and Saturn, Uranus was discovered in 1781. It had been seen previously, but observers incorrectly classified it as a star. Uranus is named for the Greek god of the sky and heavens, who was grandfather to Zeus. (The Romans knew the god as Jupiter.)

Beyond the huge gas giants Jupiter and Saturn lie the ice-cold outer reaches of the solar system. Uranus is nearly 1.8 billion miles (3 billion km) away from the Sun and visible in binoculars. Only one probe has passed close to Uranus: *Voyager 2*. It flew past the planet in 1986, eight and a half years after its launch. Images from the probe showed a nearly featureless surface. The upper atmosphere was opaque to *Voyager*, and only very slight color differences revealed any kind of structure resulting from storms. The atmosphere is composed of 83 percent hydrogen, 15 percent helium, and 2 percent methane. The methane absorbs red light, giving the planet its characteristic blue-green color. The core of Uranus is assumed to be rock and various types of ice, with hydrogen making up only 15 percent of the total composition.

Uranus is unusual in that its axis of rotation is almost parallel to the ecliptic plane (the plane of the orbits of the planets), as if it has been tipped over. The most likely explanation is that it was hit very hard by another planet-sized body not long after it formed, much like the current theory behind the formation of Earth's Moon. When *Voyager* visited Uranus, the south pole of the planet was facing the Sun. Despite this, the equatorial region was hotter than the point closest to the Sun. The reason for this is unknown. As Uranus orbits, the side facing the Sun moves around to the north pole. Currently the Sun is over the equator. This huge tilt is expected to produce unusual seasonal changes.

Iapetus

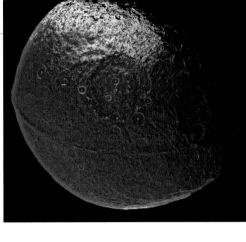

Iapetus was discovered by Giovanni Cassini in 1671. Although he could see it clearly on one side of Saturn, he could not find it at all on the other. He came to the conclusion that one side of the moon was much darker. His theory was proved right over 300 years later by images from the probe bearing his name.

Iapetus's orbit is more than twice as far away from Saturn as Titan's orbit. The orbit is tilted at an angle of 15 degrees from Saturn's equator, making Iapetus the only large moon to see the rings from a position other than edge-on. As of early 2006 the *Cassini* probe had not viewed the entirety of Iapetus, but it had shown that its shape was not a standard spheroid. Parts of it appear to be flattened, and there is a large equatorial ridge on the dark side. The ridge is massive, reaching up to 8 miles (13 km) above the surrounding plains and distorting the shape of the satellite. It is old, having just as many craters as the rest of the moon. It does not appear to have been created by tectonic processes. One theory is that Iapetus may have been closer to the rings in the past and actually collided with them, accreting the material that resulted in the ridge.

The two-tone coloration of Iapetus's surface is not fully understood, either. The leading hemisphere appears to be coated in a black carbon-based dust. Craters on the dark side do not reveal brighter material, suggesting that dust is being constantly added to the surface. It is not yet known if the material has come from within the moon as a result of volcanism or brought to Iapetus from space.

Iapetus, showing its heavily cratered surface.

DATA

Diameter: 892 mi (1,436 km)

Mass: 1.97 x 10²¹ kg

Density: 1.27 gcm³

Orbital radius: 2,212,694 mi (3,560,820 km)

Orbital period: 79.3 Earth days

Rotational period: 79.3

Eccentricity: 0.029

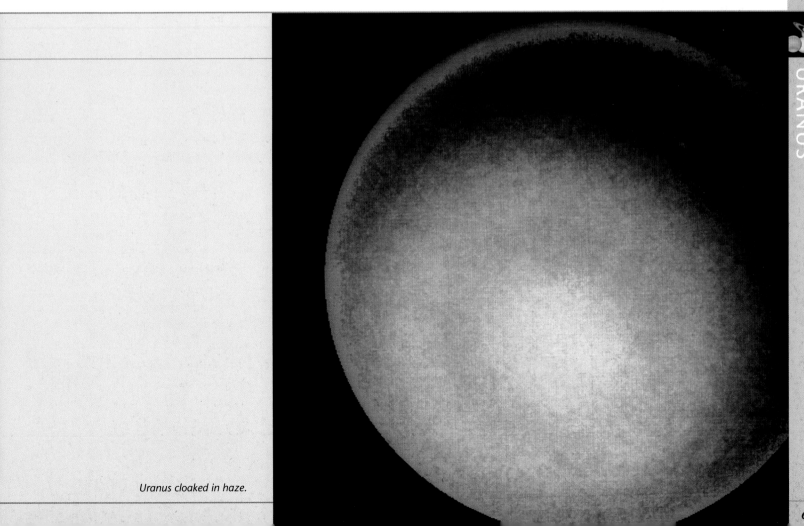

Uranus cloaked in haze.

Uranus's Rings

Like Saturn, Uranus has a ring system. The rings are oriented along the planet's equator, making them perpendicular to the plane of its orbit. They are made of similar materials—rock and ice—ranging in size from dust to fragments 30 feet (10 m) in diameter. The rings were first discovered by a team of astronomers who were hoping to research the makeup of Uranus's atmosphere. They were waiting for Uranus to eclipse a background star (an event known as occultation). By examining the nature of the light as the eclipse began and ended, they hoped to see what frequencies were absorbed in the atmosphere. When performing the observation, they found that the star disappeared from view several times before and after it passed the planet, indicating that there were rings blocking the light. The probe *Voyager* was able to confirm the findings, examining the 11 named ring structures. Two of the inner moons of Uranus appear to act as shepherd satellites for the epsilion ring, the brightest and outermost ring.

Uranus's strange tilt is not fully reflected in its magnetic field. It is off center by one-third of a Uranian radius and is at an angle of 60 degrees to the axis of rotation. (Neptune has a similarly skewed magnetic field.) The tail of the magnetosphere, which is created by interaction with the solar wind, is twisted into a corkscrew shape. It is not known what generates Uranus's magnetic field. Typically, an internal conducting liquid core or layer is required to generate the currents, but there is no evidence that Uranus has such a system.

Miranda and Other Moons

The moons of Uranus are named for characters from works by Shakespeare and the English poet Alexander Pope. There are 13 small inner moons, five large major moons, and nine small outer moons.

The innermost large moon is Miranda, named for the daughter of the central character Prospero in the Shakespeare's *The Tempest*. When *Voyager 2* needed a gravitational slingshot to give it the momentum to reach Neptune, Miranda was the closest satellite to its planned path. The probe had to use careful maneuvers to keep the moon steady in its view, and long exposures were necessary owing to the great distance from the Sun. The images are some of the best of *Voyager*'s entire mission. Features as small as 2000 feet (600 m) across can be picked out.

Miranda was expected to be a simple cratered ball of ice and rock, much like Callisto, but the pictures show a chaotic surface. Old, heavily cratered terrain is present as expected, but is interspersed with younger terrain characterized by long parallel grooves. The grooves are linear or, in some cases, appear to go around corners, giving the appearance of a racetrack. Huge faults exist, creating canyons up to 12.5 miles (20 km) deep. The terrain appears so jumbled that initial thoughts were that Miranda had been heavily impacted and shattered several times in its history. In this scenario the fragments reformed over time in a similar orbit. Recently, however, a slightly less violent theory has been put forward: that tectonic action created the observed features. Miranda could be subject to the same tidal forces that make Jupiter's moon Io such a volcanic object. Warm ice heated by the interior of the moon may well up and flow onto the surface, much like lava.

Above: *Detail of Uranus's rings.*
Right: *The faint rings of Uranus have been enhanced in this image. Also seen are some of Uranus's satellites.*

Miranda's surface has terraced layers, indicating the mixing of ancient and recent surfaces.

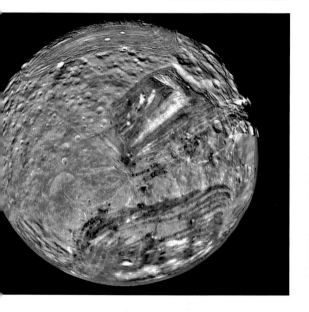

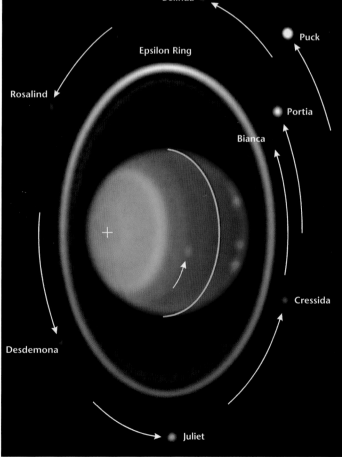

Belinda

Puck

Epsilon Ring

Rosalind

Portia

Bianca

Cressida

Desdemona

Juliet

DATA

Mass: 6.3 x 10¹⁹ kg

Magnitude: 16.3

Equatorial diameter: 293.1 mi (471.6 km)

Density: 1.15 g/cm³

Rotation period: 1.41 days

Orbital period: 1.41 days

Radius of orbit: 81,000 mi (129,780 km)

Average surface temperature: 86 K (–305°F/–187°C)

False-color image of the clouds of Uranus, its rings, and eight of its moons. The arrows indicate the orbits of the moons.

Ariel and Titania

TITANIA

Mass: 3.5 x 10²¹ kg

Magnitude: 13.73

Equatorial diameter:
980.5 mi (1577.8 km)

Density: 1.72 g/cm³

Rotation period:
8.7 days

Orbital period:
8.7 days

Radius of orbit:
272,300 mi
(436,300 km)

Average surface temperature: 60 K (−352°F/−213°C)

Beyond Miranda are the major moons of Uranus—Ariel, Umbriel, Titania, and Oberon. All four are similar in size. Titania and Oberon (king and queen of the fairies in Shakespeare's *A Midsummer Night's Dream*) are slightly larger than Ariel and Umbriel (two sylphs in a poem by Alexander Pope, "The Rape of the Lock"). Geologically speaking, the moons split into two distinct groups.

Ariel is the brightest of Uranus's satellites, with a reflectivity of 39 percent. By comparison, Earth has a reflectivity of 38 percent, and the Moon 12 percent. The surface of the satellite is water ice, which makes up nearly half of the total mass, the rest being rock. The main features on Ariel are the huge rift valleys that stretch for hundreds of miles. They can be as deep as 6 miles (10 km) and often intersect. This points to a period of tectonic activity that created great cracks in the surface before the water ice froze solid. The relative scarcity of impact craters suggests that the surface is young and that the tectonic events eradicated the old craters.

Titania is larger and darker than Ariel, but it has the same characteristic features. One chasm imaged by the *Voyager* probe stretches for 994 miles (1,600 km) around the moon.

Color view of Titania.

Uranus's outermost moon, Oberon.

Oberon and Umbriel

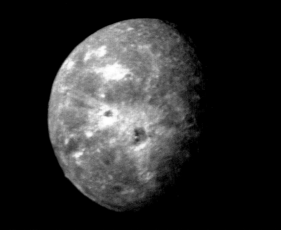

Although Umbriel is the twin of Ariel in terms of size and mass, it is the darkest of Uranus's moons. The surface is old, with large numbers of impact craters. There is no evidence of recent tectonic activity, no cracks or canyons visible in images taken by *Voyager*. Across the southern hemisphere imaged by the probe the color of Umbriel is a consistent dark gray, except for one particular patch. Near the equator there is what appears to be a ring of material that is particularly reflective. It could be frost—clean water ice deposited in an impact crater.

Oberon is similar in that it appears to have had no geological movements since it was originally formed. Its surface is also pockmarked with a lot of impacts, but in this case the largest impacts have bright ray structures associated with them. These features are similar to those found on Callisto. They are most likely to be patterns formed by debris excavated from below the surface by the huge amount of energy generated by the impacts. The crater floors are unusually dark—they are possibly flooded with less reflective material from within the crust.

Voyager 2 was the only probe to visit these moons and it did not remain long. There are no images of the northern

OBERON

Mass: 3.03 x 10²¹ kg

Magnitude: 14.2

Equatorial diameter:
950 mi (1,526 km)

Density: 1.64 g/cm³

Rotational period:
13.463 days

Orbital period:
13.463 days

Radius of orbit:
362,000 mi
(582,600 km)

Mass: 1.35 x 10²¹ kg

Magnitude: 14.16

Equatorial diameter:
719.5 mi (1157.8 km)

Density: 1.67 g/cm³

Rotation period:
2.52 days

Orbital period:
2.52 days

Radius of orbit:
118,600 mi
(190,900 km)

Average surface temperature: 58 K (–355°F/–215°C)

Scars on Ariel suggest tectonic activity, craters indicate bombardment by meteorites, and smooth areas are evidence of material deposited by some geological process over the older terrain.

UMBRIEL

Mass: 1.2 x 10²¹ kg

Magnitude: 14.81

Equatorial diameter:
726.7 mi
(1,169.4 km)

Density: 1.40 g/cm³

Rotation period:
4.1 days

Orbital period:
4.1 days

Radius of orbit:
165,000 mi
(266,000 km)

Average surface temperature: 61 K (–350°F/–212°C)

The heavily cratered surface of Umbriel.

hemispheres of any of the four largest moons, because they would have been pointing away from the Sun. No other probes are planned in the near future. The best telescopes on Earth have been able to pick out the moons of Uranus but are unable to see any kind of detail on them. It may be a long time before we get to see the other sides of these moons.

NEPTUNE

DATA

Type:	Gas giant
Mass:	1.02×10^{26} kg
Equatorial diameter:	30,754 mi (49,492 km)
Density:	1.64 g/cm³
Surface gravity:	1.14
Mean cloud temperature:	50 K (–370°F/–223°C)
Magnitude:	7.84
Mean distance from the Sun:	2,800,000,000 mi (4,504,300,000 km)
Orbital eccentricity:	0.0097
Neptune year:	164.8 Earth yrs
Neptune day:	16.1 hrs

Neptune is the last gas giant in the solar system. It is named for the Roman god of the sea, known to the Greeks as Poseidon. Some years after the discovery of Uranus, astronomical tables of its location were published to help astronomers find the planet in the sky. However, the orbit of Uranus did not agree with the calculated path. The error could be explained by the presence of an undiscovered planet beyond Uranus's orbit. Independently, English astronomer John Couch Adams and French mathematician Urbain Le Verrier calculated the unknown planet's orbit. The new planet was discovered in 1846 using Le Verrier's numbers. It turned out that it had been observed but not recognized by English astronomer James Challis, using Adams's calculations.

Neptune's composition is similar to that of Uranus. It is smaller but heavier, indicating a larger proportion of dense material. The atmosphere consists of hydrogen, helium, and methane, with trace elements that give the surface a darker shade of blue than Uranus. Well below the outer layers the amounts of methane, ammonia, and water increase, until the gases are compressed into a liquid phase at 20 percent of the distance to the core. The core is small (no larger than Earth) and made of rock and metal.

Neptune's Atmosphere

Neptune radiates roughly twice as much energy as it receives from the Sun, indicating that there is a heat source in the interior. Many other planets share this feature. Earth's internal heat is generated by radioactive decay of elements trapped in the core and mantle. Jupiter and Saturn both have internal heat sources driven by the release of gravitational energy as mass falls toward the center of the planet. As a fellow gas giant, Neptune is expected to be similar to Jupiter and Saturn in this regard. The internal energy drives the most powerful winds in the solar system—up to 1,250 miles per hour (2,000 km/h).

Right: *The Great Dark Spot, a huge storm in Neptune's atmosphere.*

Full view of Neptune with the darker Great White Spot and surrounding brighter areas.

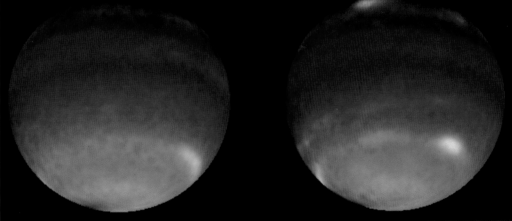

Right: False-color images of Neptune's opposing hemispheres, showing atmospheric conditions. The dark blue areas indicate Neptune's lower level methane atmosphere, with the higher clouds showing white and the highest yellow-red.

Voyager 2 is the only probe to have visited Neptune, arriving in 1989 after examining Uranus in 1986. It was able to confirm the existence of a ring system, which had been thought previously to consist only of incomplete arcs. The rings were, in fact, complete—the arcs were simply areas of greater brightness. In general the rings are thin and composed of dark dust. The areas of greater brightness were seen by *Voyager* as clumps of matter, possibly held in place by small shepherd moons.

Because of the internal heat, there is more meteorological activity on Neptune than on Uranus. White methane-cloud streaks across the planet cast shadows on the opaque blue atmosphere below. A huge hole in the planet's atmosphere was named the Great Dark Spot because it looked similar to the Great Red Spot storm on Jupiter. The fastest wind speeds were recorded in this dark patch, which is roughly the size of Earth. It disappeared in 1994 and has not yet returned, although the Hubble Space Telescope has seen other dark spots since.

Triton

Neptune has 13 known moons. Four orbit within the rings and are likely to be responsible for the observed clumping effects. Triton was discovered shortly after Neptune. It is the coldest recorded object in the solar system, with a surface temperature of only 35 K. At this temperature nitrogen, methane, and carbon dioxide all freeze solid, forming an ice cap over the moon. The low temperature is caused by Triton's high albedo, reflecting three-quarters of the Sun's energy back into space. It has a very thin atmosphere of nitrogen.

Triton's surface has very few impact craters, suggesting that some process renews it. Despite being so cold, Triton is volcanically active, although not in the same way as Earth.

Heat from the interior melts the frozen gases, causing them to erupt through the ice cap. These ice volcanoes spew liquid nitrogen and dark rocky dust up to 5 miles (8 km) high.

Triton is considered to be a captured satellite rather than one that formed alongside Neptune. Its orbit is retrograde, moving around Neptune in the opposite direction to the planet's spin. No other large moon in the solar system has that property. Although its orbit is currently a near-perfect circle, as a captured body its initial orbit would have been very eccentric. The tidal forces produced by being so close to Neptune that heated up the interior of the moon also eroded its orbit to a circular shape.

PLUTO AND CHARON

Discovered in 1930, Pluto is named for the Roman god of the underworld. Although historically regarded as a planet, Pluto is now classified as a Dwarf planet (under the new IAU solar system classification) more closely associated with Kuiper Belt objects or Trans-Neptunian objects.

Our knowledge of Pluto is much more limited than that of the planets. It has not yet been visited by a probe, but NASA's *New Horizons* spacecraft is expected to arrive in 2015. Its great distance from the Sun means that even our most powerful ground- and space-based telescopes can discern very little detail. Its size and mass are not known with any great accuracy. It is known to be smaller than all the planets, even smaller than the seven largest moons—in size order Ganymede, Titan, Callisto, Io, Earth's Moon, Europa, and Triton.

Pluto's orbit is unusual. It spends most of its year outside the orbit of Neptune, but its eccentricity takes it inside for about 20 years per orbit. This occurred for the years between

1979 and 1999 and will not happen again until 2226. Like Uranus, Pluto has an axis of rotation that is nearly at right angles to its plane of orbit. The orbit is inclined by 17 degrees to the plane in which all the other planets (except Mercury) lie.

Because of the eccentric nature of its orbit, the temperature on Pluto varies considerably. When it is near the Sun, it has a tenuous atmosphere of nitrogen with traces of methane, carbon monoxide, and carbon dioxide. As it moves away, the atmosphere freezes onto the surface. A lot of the gases have been frozen onto the southern polar cap, making it appear bright in the low-resolution images we have of it.

Initial estimates suggested that Pluto was larger than Mercury. That was because images of the distant planet could not distinguish between Pluto and its relatively large moon. It was discovered to be a separate body in 1978 by James Christy, after close examination of images showed a bulge appearing and disappearing regularly on Pluto's flank. Charon is named

PLUTO	
Type: Kuiper Belt Object or Planet	
Mass: 1.25×10^{22} kg	
Equatorial diameter: 1,585 mi (2,390 km)	
Density: 1.75 g/cm³	
Surface gravity: 0.059	
Surface temperature: 33 K (–400°F/–240°C) to 55 K (–361°F/ –218°C)	
Magnitude: 15.12	
Mean distance from the Sun: 3,674,660,000 mi (5,913,520,000 km)	
Orbital eccentricity: 0.2482	
Pluto year: 248.5 Earth yrs	
Pluto day: 6.39 hrs	

DATA

Mass: 2.15 x 10²² kg

Magnitude: 13.47

Equatorial diameter:
1,700 mi (2,700 km)

Density: 2.07

Rotation period: 5.88 days

Orbital period: 5.88 days

Radius of orbit:
220,500 mi (354,800 km)

**Average surface tempera-
ture:** 35 K (−397°F/−
238°C)

Triton with Neptune behind.

CHARON

Diameter: 749 mi
(1,205 km)

Mass: 1.58 x 10²¹ kg

Density: 1.73 g/cm³

Orbital radius:
12,060 mi
(19,410 km)

Orbital period:
6.39 Earth days

Rotational period:
6.39 Earth days

*Nearest view ever
of Pluto and its
moon, Charon.*

after the boatman in Greek mythology who ferried souls to the underworld over the Styx River. The mythological figure's name is pronounced with a hard "ch" sound, but the moon was named with a soft "ch." The name therefore sounds like Sharon (which was the discoverer's nickname for his wife).

Charon is not the only confirmed moon of Pluto—two other tiny satellites have been discovered. The presence of Charon has allowed astronomers to make better estimates of Pluto's mass by examining the orbit of the two bodies around each other. Some would classify the Pluto–Charon system as a

double planet, since they are similar in size. Charon is tidally locked to Pluto in the same way as many moons are to their parent planet, but Pluto is also locked itself. Both objects therefore keep the same face to each other at all times.

It is not known exactly how the Pluto–Charon system formed. The two bodies have different densities, suggesting that they were not created in the same way. One possibility is that Charon formed when a huge collision disrupted Pluto, in a similar fashion to the way in which Earth's Moon was created.

OUTER BELTS AND COMETS

The solar system is more than a group of eight planets clustered around a central star. Outside even Pluto's orbit countless chunks of ice orbit the Sun. They are ancient objects, flung out of the solar system as it was forming. Occasionally one is flung back in as a comet.

Kuiper Belt

The Kuiper Belt is a region of space at the edge of the solar system, similar to the asteroid belt between Mars and Jupiter. It begins just within Neptune's orbit and extends to 50 astronomical units from the Sun (an astronomical unit is equal to the distance between the Sun and Earth). It is confined to the plane of the solar system. The region has a sharp outer edge where the number of objects drops off rapidly. This has led to speculation that a body roughly the mass of Earth is cleaning up anything straying from that outside edge, much like the shepherd moons in Saturn's rings. If this hypothetical object exists, it has not yet been discovered.

The objects in the Kuiper Belt are all made of dirty ice. They tend to be very dark, covered in a carbon-based substance that has a slight red coloration. Since 1992 more than 1,000 Kuiper Belt objects (KBOs) have been recorded, of which 70,000 are more than 62 miles (100 km) in diameter. Pluto is on the verge of being reclassified as a KBO, along with its moon Charon. After them, there are several other KBOs currently known that are of a similar order of magnitude in terms of size. Quaoar, which orbits just outside Pluto, has a diameter of 745 miles (1200 km). Computer-based predictions suggest there may be several bodies of this size, each with a volume greater than all the asteroids in the asteroid belt put together. There are thousands upon thousands of smaller bodies, some of which get thrown into the inner solar system through gravitational interaction and become comets.

Oort Cloud

In 1950 Dutch astronomer Jan Oort tried to explain the source of comets. He discovered that many were in exceptionally long orbits around the Sun—some had the farthest point of their orbits nearly a light-year from the Sun. Oort imagined a spherical cloud of millions of comet nuclei, surrounding the solar system.

Halley's Comet

Halley's comet is named not for its discoverer but for the person who realized that the blurry apparition in the night sky was an object in an orbit that would bring it back to Earth every 76 years. In 1705 Edmund Halley studied many historical reports of comets. He noted that three sightings, many years apart, were objects on a very similar orbital path. He stated that they were the same object and that it would return in 1759. Unfortunately, he did not live to see his hypothesis confirmed.

The nucleus of Halley's comet is 10 miles (16 km) long, 5 miles (8 km) wide, and 5 miles (8 km) deep. Because it has a very low density, its structure is probably a loose rubble pile. The surface is covered in a carbon-rich substance, making it blacker than soot. Like all comets, its nucleus is surrounded by a coma of boiled-off volatile compounds produced by heat from the Sun. Such compounds also make up its pair of spectacular tails, each of which is millions of miles long. A jet of ions is ripped from the coma by the magnetic effect of the solar wind. This tail can appear blue, and points directly away from the Sun. A second tail consists of dust particles shed by the coma through radiation pressure. Data gathered during the 1986 visit of the comet to Earth show that four-fifths of the material that comes off the comet in jets is water. Carbon monoxide, methane, and ammonia make up the rest.

Halley's comet has featured several times in recorded human history. Sightings of comets were often viewed as omens. Its visit in 1066 is depicted in the Bayeux Tapestry, which recorded the death of King Harold II of England at the hands of Norman invaders. The *Giotto* probe was named for an Italian painter who, having viewed the 1301 visit, depicted the Star of Bethlehem as a comet in his painting *Adoration of the Magi.* It is mentioned in records of the observations of Chinese astronomers as far back as 240 B.C. The next visit will be in 2061.

Halley's comet, showing its spectacular tail.

Artist's view of the Kuiper Belt Object called 1998 WW31. Its companion is in the background. They orbit the Sun every 301 years.

The Oort Cloud is thought to reach halfway to the star nearest the Sun. It is affected by giant molecular clouds (accumulations of cold hydrogen) that inhabit the regions between the stars, as well as by the gravitational tidal effects from the galactic core. These influences can strip the comets off into interstellar space or send them into a trajectory that passes into the inner solar system. Once they get close to the Sun, comets either smash into a solid body or eventually evaporate. If there are comets still in the Oort Cloud, there must have been a huge number to begin with. Some estimates put the number of objects in the Oort Cloud as high as 10^{12}.

No direct observation of the Oort Cloud has been made, since the objects are too small and too far away. Only one object that may be a part of the cloud has been seen—Sedna is slightly smaller than Pluto and takes nearly 12,000 years to orbit the Sun once. Its farthest point is estimated at 942 times the Earth–Sun distance. But it is still too close to be within the predicted area of the Oort Cloud. Arguments continue as to whether Sedna is truly a part of the Oort Cloud or a distant KBO.

Hale-Bopp

The comet Hale-Bopp was discovered simultaneously by Alan Hale and Thomas Bopp in 1995. At that time it was between Jupiter and Saturn and was already showing a coma of ice and dust. No other comet has been spotted at such a great distance.

With a nucleus 31 miles (50 km) across, Hale-Bopp is the largest known comet. As it reached the point of closest approach to the Sun in 1997, it was brighter than all the stars in the sky except Sirius. Its two tails reached nearly 40 degrees across the sky, although the naked eye could only see part of their length. Had it passed Earth as closely as some comets, it would have been brighter than the full Moon. However, it never came within 100 million miles (160 million km) of our planet. Few comets would be visible to the naked eye at such a range, but Hale-Bopp was bright enough to be seen unaided for 18 months. It faded from view at the end of 1997 and will return in the year 4380.

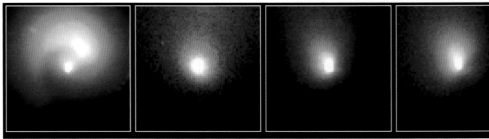

Tempel 1 and *Deep Impact*

Comet Tempel 1 was discovered by French astronomer Ernst Tempel in 1867. It orbits the Sun in the inner solar system every 5.5 years. Like all comets, it is composed of ice, gas, and dust. In January 2005 a unique space probe was launched to perform one of the most complex experiments ever in space. The *Deep Impact* spacecraft was sent on a journey to meet comet Tempel 1 and to fire a smaller "impactor" spacecraft into its path and surface. On July 3, 2005, the spacecraft bullet achieved its mission of colliding with its cometary target.

The parent spacecraft recorded images of the impact, while the impactor captured images of the comet's nucleus up to the time of the collision. The impact occurred on the sunny side of the comet, which illuminated the vaporized ice and dust that was ejected from the crater.

The ejected material was analyzed by telescopes and found to contain unexpected compounds, such as carbonates and aromatic hydrocarbons. How these chemicals came to exist in the primordial solar system is still a mystery.

A second flyby of the comet was made February 14, 2011, by the NExT mission, renamed from Stardust, to obtain a current image of the impact-site from the Deep Impact mission.

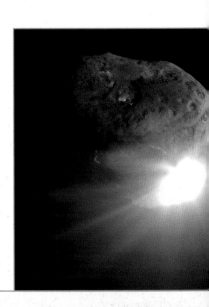

Far right: *Artist's impression of the point of impact with the* Deep Impact *spacecraft above the surface of Tempel 1.*
Right: *A bright splash of light seen by* Deep Impact *67 seconds after the impactor crashed into Tempel 1. Features on the comet can be seen clearly.*

Comets are believed to be nuggets of ancient material that date back to the formation of the solar system. Hale-Bopp's composition was determined by Earth-based observations. Astronomers were looking for the ratio of water ice to carbon-monoxide ice, which would provide clues as to where the comet was formed. They now estimate that Hale-Bopp's birthplace was between Jupiter and Neptune. After its formation it would have been flung out of the inner solar system through gravitational interaction with a planet, probably Jupiter. It would then have resided in the Oort Cloud for some time until another chance push sent it plummeting back toward the Sun.

Left: *Comet Hale-Bopp in the sky behind space shuttle* Columbia *as it awaits launch.*

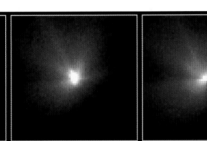

Left: *Series of images of Hale-Bopp from (far left) just beyond the orbit of Jupiter to its closest position to Earth (far right), showing changes as it is warmed by the Sun.*

Other Comets

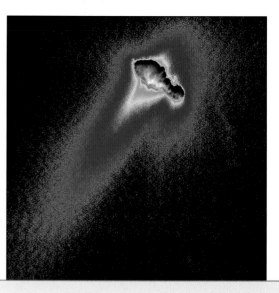

False-color image of comet Borrelly, showing typical comet structure. Dust jets are seen escaping behind the comet's solid nucleus, with a cloudlike coma of dust and gases also surrounding the nucleus. Borrelly's nucleus is about 25 miles (40 km) in diameter. The comet orbits the Sun every 6.9 years in an elliptical orbit between Earth and Mars.

Infrared view of comet Encke riding along its trail of debris of larger particles (long diagonal line). The trail encircles the solar system, and Earth passes through it periodically, resulting in the Taurid meteor shower. Encke is a short-period comet that orbits the Sun every 3.3 years. Its return in 2005 was the 59th time it had been recorded. Twin jets of smaller material can be seen spreading horizontally from the comet.

SPACE TELESCOPES

Earth's atmosphere is a shield that has protected life from dangerous amounts of radiation as well as from minor meteors and the solar wind. However, the shield also blurs and in some cases blocks our view of the external universe. Telescopes outside of the atmosphere are required in order for astronomy to advance.

Kepler's Supernova Remnant seen through the "eyes" of different telescopes: Chandra X-ray Observatory (below left), Spitzer Space Telescope (below center), and Hubble Space Telescope, viewed in visible light (below right).

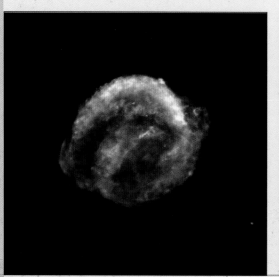

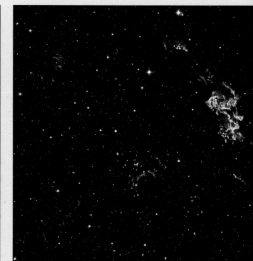

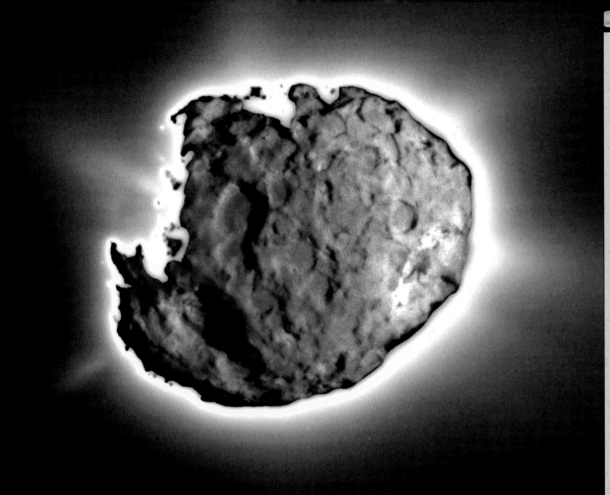

Close-up view of comet Wild 2, showing several large depressed areas. The intensely active surface is jetting dust and gas streams into space and leaving a trail millions of miles long. Wild 2 formed near Pluto and only recently reached the inner solar system. It is about 3.1 miles (5 km) in diameter. The image was taken at a distance of 147 miles (236 km) by the Stardust spacecraft, which survived a bombardment of millions of high-speed dust particles, some of which it collected and returned to Earth.

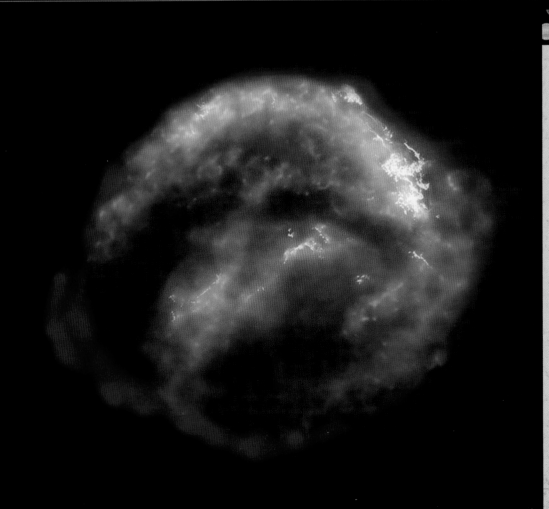

Right: *Composite image of X-ray, infrared, and visible-light observations.*

Hubble Space Telescope

The Hubble Space Telescope (HST) is one of the most important tools in the history of astronomy. It allows us to see farther and with more detail than any earlier telescope. The amazing pictures it has produced have arguably done more to raise public awareness of and interest in the universe outside Earth than any individual or any professional or scientific organization could have done.

Astronomers aiming for clarity in their observations have to cope with the distorting effect of the atmosphere. Stars in the night sky appear to twinkle because the light from them is being bent slightly by the air, in much the same way as water changes the perceived angle of a half-sumerged stick. The atmosphere is turbulent, and the distorting effect changes continuously, making it impossible to get a sharp image. Additionally, infrared, ultraviolet, and X-ray light that would provide much useful information is absorbed by Earth's protective atmospheric layers. The HST avoids the problem by orbiting on the very edge of the atmosphere but close enough to be serviced by the space shuttles.

Launched in 1990, the HST was not an instant success. Images returned from the telescope were no clearer than those achieved by ground-based telescopes. The light from stars was still being spread out across wide areas of images. The cause was discovered to be the telescope's primary mirror. This 7.8-foot- (2.4-m-) wide circular sheet of glass had taken over two years to shape and polish into a slightly concave shape. To work properly, the mirror could not be more than 30 nanometers away from the planned shape. Analysis of the flawed pictures showed that although the mirror was smooth, it had been polished to an incorrect shape. The focus of imagestherefore did not match up with the other components of the telescope.

The HST was in danger of being a very expensive failure, but the optical aberration was a solvable problem. The first service mission took place in 1993, when the space shuttle *Endeavour* docked with the HST. Astronauts installed corrective systems, known collectively as COSTAR (Corrective Optics Space Telescope Axial Replacement). At the start of 1994 the first sharp images were released. Galaxies and nebulae were revealed in images with high resolution and stunning color.

COSTAR has since been used to correct the aberration of the primary mirror for all of the remaining instruments on

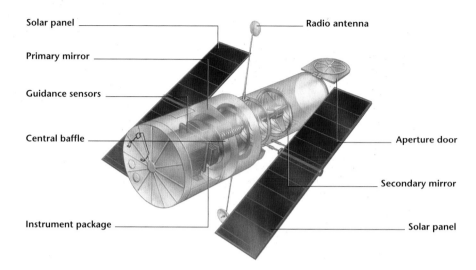

Solar panel — Radio antenna
Primary mirror
Guidance sensors
Central baffle — Aperture door
Secondary mirror
Instrument package — Solar panel

Hubble, with the exception of WFPC2, which has its own corrective optics. After the first servicing mission, most of the images from Hubble were made with the WFPC2 instrument, including the significant S/L-9 images as well as those of the Eagle Nebula. This continued until the installation of ACS in 2002. When ACS failed in 2006 WFPC2 again took over until its replacement by WFC3 in 2009.

The HST has not just been used for capturing pretty pictures. It has been gathering huge amounts of data for cutting-edge science. Shortly after the COSTAR package allowed the telescope to see clearly, comet Shoemaker-Levy 9 crashed into Jupiter.

Despite the impact being on the far side, the HST was able to view the breakup of the comet and the huge bruises it left on Jupiter's surface. This event provided information about the planet and the nature of comets. One of the planned missions for the HST had been to measure the distance to Cepheid variable stars, an important yardstick in determining the Hubble constant—the key to the age of the universe. (American astronomer Edwin Hubble had found evidence that the universe was expanding and that galaxies outside our own existed.) The value for the Hubble constant was found to be about 73 km/s/Mpc, and scientists have calculated the age of the universe as 13.7 billion years. Before the HST, it had not been possible to determine either figure with any degree of accuracy.

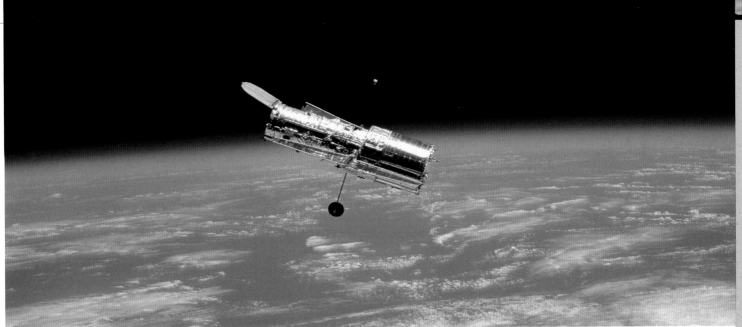

*The Hubble Space Telescope
seen from the space shuttle
shortly after its redeployment
during the second servicing
mission in 1997.*

*Attached to a shuttle's robot arm,
the Hubble Space Telescope is
prepared for redeployment after the
second servicing mission in 1997.*

Spitzer Space Telescope

The final mission in NASA's great observatories program, the Spitzer Space Telescope, named for Lyman Spitzer's (1914–1997) contributions to infrared astronomy, was launched in 2003. Spitzer, which would later be instrumental in the development of Hubble, was designed to view the universe in the infrared part of the spectrum. All objects whose temperature is above absolute zero emit some infrared light as heat, including the telescope itself. Obtaining clear infrared images is therefore a challenge. The imaging section of the telescope must be as cold as possible to minimize "background noise" from the heat of the instruments. Liquid helium is used to reduce the temperature to 5.5 K (–450°F/–268°C). The spacecraft itself is not in orbit around Earth, because reflected heat would overwhelm its viewing ability. Instead, it shares Earth's orbit around the Sun, but it lags behind our planet at a safe distance. The outer shell is designed to minimize the amount of heat absorbed from the Sun. The internal telescope structure is made primarily of beryllium, a lightweight but stiff metal that cools easily.

Infrared light passes straight through the interstellar dust clouds that can hide details from visible light observatories such as the Hubble Space Telescope. New stars are born in dust clouds, and Spitzer can peer inside, viewing the formation of a distant planetary system. Greater details can be seen in the hearts of galaxies when the lanes of dust between spiral arms are transparent.

Some objects in the universe that are interesting to astronomers are not hot enough to emit visible light. Brown dwarfs are a good example. They do not contain enough mass to initiate the fusion reactions that drive a star like our Sun—the temperature and pressure are not high enough. With surface temperatures of up to 3000 K they are, however, hot enough to shine in the infrared and can be observed by Spitzer.

Infrared spectroscopy is used in chemistry to determine the

Chandra Observatory

The Chandra X-ray Observatory is named for the Indian physicist Subrahmanyan Chandrasekhar. He calculated the maximum mass of a white dwarf, above which gravitational pressures would crush it into a neutron star. X-rays are high-energy photons and are emitted from some of the most violent events in the universe. One of the processes that generates them is when gas is heated up to millions of Kelvin. This can occur when material falls onto the surface of a neutron star or into a black hole. The gravitational energy released can be huge, far more than that generated by a normal star. Such events are not always very bright in the visual region of the spectrum, and so a dedicated X-ray telescope is required to view them.

X-rays are blocked by Earth's atmosphere. They are also very difficult to focus onto a viewing surface. Lower-energy light, such as visible or infrared, will reflect off metal mirrors perfectly. X-rays are so energetic that they can be absorbed by even the best reflecting surfaces. Imagine a brick wall as the reflector. A visible light ray is like a tennis ball that will bounce easily and predictably, but an X-ray is like a bullet that will smash straight into the brick (and potentially out the other side). For an X-ray to alter its path, it is necessary for it to hit the reflector at a very shallow angle, allowing it to ricochet off in the correct direction. The mirrors in Chandra are four nested circles, their edges almost parallel to the incoming radiation. They are the smoothest and cleanest mirrors that have ever been built.

DATA	
Altitude:	9,942 mi (16,000 km) to 82,646 mi (133,000 km)
Orbital period:	64 hrs 18 mins
Mass:	10,584 lb (4,800 kg)
Length:	45 ft (13.8 m)
Launched:	July 23, 1999
Power usage:	2350 watts

Artist's impression of Chandra in orbit.

Artwork of Spitzer Telescope seen against the infrared sky with a band of light depicting the Milky Way.

DATA

Orbit: Earth-trailing helio-centric

Mass: 2,095 lb (950 kg)

Length: 13 ft (4 m)

Launched: August 25, 2003

Power usage: 427 watts

compounds present in samples. Bonds between atoms and molecules absorb and emit specific frequencies of light. By analyzing the light that passes through a sample or is emitted by an object, its chemical composition can be determined. These light frequencies lie in the infrared, and Spitzer is the perfect tool for examining the chemical makeup of astronomical objects. It has been able to discover the compounds that make up a protoplanetary disk, giving us insight into how our own solar system was formed.

Spitzer also joined the Deep Impact mission in examining the chemistry of comets. The *Deep Impact* space probe sent a 816-pound (370-kg) weight into the comet Tempel 1 at a speed of 23,000 miles per hour (10 km/s), throwing up a cloud of debris. Spitzer examined the debris and discovered surprising ingredients, such as clay, carbonates, and aromatic hydrocarbons.

International Space Station

In 2001 the Russian space station Mir was taken out of orbit after 15 years of providing a platform for humanity in space. It was the first modular space station, made of components put into orbit by multiple launches. That modular system was reused for Mir's successor, the International Space Station (ISS). Russian plans for Mir-2 and American plans for a new station named Space Station Freedom were combined—with assistance from other countries—to build the ISS.

The ISS is the largest artificial satellite and it is still only half built. The first section was launched in 1998. Since then another 12 sections have been added. More laboratories are planned, along with extensive solar panel arrays that will power experiments. When it is completed, the station will have a mass of 816,000 pounds (370,131 kg) and will measure 356 feet (108.4 m) long. Today the ISS can support a crew of three people, but that will increase to six.

The ISS is used as a floating science lab, capable of performing experiments in a zero-gravity environment. Growing a pure crystal of a protein, for example, is easier without the influence of gravity. Crystals are helping scientists understand the structures of viruses and enzymes. The effect of weightlessness on living cells and humans is another important area of research for future crewed space exploration. Alongside the scientists, three paying customers have visited the ISS so far, marking the start of space tourism. The ISS will be completed in 2011.

Altitude: 219 mi (353 km)

Orbital period: 92 mins

Mass: 816,000 lb (370,131 kg)

Internal volume: 12,700 ft³ (425 m³)

Launched: (First module) November 20, 1998

International Space Station seen from space shuttle Discovery *after a crew change in 2001.*

UNCREWED PROBES

Humankind has set foot on only two heavenly bodies: Earth and its moon. But machines—the tools of humans—have reached much farther. Space probes, both crewed and uncrewed, have revolutionized the study of the Moon as well as planets and other bodies in the solar system, which until the 1960s could be studied only through terrestrial telescopes. The Apollo missions of the 1960s and 1970s returned space-craft to Earth and brought rock samples from the Moon. In 2006 the Stardust mission capsule returned to Earth and crash-landed in Utah. It brought back interstellar dust from outside the orbit of the Moon.

Left: *After a 2.88-billion-mile (4.5-billion-km) mission in space, the Stardust capsule crash-landed as planned in Utah in January 2006. Launched in 1999, the spacecraft collected particles from comet Wild 2 in 2004 and interstellar space on the way. The capsule was released from the spacecraft while in orbit around the Earth and parachuted down to the ground.*
Right: *An artist's depiction of Stardust's encounter with comet Wild 2.*

Transporting a package of scientific instruments to another world is difficult. Landing them safely is even harder. For each successful landing there have been many landers that have been destroyed in the atmosphere, crash-landed, or even missed the destination planet entirely. Controlled soft landings have been achieved on other planets and surface measurements have been taken.

Orbital probes are the first stage in exploring another planet. They can scan the surface looking for clues as to the history of the planet and search for good landing spots for subsequent probes. Close-up images from orbiters surpass anything that can be achieved using Earth-based telescopes. Photography at visible and invisible wavelengths, imaging radar, magnetic field instruments, spectroscopy, and other forms of instrumentation have provided us with vital data from orbiting space probes and flybys.

As well as adding to our knowledge of the planets themselves, the Pioneer series of probes in particular provided a great deal of information about interplanetary space. Other probes have given us important facts about the Sun. Since 1991, the Hubble Space Telescope has sent back vital data on the planets and provided images of objects in deeper space.

Several probes have already left our solar system. It is hoped that useful data will be received from *Voyager 2*, which is well beyond the orbit of Pluto. In 2006 NASA launched the New Horizons mission, which is expected to reach Pluto in 2015 and the Kuiper Belt five to ten years later.

Left: *Artist's concept of the European deep space probe* Giotto, *which took photographs of Halley's comet in 1986.*

Above: *Artist's concept of* Deep Space 1, *launched in October 1998 primarily to test new technologies for future deep space missions. It took photos of asteroid Braille and later of comet Borrelly.*

Left: *Image of the impact site (far right) of Mars Exploration Rover Opportunity's heat shield. The inverted heat shield is seen far left, with debris toward the center.*

Right: *A U.S. Viking lander, two of which arrived at Mars in 1976.*

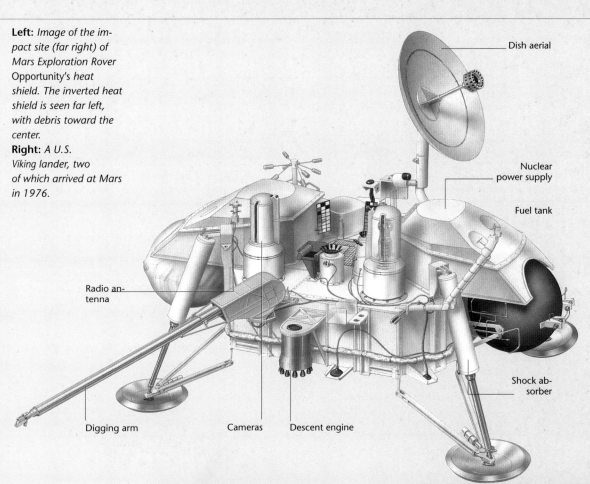

Dish aerial

Nuclear power supply

Fuel tank

Radio antenna

Shock absorber

Digging arm

Cameras

Descent engine

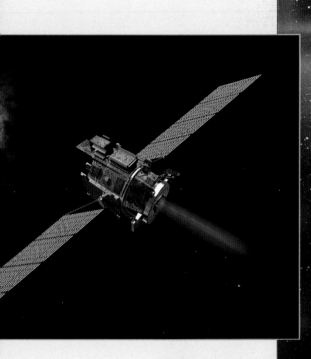

Right: *Artist's concept of the* New Horizons *spacecraft, launched in January 2006 on a mission to Pluto and its moon Charon.*

Mars Lander—*Viking*

The first successful soft (controlled) landing on the surface of Mars was the Soviet-made *Mars 3*. It landed in 1971, a few days after its sister craft *Mars 2* had crash-landed onto the surface after a malfunction. Unfortunately, *Mars 3* returned data for only 20 seconds before contact with the probe was lost, so very little information was received.

The first useful information from the surface of Mars was sent by the two U.S. Viking probes. Launched in 1975, each probe consisted of an orbiter and a lander. After arriving in 1976, the two landers were released. Parachutes were used to slow their descent through the Martian atmosphere, and retrorockets were fired to slow them to a speed of just a few feet per second as they touched down.

Viking 1 and *Viking 2* sent back color pictures of their surroundings. They analyzed the atmosphere and planted seismometers on the surface of Mars. Diurnal temperatures at the *Viking 1* landing site ranged from 184 K (−89°C/−129°F) to 242 K (−30°C/−22°F). The landers survived long enough—nearly four years in the case of *Viking 2*—to record seasonal variations in the temperatures and dust-storm activity.

The most famous part of the Viking program was the search for life on Mars. Both landers carried biological experimentation packages for testing samples of Martian soil. The purpose was to look for gases that might be consumed or released by bacteria. The tests came up with potentially positive results, but they are likely to have been simple chemical reactions within the soil itself. The final conclusions are still being debated.

A photo taken by Viking 2 *from its position on Mars's Utopian Plain reveals that the surface is strewn with boulders as far as the horizon 2 miles (3.2 km) away. The colors of the surface and the sky are thought to be true colors.*

Mars Lander—*Pathfinder*

Before 1996 fewer than one-third of the landers and orbiters sent to Mars had fulfilled their missions. So many probes were lost or destroyed that people began to talk of a "Mars curse" that prevented us from exploring our nearest planetary neighbor. The curse was given the nickname "the Great Galactic Ghoul"—a monster that devoured millions of dollars' worth of spacecraft with no warning. NASA therefore began to focus on simpler, lower-cost missions. The Mars Pathfinder program was one of the first examples, costing relatively little—$150 million—to develop.

In order to land *Pathfinder* safely on the Martian surface, parachutes and retrorockets were used as in the Viking missions. But *Pathfinder* had a new trick. At about 33 feet per second (10 m/sec), just before impact, 24 large airbags inflated and cushioned its landing. After the craft had bounced several times, the airbags deflated and the probe unfolded itself to reveal the rover *Sojourner*. It was the first time a rover had been used on the surface of another planet. The tiny robot, weighing just 22 pounds (10 kg), was designed to drive up to Martian rocks and analyze them.

Rocks were named after cartoon characters: Barnacle Bill was a volcanic rock similar to andesite on Earth, Yogi was basaltic. *Sojourner* accumulated evidence that the rocks in the region had been deposited in a huge flood, leading to the conclusion that Mars had been wetter and warmer in the past.

Despite being expected to last only a week on the cold and dusty surface of the planet, the rover returned data for three months in total. Its success proved that the technologies and techniques it used were sound and enabled NASA to develop and launch two successors to *Pathfinder*—named *Spirit* and *Opportunity*.

A color-enhanced photo taken by the Imager for Mars Pathfinder (IMP). The large rock being investigated by the rover Sojourner at the top right of the photo has been named "Yogi."

Mars Landers—*Spirit* and *Opportunity*

Following on from *Pathfinder* came the Mars Exploration Rovers, *Spirit* and *Opportunity*. The new rovers used the technology tested successfully by the Pathfinder mission, but they were even larger and more capable. Their mission was to examine the geology of Mars, to classify the rocks and dust, and potentially to provide scientists with greater understanding about the presence of water.

The probes were launched in 2003 and landed on Mars in 2004. *Spirit* came down in Gusev Crater, and *Opportunity* landed on the other side of the planet in Meridiani Planum. Both sites were chosen for their possible associations with water in the past. Gusev Crater was once thought to have been a lake. Meridiani Planum has a large amount of the mineral hematite, which on Earth is often formed in water.

The rovers are twins, with identical hardware and capabilities. They have several cameras for panoramic and close-up images as well as for navigating the rock-strewn surface. An instrument mounted on the robotic arm of each vehicle is able to grind away the surface of rocks to allow the spectrometers to analyze their composition. Power for such operations comes from solar panels that can provide 140 watts for about four hours a day when the Sun is overhead. The Martian surface is a very dusty place, and there is a great deal of dust in the atmosphere, which limits the amount of energy that the rovers can receive. Eventually the dust is expected to damage the rovers, coating the panels to the point where they will no longer be able to function. This has not happened as early as predicted, however, because dust is being blown off the solar panels by gusts of wind, giving the rovers an extended life span. The mission was expected to last for 90 days, but *Spirit* and *Opportunity* were still moving and working in January 2006, more than 650 days after they first reached Mars.

Venus Landers

Landing a probe on the surface of Venus softly is only half the battle. Before it can even land, it must be able to withstand temperatures capable of melting lead and atmospheric pressures 90 times greater than that on Earth. Several probes have been lost as they descended through Venus's thick corrosive clouds.

The only attempts to land on Venus were part of the Soviet Venera program. It began in 1961 with *Venera 1*, which was able to confirm earlier theories about the existence of the solar wind, but it stopped responding well before it reached Venus. *Venera 3* was the first human-made artifact to reach another planet, crash-landing in 1966. Later probes gathered information from within the atmosphere by descending on parachutes. Each one was crushed by the tremendous pressure as it neared the surface. In 1970 the first probe to survive the descent and return data to Earth was *Venera 7*. It lasted for just under an hour.

Right: *Image of the surface of Venus taken by* Venera *14's lander. The area is made up of basaltic rock. Part of the lander can be seen near the bottom of the image, and near the center is a lens cover.*

Left: *An artist's impression of a Mars Exploration Rover on the surface of Mars.*
Right: Opportunity *looks back over its wheel tracks and discarded descent cocoon lying in Mars's Eagle Crater.*

Left: Venera 13*'s lander.*

The first pictures were returned by *Venera 9*. They provided a close look at the Venusian surface. Problems with the design of the lens caps on the cameras, however, meant that only one picture each was taken by *Venera 9* and *Venera 10*. *Venera 11* and *Venera 12* were not able to get any images because the lens caps failed to come off, although they were able to detect lightning in the cloudy atmosphere. The record for the longest operational life of any lander on the surface of Venus—2 hours and 7 minutes—was set in 1982 by *Venera 13*. It was also the first probe to take color images of the surface. Sampling of the terrain by *Venera 13* and *Venera 14* revealed that the surface was generally basaltic rock, which typifies volcanic regions. The last probes to touch the surface were dropped by the international *Vega* spacecraft (on a mission to study Venus and Halley's Comet) in 1985.

LANDERS
Titan Lander—*Huygens*

Christiaan Huygens was a Dutch astronomer who lived during the 17th century. His contribution to the scientific revolution of the period was significant, and he made many discoveries including Titan, the largest moon of Saturn.

Titan is unlike any other moon in that it has a dense organic atmosphere. It is considered one of the potential sites for life outside Earth, which is why a probe was sent there. The *Huygens* lander was brought to the Saturnian moons by the *Cassini* probe. It was detached from *Cassini* on December 25, 2004, and it coasted toward Titan for 20 days. It was aimed at a point where preliminary images suggested the shape of a shoreline. The probe design had taken into account the possibility of landing on liquid as well as on land. After a descent through the atmosphere lasting two and a half hours, *Huygens* landed on a solid icy surface. Its lifetime was short—the battery it carried was good for only another 90 minutes.

An employee in the Payload Hazardous Servicing Facility at NASA sews thermal insulation material onto the back cover and heat shield of Huygens.

ORBITERS
Sun Orbiter—SOHO

The Solar and Heliospheric Observatory (SOHO) was launched in 1995. It carries 12 different instruments, all gathering data on the star that is at the center of our solar system. The spacecraft sits at the point between Earth and the Sun where the gravitational pull of both bodies cancels out. This is 930,000 miles (1.5 million km) from Earth, and its position ensures that SOHO has an uninterrupted view of the Sun.

Unlike the other bodies in the solar system, the Sun is not a solid, slowly changing object. It has many active phenomena that can be studied. SOHO can measure the oscillations that occur, thereby providing information on the state of the Sun's energy-generating core. Its cameras can look at the surface of the Sun where magnetic fields produce prominences and sunspots. (Such images require a specially built telescope. If the Hubble Space Telescope, for example, looked at the Sun, its delicate internal systems would be irrevocably damaged.)

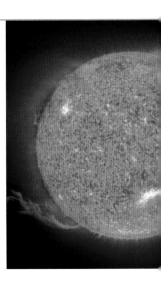

SOHO operates an early-warning system for solar flares that can damage power grids and communications systems on Earth. It has instruments on board for sampling the particles that make up the solar wind, determining their composition and density. One particular system examines the extent of the solar wind and its interaction with the cloud of interstellar hydrogen through which our solar system is moving. By blocking out the Sun's disk, SOHO can view its faint corona. This region of incredibly hot tenuous gas is not well understood.

SOHO's scientific investigations are of enormous value because they help us understand our solar neighborhood and they also provide useful data for our theoretical models of stars. The structures and life cycles of stars are important measuring sticks in many models of the galaxy as well as of the universe itself.

A radar image taken by Cassini of the surface of Titan reveals a complex geologic surface, thought to be made up of icy materials and hydrocarbons.

As the lander descended, its instruments examined Titan's atmosphere. Accelerometers measured the wind that buffeted the probe. A microphone recorded wind sounds and listened for lightning. Cameras took pictures of the landscape and saw shapes reminiscent of the erosion caused by water on Earth. Analysis of the atmosphere was carried out using a mass spectrometer, a device that identifies the type and proportions of chemicals. The speed of sound was measured, giving more information on the nature of the atmosphere.

One experiment designed to measure wind speed seemed to have been lost after one of the channels *Huygens* used to relay data back to *Cassini* failed. However, Earth-bound radio telescopes were able to pick out the faint signal from the lander and reconstruct the data. Incredible sensitivity was required to tune in to a tiny radio station broadcasting more than one and a quarter billion miles away!

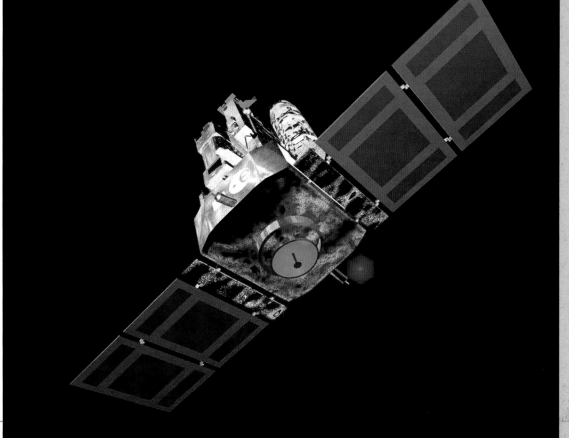

Left: *This ultraviolet image from SOHO reveals clouds of plasma suspended in the Sun's corona. The hottest areas in the image are white, while the darker red areas indicate cooler temperatures.*

Right: *With its solar panels extended, SOHO measures about 25 feet (7.6 m) across.*

ORBITERS
Mercury Orbiter—*Mariner 10*

The only probe to reach Mercury was *Mariner 10*. It was significant because it was the first spacecraft that used the mass of a planet to give it a gravitational slingshot toward its destination. This technique allows interplanetary probes to travel much farther into space, using the same initial boost from Earth. When a slowly moving probe gets near to a planet, it speeds up as gravity pulls it closer. This accelerates the probe up to the planet's speed, radically changing the shape of the probe's orbit.

Mariner 10 was launched in 1973 and passed close to Venus in 1974. As well as examining Venus's atmosphere, *Mariner 10*'s path was bent closer to the Sun, putting it on a flyby trajectory for Mercury. Slingshots require extreme precision in the initial direction of the probe. In order to reach Mercury, *Mariner 10* had to aim for a region only 250 miles (400 km) across at Venus. Several small thrusters onboard were able to correct the probe's course en route. In addition to the thrusters, the probe's solar panels were used as sails against the light outward solar wind and the radiation pressure exerted by the Sun, thus saving valuable thruster fuel for maneuvers near Mercury.

Mariner 10 was able to make three flybys of Mercury, and mapped nearly half of its surface. It had instruments designed to analyze the solar wind in order to observe any change in Mercury's locality. These instruments discovered that the planet generated a weak magnetic field. *Marnier 10*'s own orbit allowed scientists to determine the mass of Mercury with great accuracy.

Mariner 10 had to be engineered to withstand the cold of space and the heat emitted from the Sun. It had a sunshade made of Teflon-coated glass-fiber cloth and protective blankets made from the same material.

ORBITERS
Venus Orbiter—*Magellan*

The first planetary probe to be launched by a space shuttle was *Magellan* (named for the famous 15th-century Portuguese explorer and navigator Ferdinand Magellan). In 1989 the space shuttle *Atlantis* placed *Magellan* into Earth orbit. *Magellan* then fired its own engine, which sent it spiraling toward Venus.

Although spacecraft such as the *Pioneer* series and *Venera* series had studied Venus's surface during the 1970s and 80s, they could only view large-scale features. The thick clouds that cover the planet mean that the only way to get information about the surface from orbit is by radar. *Magellan* used several new radar techniques to improve on the resolution of images obtained by previous probes.

In August 1990 *Magellan* was placed into an unusual orbit that took it over both poles of Venus. The orbit was also highly elliptical, bringing the probe close to the surface of the planet for less than half the orbital time. When *Magellan* was close to the surface, it scanned a 15-mile- (25-km-) wide strip of the planet from pole to pole. As its orbit took the probe away, it beamed back the images to Earth for processing. Using this method, it mapped 98 percent of the surface over four years.

Although the radar imaging system was the only scientific instrument onboard *Magellan*, it could be used to determine other features. The nature of a radar reflection means that it can provide information on the type of material it bounces off. The radar antenna can also look at temperature variations across the surface. *Magellan*'s own orbit was analyzed to reveal any slight deviations that would indicate a density variation within Venus.

Magellan's final experiment was to test the technique known as aerobraking. It involved dipping the orbit of the spacecraft into the planet's upper atmosphere to slow it down—something that had not been tried before. In 1994 *Magellan* fell into the Venusian atmosphere and finally lost contact with Earth.

Above: *View of Mercury from Mariner 10.*
Right: *Technicians prepare Mariner 10 for encapsulation.*

Three days before launch the space shuttle Atlantis stands with its payload doors open to reveal the Magellan space probe inside.

Mars Orbiter—*Global Surveyor*

After the success of the Viking landers in the 1970s NASA did not return to Mars until the launch of the Mars Global Surveyor (the MGS) in 1996. Its primary mission was to take detailed images of the entire Martian surface. The information was used to plan later missions, including the launch of the Pathfinder, Spirit, and Opportunity rovers.

The *MGS* began orbiting Mars with a highly elliptical orbit, ranging from 163 miles (262 km) above the surface at closest approach to 34,000 miles (54,000 km) half an orbit later. Aerobraking was used to bring the probe into a closer, more circular orbit. The technique, which had been pioneered by the Venus orbiter *Magellan*, required careful positioning of the spacecraft at the top of the planet's atmosphere. The drag on the *MGS* bent one of the solar panels, so the probe was moved to a slightly higher orbit to prevent any more damage. The final orbit is a near-circular one passing over the poles of the planet. Its movement across the face of the planet coincides with the time at which the Sun shines on the surface below, ensuring that it is always illuminated.

As well as mapping the entire surface, the *MGS* has kept an eye on Mars's changing weather systems. Dust storms and the state of the ice caps during different seasons have been closely monitored. It has also acted as a data relay for the three rovers that have landed on Mars in recent times. The high-resolution camera has even been able to take pictures of the tracks left in the Martian dust by *Spirit* as well as imaging the two other recent orbiters, *Mars Express* and *Odyssey*.

The last communication was on November 2, 2006 in Safemode and the mission ended in January of 2007 with a final orbit of 11,060 miles (17,800 km) (apoapsis) and 106 miles (171 km) (periapsis).

Mars Orbiter—*Odyssey*

Launched in 2001, *Odyssey* was named for the famous science fiction movie *2001: A Space Odyssey*. Its mission was similar to that of the *Mars Global Surveyor* that was launched five years earlier. It mapped the surface, this time in infrared light, revealing the composition of the surface material. Minerals can be identified by the different frequencies of infrared light they emit when heated by the Sun. Particular attention has been paid to those minerals that are known to form in water here on Earth. Large deposits of such minerals may point to regions of Mars where lakes once existed. Such places would be the best starting points to search for life on the planet.

More information regarding the chemical makeup of Mars can be gleaned by detecting gamma-ray emissions. Cosmic rays that bombard the planet collide with atoms on or near the surface. The atoms then emit radiation that can be identified as coming from a particular element. Using this technique, *Odyssey* was able to detect a large amount of hydrogen just 3 feet (1m) below the surface of Mars. Hydrogen exists in water molecules, and in this case it has been frozen into permafrost just below the surface dust. Mars was thought to have had water at some time in the past, and *Odyssey* has discovered what happened to it.

Odyssey also measured the radiation in the area around Mars. The planet lacks Earth's protective features, such as a dense atmosphere and a strong magnetic field. Cosmic rays from the Sun and outer space could therefore pose a serious health risk to any crewed expedition to Mars. The instrument that measured the radiation was damaged by a particularly strong outburst from the Sun, underlining the potential effects of space-based radiation.

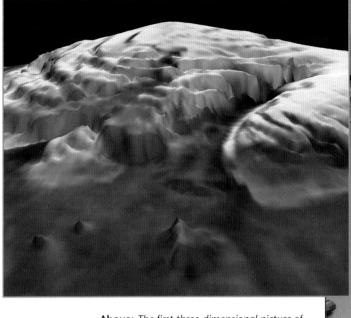

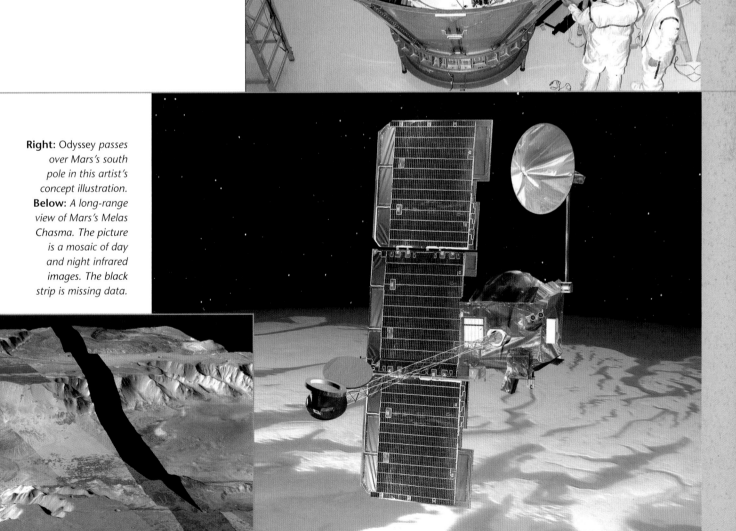

Above: *The first three-dimensional picture of Mars's north pole from the MGS in 1999.*
Right: *The MGS is placed in a protective canister for transporting to the launch pad.*

Right: Odyssey *passes over Mars's south pole in this artist's concept illustration.*
Below: *A long-range view of Mars's Melas Chasma. The picture is a mosaic of day and night infrared images. The black strip is missing data.*

ORBITERS
Mars Orbiter—*Mars Express*

The European Space Agency (ESA) launched its own Martian orbiter in 2003. The probe was designed to be built as inexpensively and quickly as possible, leading to the name *Mars Express*. The design of many of the instruments on *Mars Express* originated from an ill-fated Russian mission—*Mars 96*—which burned up in Earth's atmosphere after a malfunction in the rocket that launched the probe into its initial orbit.

Like the other orbiters that were active around Mars at the same time, *Mars Express* was designed to map and classify the nature of the planet's surface. The probe is able to take high-resolution images in color and in stereo. This provides three-dimensional views of the surface—useful for spotting large-scale features that may have been formed by the action of water.

The distribution of minerals is mapped by another instrument that uses the infrared part of the spectrum.

Ultraviolet light is typically absorbed by atmospheric components such as water vapor and ozone. A camera on the probe that is sensitive to this light has been able to map out the nature of the atmosphere, measuring variations relating to position and season. The orbiter's instruments have found small amounts of methane in the atmosphere, concentrated in areas where other observations have found high levels of water vapor and underground water ice. Using low-frequency radio waves, *Mars Express* can look several feet below the surface and reveal information about possible permafrost and the structure of Mars's sand dunes.

When it approached Mars for the first time, the orbiter released a lander called *Beagle 2*. Its mission was to search specifically for signs of life, past or present. However, the lander failed to respond to communications after its descent, and is now lost.

ORBITERS
Jupiter Orbiter—*Galileo*

In 1610 the Italian astronomer Galileo Galilei observed the four largest moons of Jupiter—Io, Europa, Ganymede, and Callisto—with one of the first telescopes. Nearly 400 years after this important event in the history of astronomy the probe bearing his name made close-up, long-term observations of the planet and its moons.

Reaching Jupiter directly from Earth is difficult, since it requires a very large amount of energy. When NASA planned the launch of *Galileo*, the decision was made to use only a small initial thrust before sending the probe on a roundabout path through the inner solar system in order to build up enough speed to reach Jupiter. After launch in 1989, its first encounter was with Venus. It took readings of the space environment around the planet as well as infrared pictures of lower cloud layers that had not been imaged before. After looping past Venus, it returned to Earth for a flyby visit during which it also observed sections

of the Moon. *Galileo* was then flung into the asteroid belt for a short time and made the first flyby of an asteroid, Gaspra. The final gravitational boost came from Earth in 1992, increasing the probe's speed by 2.3 miles per second (3.7 km/s). The flight path to Jupiter came close to another asteroid, Ida. *Galileo*'s photos revealed Ida's companion, a tiny asteroid moon named Dactyl. Finally, in 1995 the probe reached Jupiter and began its primary mission.

The *Galileo* probe was large, weighing over 5,500 pounds (2,500 kg). Its power-consumption requirements and distance from the Sun meant that it could not use solar panels. Instead, it was powered by the heat generated by the decay of plutonium. This source of power is reliable in the dark and severely cold environment and in the presence of the damaging radiation bands created by Jupiter's magnetic field.

An artist's concept shows the Galileo *space probe as it encounters the moon Io during its approach to Jupiter.*

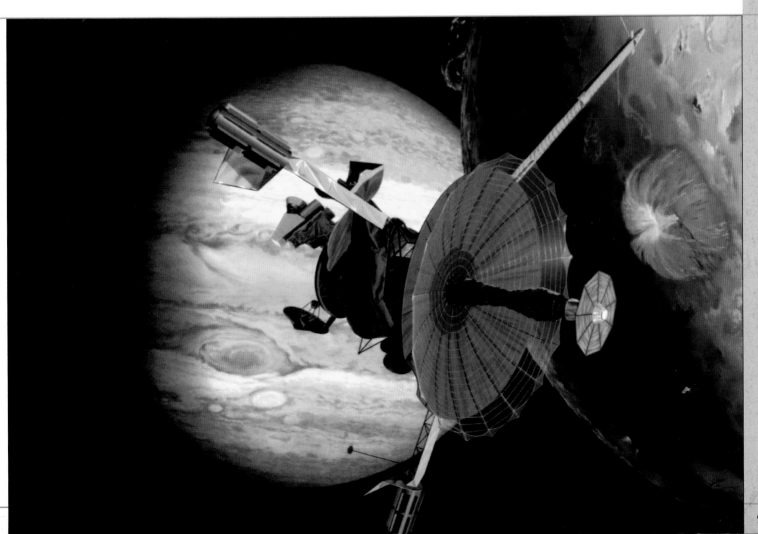

Above: *An artist's impression shows* Mars Express *orbiting Mars.*
Left: *Relayed to Earth via* Mars Express, *this false-color image of
the inside of Mars's Endurance Crater was taken by* Opportunity.

ORBITERS

Left: *the four Galilean satellites are shown to scale. In increasing distance from Jupiter they are (left to right) Io, Europa, Ganymede, and Callisto.*

Mission Details

Galileo's orbital path around Jupiter was carefully chosen to maximize the number of close-up flybys of the major moons. It was able to get highly detailed images of the surfaces of the Galilean satellites, demonstrating the rich variety in the Jovian system.

On the whole, the mission was a success, although it experienced some major problems along the way. The main antenna, through which most of the data was to be sent back to Earth, failed to unfold correctly. It had been kept stowed away throughout the trip to Venus in case it was damaged by the proximity to the Sun. The data from Venus had been transmitted through a nondirectional antenna with a lower bandwidth. The low gain antenna had to be used for the rest of the mission, resulting in a reduction of the amount of data that could be passed to listening stations on Earth. Imaging Io was expected to be dangerous, because the magnetic field of Jupiter channeled large amounts of damaging radiation in the area around the moon. But despite problems caused by the degradation of electrical components as a result of constant bombardment, the instruments survived.

ORBITERS
Saturn Orbiter—*Cassini*

The joint mission to Saturn developed by NASA and the European Space Agency (ESA) was launched in 1997. The *Cassini* probe was named for the Italian astronomer Giovanni Cassini, who made many observations of Saturn in the 17th century. It carried with it the *Huygens* descent probe, designed to fall onto the surface of Titan. Together the two probes made up the largest and heaviest interplanetary probe ever launched, with a mass of over 12,000 pounds (5,600 kg). Like *Galileo*, *Cassini* was powered by radioisotope thermoelectric generators, since solar panels would be impractical at such a distance from the Sun. The distance also posed problems in sending commands to the probe. Radio signals take an average of 75 minutes to reach the orbit of Saturn, meaning that all maneuvers must be planned meticulously in advance.

In order to reach Saturn, *Cassini* had to use multiple gravitational slingshots. Its route took it past Venus twice, back past Earth, out to Jupiter, and finally into orbit around Saturn in 2004. During the flyby of Jupiter it took the most detailed image of the planet to date. At Saturn, *Cassini* observed the planet's magnetic field and its effects on the solar wind at such as distance from the Sun. One of *Cassini*'s main functions was to examine the ring system for clues as to how it formed. In doing so, *Cassini* discovered a few new moons, including Daphnis, which orbits within a gap in the rings (Keeler gap). Unexplained spokes in the ring structures first seen by the Voyager probes were also spotted.

Cassini's orbits took it close to Saturn's moons, especially *Huygens*'s target, Titan. The surface is obscured by the opaque atmosphere, so *Cassini* was equipped with radar to image the surface through cloud.

Right: Cassini *captures an image of the moon Pandora skimming Saturn's F ring.*
Far right: *Artists's concept of* Cassini *as it moves into Saturn's orbit.*

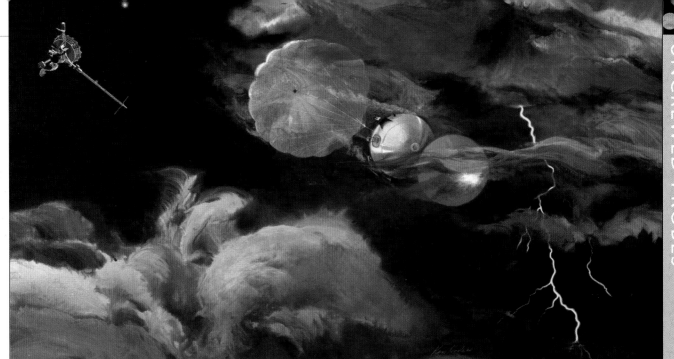

Right: *Seen here in an artist's impression, the* Galileo *probe penetrated 120 miles (200 km) into Jupiter's atmosphere before being destroyed by the pressure and temperature.*

As *Galileo* neared Jupiter it released a separate probe intended to fall into Jupiter's atmosphere and return data about the weather beneath the clouds. This probe lasted for an hour in the atmosphere before the heat proved too much for its electronics. It measured temperature and pressure to a depth of 87 miles (140 km) and wind speeds of over 435 miles per hour (700 km/h). At the end of its mission in 2003, *Galileo* was sent to burn up in Jupiter's atmosphere to prevent any possible contamination of the moons. Microbes from Earth carried by the probe could confuse any future missions searching for life.

ORBITERS
Pioneer 10 and *Pioneer 11*

During the 1970s the outer planets of the solar system lined up in such a way that a spacecraft could take advantage of the phenomenon of gravitational slingshot and travel from one to the other. Plans were made for a "Grand Tour," designed to take in all the sights that Jupiter, Saturn, Uranus, and Neptune could offer before heading out of the solar system for good. Although these plans were not fully realized, the notion resulted in the development of the Pioneer and Voyager probes.

The first spacecraft to cross the asteroid belt was *Pioneer 10*. It was launched in 1972 and aimed to make the first flyby of Jupiter. It took close-up photographs of the planet and its moons. Other instruments on the probe detected the areas of dangerous radiation near Jupiter that subsequent probes would need to survive. After visiting Jupiter, *Pioneer 10* continued to travel away from the Sun, making measurements of the environment in the outer solar system. The last signal from the probe was received in 2003. It is believed that the power has run too low for the probe to contact Earth again. By 2005 *Pioneer 10* was twice as far from the Sun as Pluto is and was heading for Aldebaran. It will take two million years to reach the star.

Pioneer 11 was identical to *Pioneer 10*. It too was sent to Jupiter, but in a trajectory that would allow it to slingshot past to Saturn. *Pioneer 11* took the first close-up pictures of Saturn in 1979 before flying on into the darkness of the outer solar system. The power that operated its scientific instruments ran out in 1995.

Both Pioneer probes are traveling away from the Sun at about 230 million miles per year (375 million km/year). However, while analyzing the probes' trajectories, scientists have noticed a tiny but unexpected acceleration toward the Sun, known as the Pioneer anomaly. It has led to speculation that the predicted trajectories are based on a flawed understanding of the theory of gravity, but there could be another explanation, such as interaction with an as yet undiscovered mass.

Above left: Pioneer 10's pictorial plaque shows, for the benefit of any other intelligent life forms, when and where it was launched and by what kind of beings.

ORBITERS
Voyager 1 and *Voyager 2*

After the Pioneer probes had proved that uncrewed missions beyond the asteroid belt were possible, the Voyager probes were launched in 1977. Their targets were the huge gas giants Jupiter and Saturn, and there was the possibility of subsequent visits to Uranus and Neptune.

As *Voyager 1* reached Saturn, the controllers decided to forego the chance of reaching Uranus and Neptune in favor of a close-up look at Saturn's major moon, Titan. The probe was able to measure the composition and density of the moon's atmosphere. Its new trajectory took it out of the plane of the planets. Because it had a greater initial velocity than the Pioneer probes, it is now the farthest human-made object from Earth, at a distance of 9 billion miles (14 billion km). The power sources on board are expected to last until 2020, and *Voyager 1* is still returning information. Messages containing data about the edge of the solar system take 13 hours to travel back to Earth.

Voyager 2 was the only probe to complete the "Grand Tour" of the outer planets (a plan that had been conceived several years earlier). It reached Jupiter in 1979 and Saturn in 1981. It was the first probe to reach Uranus (1986) and Neptune (1989). Almost all the information we have about these outer planets comes from data that was provided by *Voyager 2*. The probe left the solar system after taking a last look at Neptune's large moon, Triton. It is currently over 75 times the Earth–Sun distance away from us, traveling at 9 miles per second (15 km/s).

Both Voyager probes carry a gramophone record made from gold. These records contain messages to any extraterrestrial life forms that might recover the probes. They contain encoded images as well as greetings in 55 different languages.

A Voyager spacecraft as depicted by an artist.

Above: *A depiction by an artist of a Pioneer spacecraft heading out into interstellar space.*

THE MILKY WAY

Summer sky showing the Milky Way, seen over the Nordic Optical Telescope on the island of La Palma.

Inside the Milky Way

The hazy band of stars faintly visible in the night sky is our galaxy, the Milky Way. It is a flat disk of some 200 million stars and an unknown number of nebulae. It is 100,000 light-years across and 3,000 light-years thick. The solar system is 28,000 light-years from galactic center, on the inner edge of a spiral arm. The Milky Way is part of the Local Group of galaxies that includes the Large Magellanic Cloud.

1 *The Sun.*

2 *Sagittarius arm.*

3 *Centaurus arm.*

4 *Orion arm.*

5 *Perseus arm.*

6 *Cygnus arm.*

7 *Galactic center.*

Stylized view of the Milky Way, showing a number of key features. No matter which way we were to look from the solar system the view would always be a superimposition of spiral arms. The Milky Way appears at its most dense when we look toward the galactic center. Other views pass through different volumes of stars—some more, some less.

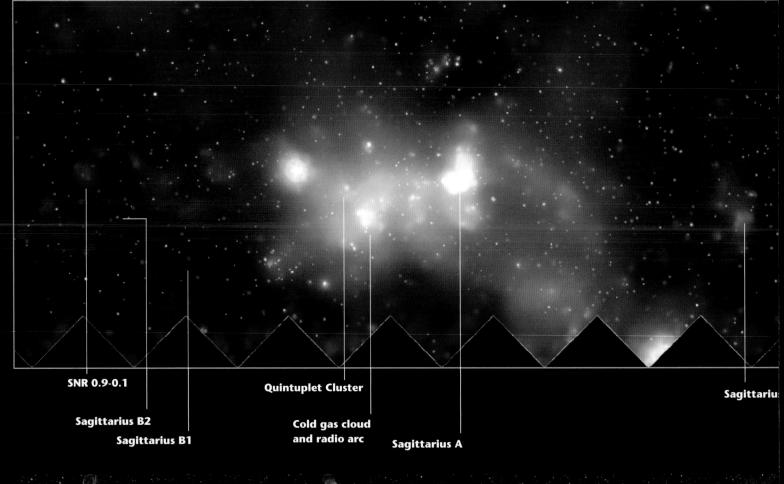

SNR 0.9-0.1

Quintuplet Cluster

Sagittariu

Sagittarius B2

Sagittarius B1

**Cold gas cloud
and radio arc**

Sagittarius A

13.2 billion years (the age of a star in the galactic halo) and not more than the age of the universe, 13.7 Gyr. It has completed roughly 21 orbits and is currently situated on the trailing edge of the Orion arm. As the name implies, this is the arm that contains most of the stars in the constellation of Orion. Some recent work on mapping the galaxy has suggested that Orion may not, in fact, be a complete spiral arm but simply an offshoot that connects the Sagittarius arm and the Perseus arm. If that is the case, our location would be described more correctly as being within the Orion bridge, or spur. The Sagittarius arm liesbetween us and the galactic center, while the Perseus arm curls around outside the Sun.

The galactic center is a place of mystery, shrouded in dust and gas clouds that cannot be penetrated by visible light. Astronomers rely on observations at other wavelengths of electromagnetic radiation. One of the brightest sources of radio emission in the sky comes from an object known as Sagittarius A. It lies at the galactic center and is thought to be a supermassive black hole. Evidence of its existence comes from the motions of nearby stars. The X-ray and radio radiation it emits and a newly discovered plume of antimatter also indicate energetic processes that can be powered only by an object as dense as a black hole.

The Milky Way is an average-sized spiral galaxy. Which type of spiral galaxy, however, is uncertain. For years it was believed to be a

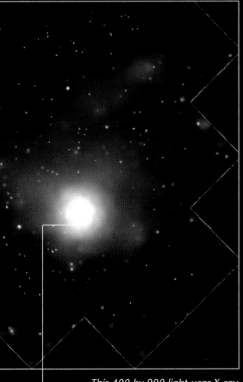

This 400 by 900 light-year X-ray image of the central region of the Milky Way galaxy contains hundreds of white dwarf stars, neutron stars, and black holes.

1E 1740.7-2942

The hazy band of stars faintly visible in the night sky is our galaxy, the Milky Way. It is a flat disk of some 200 million stars and an unknown number of nebulas. It is 100,000 light-years across and 3,000 light-years thick. The solar system is 28,000 light-years from galactic center, on the inner edge of a spiral arm. The Milky Way is part of the Local Group of galaxies that includes the Large Magellanic Cloud.

Traditionally, when people talk of the Milky Way, they are describing the misty band of light that stretches across the night sky. The Italian astronomer Galileo Galilei was the first person to look at the Milky Way with a telescope, in about 1610. He observed that it was composed of countless faint stars. Over the next three centuries astronomers came to realize that this

faint band of light is our view of our own galaxy. The reason it looks so different from the other galaxies we can see is that we are viewing it from within.

The Milky Way is a spiral galaxy and is therefore relatively flat and disklike. If we look along the plane of the disk, we see many more stars than we do by looking to each side. The Sun is not at the center of the Milky Way, but is located in one of the spiral arms. The center of the galaxy lies in the direction of the constellation known as Sagittarius.

Although the galaxy formed between 10 and 15 billion years ago, the Sun was only formed in a spiral arm some 4.5 billion years ago, and has been in orbit around the center of the Milky Way ever since. Thus, the age of the Milky Way galaxy is not less than

In visible light most of the stars in the Milky Way are hidden behind clouds of galactic dust and gas. This panoramic image of a section of the galaxy, taken by the infrared Spitzer Space Telescope, reveals distant stars and dust clouds with clarity.

standard spiral galaxy. But there is almost certainly a small bar joining the nucleus to the spiral arms, making it a barred spiral galaxy. Another interesting feature of its shape is that the disk of stars is not flat, but warped.

The Milky Way and its smaller companion galaxy (the Large Magellanic Cloud) are the main regions in which astronomers study stars and star formation. Large areas of dust and gas (nebulas) are the sites of star formation. Once formed, the life cycle of a star depends on its size. Sometimes it spans billions of years, progressing through various stages of development. Not all stars are solitary. They can be associated in pairs (known as binaries) or larger groups (clusters).

The final stages in the evolution of stars may be marked by spectacular planetary nebulas. The death of a massive star will result in a supernova and the formation of a neutron star or black hole. All these astronomical objects can be observed in the Milky Way and the Large Magellanic Cloud.

Like many large galaxies, the Milky Way has a number of smaller galaxies in orbit around it. The Magellanic Clouds are two irregular satellite galaxies, and there are a number of even smaller dwarf galaxies caught in its gravitational influence. Beyond the boundaries of its overwhelming influence, the galaxy is gravitationally bound to others in an association of galaxies known as the Local Group.

NEBULAE

The most beautiful and striking objects we have observed in our galaxy and neighboring galaxies are the nebulae. They are an essential part of the stellar life cycle, since they recycle the gas and dust from old stars and give birth to new systems.

These clouds of dust and gas are visible from Earth because they either emit starlight indirectly (emission nebulae), reflect starlight (reflection nebulae), or absorb starlight (absorption nebulae).

Right: *This superbubble of expanding glowing gas is the N70 nebula in the Large Magellanic Cloud, a satellite galaxy of the Milky Way. It measures 300 light-years across and was created by winds from hot massive stars and supernova explosions.*

Discovery of Nebulae

Charles Messier was a French astronomer at the Paris Observatory. He had been hired to search for and observe the return of Halley's Comet, expected in 1757 or 1758. As he searched for the comet along a path incorrectly predicted by his employer, he discovered an object that was cometlike in appearance. It was not a point source of light like a single star, nor did it move relative to the stars as comets should. Irritated that his observations had been sidetracked, he carefully noted the position of the object so he would not mistake it for a comet in the future. It would turn out to be the Crab Nebula, the ashes of a supernova that exploded in 1054. He decided to catalog such nebulous objects so that he and other astronomers could search for comets more efficiently. Other astronomers helped him, and by the end of Messier's life, his catalog listed 103 nebulae (the Latin word for "mist").

The nebulae were given numbers according to the order of their discovery, preceded by the letter M to mark them as part of Messier's catalog. To their discoverer they were simply "not comets," and the catalog covered a wide range of objects. The most distant are galaxies such as M31, the Andromeda Galaxy. Closer in were globular clusters, regions with a high density of stars. Smaller open clusters were also listed—the distance between the stars being too small for Messier to resolve them into separate bodies. (Later observations using better telescopes showed them to be individual stars.) Since Messier's time other catalogs of astronomical objects have been compiled, notably the New General Catalogue, that gives the prefix NGC to the objects listed.

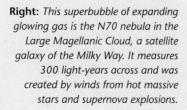

An ocean of glowing hydrogen gas and other elements such as sulfur and oxygen within the emission nebula M17 (Omega, or Swan Nebula) in the constellation Sagittarius. The wavelike patterns, revealing the three-dimensional form of the gases, are also illuminated by ultraviolet light from young stars that lie outside this image to the top left.

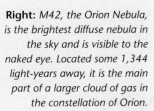

Right: *M42, the Orion Nebula, is the brightest diffuse nebula in the sky and is visible to the naked eye. Located some 1,344 light-years away, it is the main part of a larger cloud of gas in the constellation of Orion.*

Molecular Clouds

The term nebula is now used only for a subset of Messier objects. Galaxies and clusters are known to be primarily made from stars. Nebulae are made visible when interstellar dust and gas interact with the light from nearby stars.

The material in the galaxy that is not bound up into stellar systems can be found floating in giant molecular clouds in the gaps between stars. Like stars, these clouds consist mainly of hydrogen and helium, with small quantities of other elements. Stellar temperatures are so high that elements exist as individual atoms, but a molecular cloud is cold enough for atoms to combine into molecules. Hydrogen atoms pair up to form H_2, the main component of the clouds. Carbon monoxide is another common ingredient. Beyond these simple two-atom molecules many more complicated compounds have been detected, including water, alcohol, and hydrogen cyanide. The largest molecules can have up to ten atoms. The most important molecule yet discovered in these clouds is glycine. It is the simplest of the amino acids, which themselves are the building blocks of proteins. The presence of this key component of organic life in deep space provides another clue for us in our attempts to understand how life occurred on Earth.

We can detect these molecules across the vast distances of space because they emit radiation at very specific wavelengths. The peaks in the spectra received by our Earth-based and orbital instruments are a kind of fingerprint for particular molecular structures and bonds.

A central cluster of hot luminous young stars powers strong emission from hydrogen and oxygen gas in the Tarantula Nebula (NGC 2070, 30 Doradus). This powerful emission nebula is located some 170,000 light-years away in the Large Magellanic Cloud, a satellite galaxy to the Milky Way.

Interstellar Dust

Before the 1900s, interstellar space was thought to be entirely empty. This concept was challenged when it was noted that the appearance of star clusters, expected to be similar across the galaxy, differed depending on the cluster's distance. Fewer faint stars were seen in distant clusters, and the stars were consistently redder. Astronomers were able to attribute their appearance to the presence of dust scattered throughout the galaxy. Microscopic grains of silica and graphite float within the molecular clouds, scattering and absorbing light. They are believed to play an important part in the chemistry of the clouds, providing a surface on which atoms and small molecules can combine to form larger molecules. Without the dust grains to collect the chemicals, it would be exceptionally rare for atoms to approach each other closely enough to form bonds. Bonds within molecules would normally be broken by ultraviolet radiation from a star, but the presence of dust can block the radiation.

The giant molecular clouds can be from 1 to 300 light-years in diameter. The total mass of all the gas and dust can be between ten times and 10 million times the mass of our Sun. These clouds have very low density, with only 10,000 molecules per cubic centimeter. The best vacuums created on Earth still have more particles per volume than these clouds. When the cloud is not near a star, the typical temperature is only a few degrees above absolute zero.

Several thousand of these clouds exist within our galaxy. Their presence along the galactic plane blocks out starlight and prevents us from viewing the galactic center in visible light.

Cloaked in a bowl-shaped red glow of interstellar dust, a cluster of young stars is seen shining at the center of Sharpless 140, a star-forming region 3,000 light-years from Earth. In visible light the stars are completely obscured by the surrounding dust cloud, but this infrared image reveals the details within.

Cloud Collapse

Young stars are found in regions of the galaxy where the interstellar medium is densest. Nebulae are often referred to as star nurseries, and are a major part of the life cycle of stars. Collapsing clouds of dust and gas give birth to new stars. Stellar death from supernova explosions redistributes that material back into the interstellar medium. Stars are also losing mass to the interstellar regions throughout their lifetimes.

A cloud of gas and dust has two major influences on it: gravity and gas pressure. Gravity is constantly pushing the cloud inward to collapse on a point, whereas the gas pressure resists that force. Star formation requires that gravity wins in localized regions of space. This can occur thanks to the chance influence of a nearby star or by a supernova explosion sending shockwaves through the gas. The cloud internally fragments, first into threads and then into smaller, denser chunks. As the chunks get smaller, the effect of gravity grows, and the collapse is accelerated. The gravitational energy released by the infall of the gas turns into heat that will eventually ignite the interior of the cloud and give birth to a star. The radiation pressure from the newborn star blows away the surrounding gas. If the new star is large and hot enough or if there are many stars formed in the same region, a bubble will appear in the nebula. Some gas and dust will remain gravitationally bound to the new star and form a protoplanetary disk. Star clusters are formed when large nebulae give birth to a number of stars in the same neighborhood.

EMISSION NEBULAE

Emission nebulae mark the birthplaces of massive stars. The gas and dust surrounding the star begin to glow as the intensely energetic starlight ionizes the elements. Advanced telescopes such as Hubble can image the details of star formation. Giant molecular clouds in deep space can be detected only by their effect on light from distant stars and by the faint emissions from their component molecules. When such a cloud is near a star, it interacts with the starlight in such a way that we can see and classify it as a nebula.

Exceptionally hot stars—over 25,000 K—produce enough high-energy ultraviolet radiation to break up molecules and strip the resultant atoms of their electrons. This process is known as ionization and can occur as far as 300 light-years from a star. The ionized gas glows in colors specific to the elements present. Hydrogen (the most common gas and the easiest to ionize) glows red. If enough energy is pumped into the cloud by nearby stars, other elements such as helium, oxygen, nitrogen, and sulfur will add their own colors to the display. The result is known as an emission nebula. In many pictures taken by telescopes of these impressive objects, the colors displayed are not the same as the colors emitted. Each photon of light is assigned to red, green, or blue depending on which element emitted it. The combination of the three colors can create any other color, resulting in pictures that are both beautiful and scientifically useful.

Often multiple stars contribute to the ionization process. Radiation pressure from the stars, combined with gravitational interaction, creates a wide variety of nebular shapes and sizes. The largest can reach 1,000 light-years in diameter and contain hundreds of millions of times more matter than our Sun. Features such as bubbles and cometary knots are formed from the stellar winds produced by the hot stars.

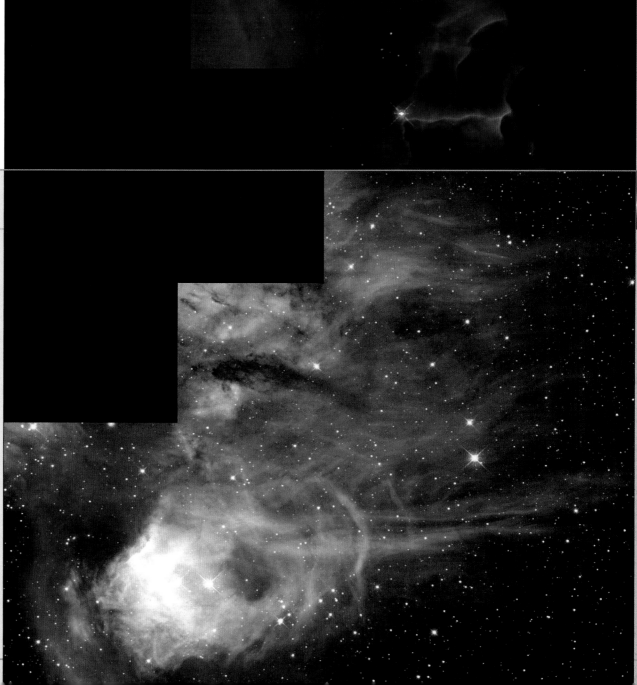

The collapsing dense cloud of dust and gas in the Trifid Nebula, located 9,000 light-years from Earth, is a stellar nursery full of embryonic stars. The source of the thin wispy projection to the upper left (a stellar jet) is a young stellar object lying buried in the cloud. Stellar jets are the exhaust gases of star formation. At the center of the nebula there is a central massive star that will continue to erode the nebula and bring to a halt the development of younger stars.

In nebula N44C, ionized gas streams from the exceptionally hot stars.

Carina Nebula—Keyhole

The Carina Nebula (NGC 3372) is a vast region of ionized hydrogen that lies roughly 8,000 light-years away. At 300 light-years wide, it covers a larger portion of the sky than any other emission nebula. It makes up part of the Milky Way as seen from Earth's Southern Hemisphere but it never appears in the sky in the Northern Hemisphere. It was discovered in 1751 by Lacaille, a French astronomer who cataloged southern stars and nebulous objects from the Cape of Good Hope.

The most massive stars in the Sun's neighborhood are in the Carina Nebula. They are 100 times heavier than the Sun and 10 times hotter on their surfaces. Stars of that size are relatively short lived and violently unstable. The most famous is the highly variable star Eta Carinae, which is just out of sight above and to the right of the Hubble image (right). It was briefly the second brightest star in the sky when it unleashed a nova in 1841. The light from this event was reflected by the surrounding nebula. John Herschel observed the nebula in 1847 and made a particular study of the central region, which he named the Keyhole Nebula. He produced detailed drawings, capturing the brighter regions around Eta Carinae. For a period of time the star dimmed and was invisible to the naked eye making its appearance very different from Herschel's original depiction. However, it has since regained magnitude.

The Hubble Space Telescope's image is of the Keyhole section of the Carina Nebula, approximately 7 light-years across. It shows the hydrogen gas filaments glowing in red, together with traces of other elements. The background dust is reflecting starlight, and appears blue. Two denser regions of gas appear toward the left and bottom edges. They are likely to be star-forming regions.

The Keyhole region within the Carina Nebula.

Carina Nebula—Star Formation

This infrared image from the Spitzer Space Telescope looks at a different part of the Carina Nebula, away from the Keyhole. Eta Carinae is above the picture. It is much too hot to be observed directly —it would overload the telescope's imaging sensors. Its effect on the nebula can be seen by the structures in the gas and dust. Most of the pillars in the picture are pointing toward the star Eta Carinae. This is because the intense radiation and the stellar winds from the star are pushing the contents of the nebula away. Some regions of the nebula are dense enough to resist the pressure, forming the tops of the pillars. The materials in the shadows of these denser regions are protected from the destructive stellar emanations.

The winds from Eta Carinae and other massive stars in the nebula make it a violent and dynamic region to observe. The clouds are moving rapidly, over 500,000 miles per hour (800,000 km/h) with respect to each other in some cases. Collisions between them generate large amounts of heat. The gas gets hot enough to emit X-rays, which have been detected all across the nebula.

In visible light a picture would show a ghostly amorphous red cloud. The light from many of the stars in the nebula itself would not reach us, and the detail in the shapes of the dust pillars would be less clear. Spitzer's infrared view sees through the intervening dust and can pick out the birth of stars in the denser regions at the top of the pillars.

The star-forming regions of Carina in infrared.

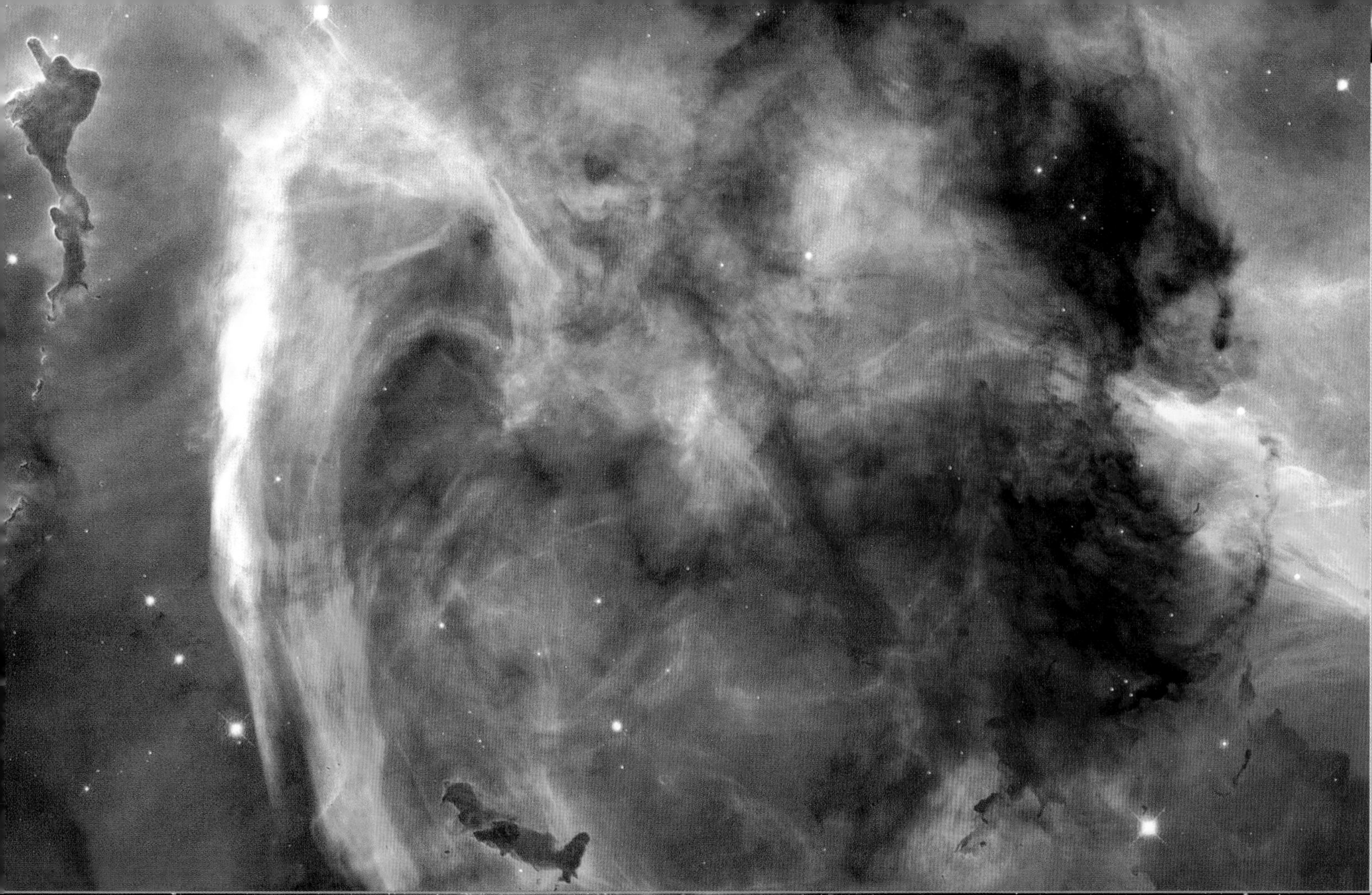

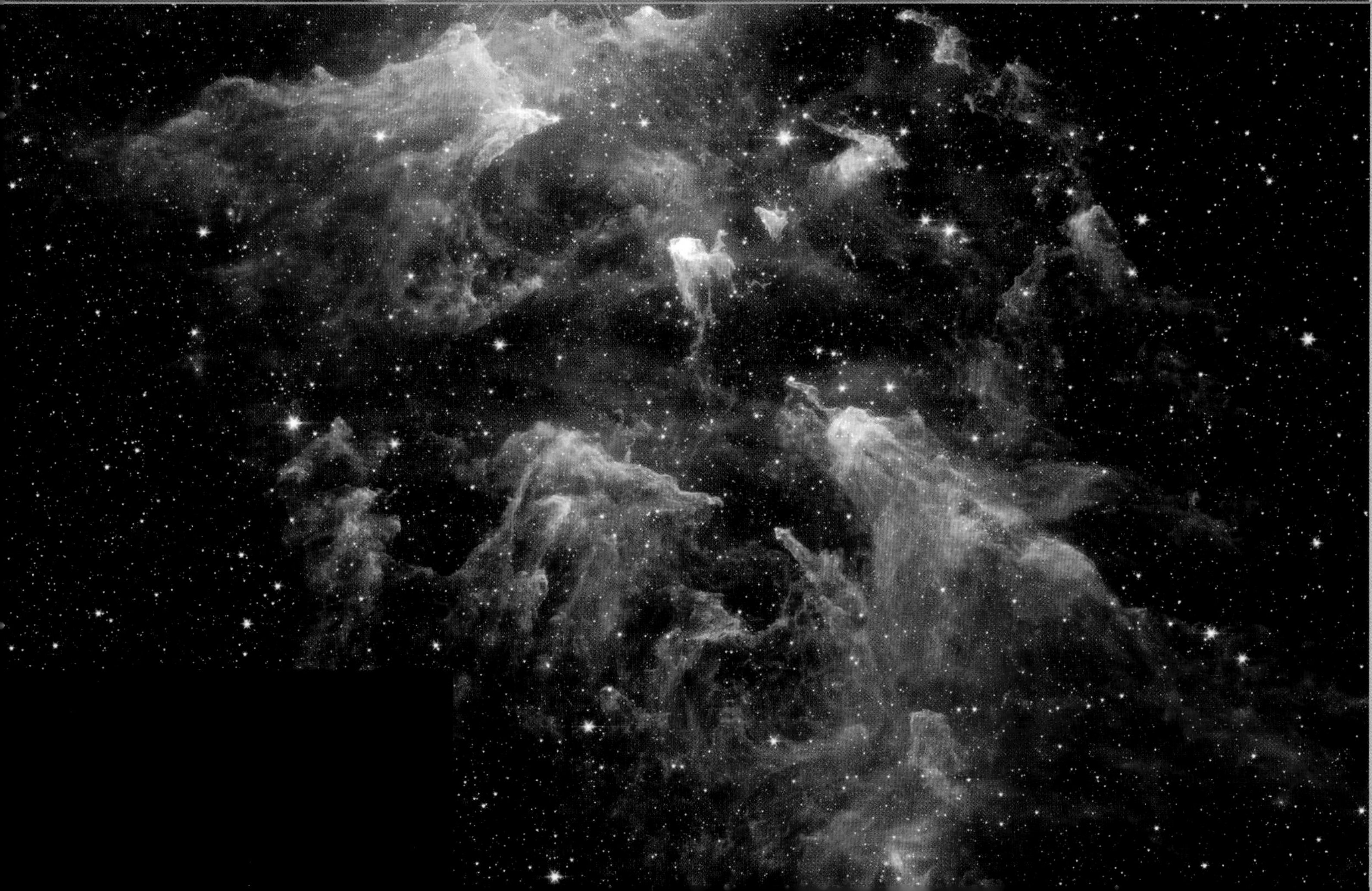

Eagle Nebula—Pillars of Creation

The Eagle Nebula and its associated cluster of newly born stars lie in the neighboring spiral arm to our solar system, 7,000 light-years toward the galactic center. They can be found in a nebula-rich area of the sky at the tail of the constellation Serpens. The open cluster is listed in Messier's catalog as M16 and also has the designation NGC 6611. The nebula is designated IC 4703 in the Index Catalogue, a publication that lists nebulae and clusters discovered between 1888 and 1905.

The full nebula is 70 light-years wide by 55 light-years tall and resembles an eagle with wings outstretched. Inside the body of the eagle lies the young star cluster, born from the clouds of dust. One of the most famous images taken by the Hubble Space Telescope shows three clouds that are known as the Pillars of Creation. The tallest one is 4 light-years from base to tip.

When observing a nebula, the Hubble Space Telescope takes a monochrome picture of the light that comes from a specific element. In this picture, Hubble imaged the Pillars of Creation in hydrogen (emits orange-red light), sulfur (red), and oxygen (green). In order to help distinguish between elements that emit light of a similar color, each monochrome picture is assigned to a primary color—in this case red for sulfur, blue for oxygen, and green for hydrogen. The three colored images are then overlayed to produce the final image. Although the picture is not what the naked eye would see when looking at the nebula, it is useful for showing the distribution of the elements.

Eagle Nebula—Hidden Stars

Stars forming in the clouds of the Eagle Nebula are hidden from view of the Hubble Space Telescope in pockets of gas that astronomers call EGGs (evaporating gaseous globules). EGGs are larger than our solar system and appear at the ends of fingerlike protrusions from the main body of the pillar. As ultraviolet radiation from neaby young hot stars heats the gas along the surface of the pillars, it dissipates, leaving behind EGGs. This process is called photoevaporation, and it is eroding the pillars. It is thought that they will not last more than another million years. The star cluster born from the Eagle Nebula consists of stars that are only five million years old themselves.

The stars in the open cluster are exceptionally hot. Their surface temperatures are about 45,000 K and they are many thousand times more luminous than our own Sun. The ultraviolet radiation from these stars causes the nebula to glow brightly. Each elemental component of the gas radiates light in a specific color, which can then be used to identify the ingredients of the cloud. The colors are not necessarily the same as the ones used in the resultant pictures.

Another pillar in the Eagle Nebula, nearly 10 light-years tall. Hydrogen appears as reddish brown and oxygen as blue. The dark regions are dust clouds that are too dense to allow much light to pass through them. Their surfaces glow when facing the bright cluster stars, but the ionizing radiation does not penetrate very far into the clouds.

Lagoon Nebula

The Lagoon Nebula is a large diffuse cloud 4,100 light-years away in the direction of the constellation Sagittarius. It was discovered by French astronomer Guillaume Le Gentil in 1747, although its associated cluster of stars had been recorded 67 years earlier. The nebula was catalogued by Charles Messier, receiving the designation M8. The star cluster NGC 6530 appears to lie just in front of the nebula, since the starlight does not pass through an appreciable amount of dust.

The Lagoon Nebula is one of the very few nebulae visible with the naked eye in the right viewing conditions. Because our eyes do not distinguish color well from low-light objects, it would appear gray. Long-exposure photography is capable of showing the rosy pink color that is characteristic of emission nebulae. The cloud is 110 light-years wide by 50 light-years tall and covers an area on the sky over three times the size of the full Moon. It contains three dark globules listed in Barnard's Catalog (the first catalog of dark nebulae) and several very large and bright stars.

The Hubble Space Telescope picture opposite shows the Lagoon Nebula's center, an area known as the Hourglass Nebula. The shape is produced by dark threads of nebula material spiraling across the bright gas cloud, powered by the star Herschel 36 (center, appears pink). The shape of the tendrils suggests that the dark material is being twisted. Astronomers have speculated that processes similar to the formation of tornadoes on Earth may be at work. The time and distance scales are massively different, however, and observations over many years would be required to examine the motion of the clouds.

The bright hourglass structure that lies in the center of the Lagoon Nebula.

N44F

The Large Magellanic Cloud is a dwarf galaxy and is part of the group of galaxies close to our own Milky Way. It lies at a distance of 160,000 light-years in the direction of the southern constellation Doradus and contains large regions of star formation. One of the brightest and busiest is N44, a superbubble that is 1,000 light-years wide. Among the many stars in the bubble are 40 white hot giants. These stars light up the gas in the region, and their radiation pressure creates the structures seen by our telescopes.

The picture shows a small section of N44, called N44F. A hydrogen gas filament leads onto a bubble structure (left of picture). The bright blue star below the center of the bubble is the sculptor of this feature. Its strong stellar winds have pushed gas and dust away, creating the wall of the bubble.

All stars lose mass over their lifetimes through stellar winds—our Sun, for example, sheds 10 million tons of material a year. The central star of N44F loses 100 million times more in the same period. The wind propagates out at speeds of 4.3 million miles per hour (7 million km/h), five times faster than our Sun's wind. The bubble is currently 35 light-years in diameter. On the inside of the bubble wall are several columns that point back toward the star. These constructions are several light-years long and are very similar to the Pillars of Creation in the Eagle Nebula. The ends of the columns probably hide collapsing dust clouds that may form stars. In a few million years the bubble will have been blown away by the radiance of a new star cluster.

The Hubble Space Telescope peers into the bubble cavity of N44F.

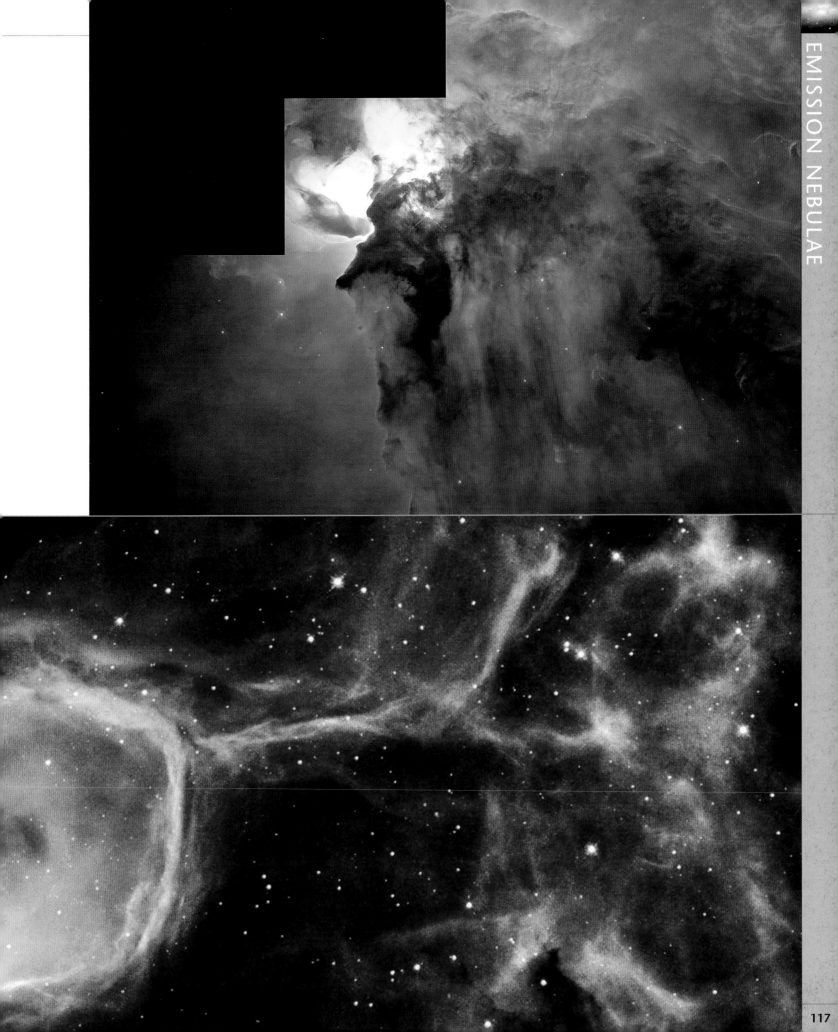

Trifid Nebula

The distance from Earth to an observed nebula is often difficult to determine. The Trifid Nebula has been assigned distances ranging from 2,000 to 9,000 light-years away. The uncertainty affects other measurements, such as the diameter of the glowing gas region (roughly 50 light-years) and the brightness of associated stars. The Trifid Nebula lies in the constellation Sagittarius. It is the 20th object in Messier's catalog, separated from the Lagoon Nebula (M8) by just two degrees.

The Trifid Nebula was discovered by Messier, who was more interested in the local cluster of stars. It appears in long-exposure photographs as a pink cloud split into three lobes by bands of dark material. Surrounding the pink region and extending above it is a blue reflection nebula. The pink color comes from the emission of radiation by ionized hydrogen, powered by a triple star system at the center of the nebula. The three massive stars have recently been born from the region where the dark bands cross, and have since pushed the cloud away due to their great luminosity and stellar winds.

The two images compare visible-light and infrared views of the Trifid Nebula. The visible-light image (main picture) shows a murky cloud with dark trails of dust. Four dense knots, or cores, of dust (circled in yellow) are "incubators" for embryonic stars. Astronomers did not believe these areas were ripe for stars until the false-color infrared image (inset) taken by the Spitzer Space Telescope showed the warmth of rapidly growing embryos inside (marked with arrows in the infrared image). They cannot be seen in the visible-light image of the same areas (but the corresponding areas are marked with arrows for purposes of comparison).

Ghost Head Nebula

This bright nebula, classified as NGC 2080, is part of the Large Magellanic Cloud, 160,000 light-years away. It has been given the nickname the Ghost Head Nebula, because the two bright white patches resemble eyes set in an ethereal face. The nebula is little more than 50 light-years from side to side.

For this picture from the Hubble Space Telescope, three different emission wavelengths were imaged and then assigned a primary color. Emission from ionized oxygen is shown in green. A star just off the left-hand side is responsible for heating up the gas and causing the oxygen component to glow. Around the edges of the nebula the red-pink color comes from ionized hydrogen. Red and green light mix to form yellow, so the yellow areas of the image indicate regions where both hydrogen and oxygen are emitting light. In the center of the nebula the stellar radiation is high enough to make hydrogen emit a second wavelength of light. This is given the color blue. When added to red and green, it produces white light. The white regions are concentrated around two centers (the eyes), designated A1 (left) and A2 (right).

A1 contains a single massive star that has already blown a bubble-shaped cavity within its dust cocoon. It can just be seen in the picture as a bright white arc. A2 has a few stars in a cluster, but they are not visible to Hubble because of the thickness of the dust. These stars must have formed only recently, since they have not yet dissipated the dust in their immediate vicinity.

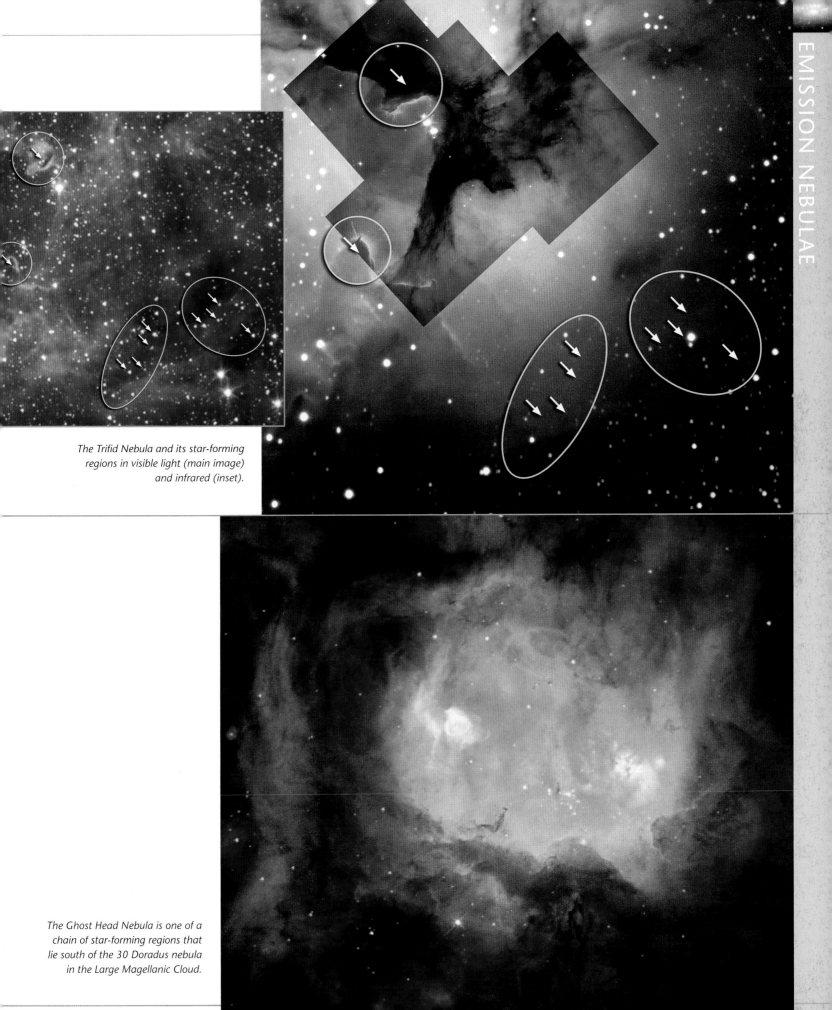

The Trifid Nebula and its star-forming regions in visible light (main image) and infrared (inset).

The Ghost Head Nebula is one of a chain of star-forming regions that lie south of the 30 Doradus nebula in the Large Magellanic Cloud.

REFLECTION NEBULAE

Reflection nebulae are prime examples of the interaction between light and dust. They range from sections of vast molecular clouds, reflecting the light from nearby stars, to light echoes produced by variable stars, as in V838 Monocerotis.

If a star is close to a nebula but its radiation is too weak to ionize the gas, it can still light up the cloud as a reflection nebula. Just as the Moon reflects the Sun's light toward Earth, the star's light is scattered toward us by the dust. The dust can be behind the star, reflecting the light back past it to us, or it can be between us and the star, creating a haze around the star comparable to the effect of fog on car headlights. For that to happen, the nebula must be optically thin, in other words, not so dense that light cannot penetrate it. Often reflection nebulae are simply part of the same huge dust clouds as dark and emission nebulae. They are similar in composition, but it is their situation that creates the distinction.

The nature of reflection nebulae was discovered by American astronomer Vesto Slipher in 1912. He noted that the light from the nebulae around the Pleiades star cluster had the same features in its spectrum as the light from the stars themselves. This indicated to him that the light was being reflected from the dust cloud.

Reflection nebulae are often blue in color, because the scattering process preferentially selects light at the blue end of the spectrum. Radiation at the red end of the spectrum is less likely to be scattered. This is the same physical phenomenon, known as Rayleigh scattering, that makes the sky blue. The Sun appears redder during sunrise or sunset because the path of light from the Sun passes through a larger amount of the atmosphere. When the blue light from the Sun's rays is removed through scattering, the light that remains is red.

The reflection nebula NGC 1999
in the constellation Orion.

V838 Monocerotis

From May to December 2002 Hubble captured a series of images of the reflected light pulse resulting from the sudden brightening of the variable star V838 Monocerotis. The star is seen at the center of the "eyeball" feature, and the light echo is caused by reflection of the light off the surrounding circumstellar dust.

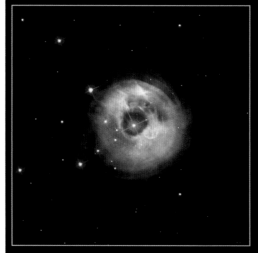 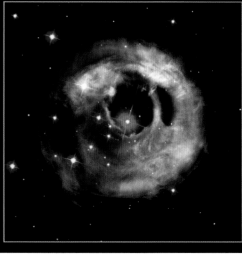

May 2002   *September 2002*

On the edge of the Milky Way, 20,000 light-years away from Earth, lies the unusual variable star V838 Monocerotis. It was discovered in January 2002 when it suddenly increased in brightness. Astronomers kept an eye on the star and in February of the same year they noticed a further increase in brightness of 10,000 times in just one day—an incredible outburst. The red supergiant star became one of the brightest in the galaxy. It swelled to a size where its radius was greater than the distance from Jupiter to the Sun. After a few weeks it faded, the luminosity dropping back to more conventional levels. A short time later, the Hubble Space Telescope was used to view the aftermath. It saw a rare but not unexpected event—a light echo.

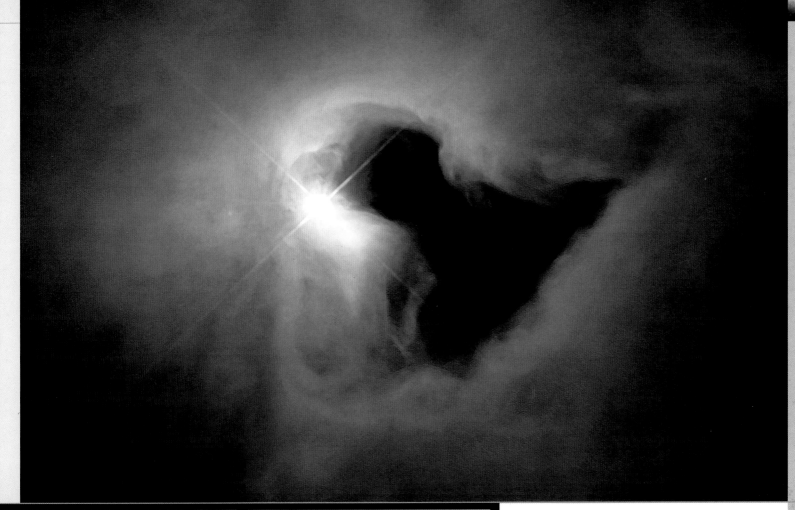

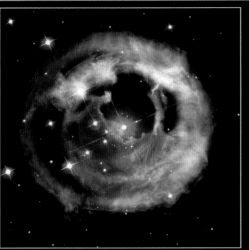

October 2002

December 2002

The light that came directly from the star reached Earth first because it traveled along a straight path. Subsequently, light that had been projected at an angle from our line of sight began to be scattered by interstellar dust. Because this light took an indirect path, it took longer to get to us. As the months passed, light that had been reflected through larger angles began arriving. The resultant images from Hubble shown here look like a bubble of material exploding from the central star. This is misleading—in reality, the "explosion" is simply that of light reflecting off dust that was already present in the space around the star. Swirls and other structures are being revealed by the expanding light wave, which is expected to fade away.

Barnard's Merope Nebula

The well-known Pleiades star cluster is currently moving through a dark molecular cloud. Strands of dust glow blue with the reflected starlight across the whole cluster. The picture shows just a tiny part of the cloud, which has broken off and is heading toward one of the stars in the cluster Merope.

Barnard discovered the nebula in 1890, and it was cataloged as IC 349. Observing it is difficult because of the close proximity of Merope, which is only 0.06 light-years away. In the picture opposite, the radial lines coming from the upper right are optical artifacts created by the relative brightness of the star. Spectral measurements show that the light coming from the star and the cloud is essentially the same, indicating that the cloud is reflecting rather than being heated and glowing as in an emission nebula.

The nebula is being slowed down by the radiation pressure exerted by the star, causing the cloud to bunch up. Larger, heavier particles of dust are less affected by the pressure and are leading the cloud in the form of streamers that can be seen pointing at the top right-hand corner of the image. Lighter particles are pushed back, and they form the main body of the nebula in the bottom left quarter. The cloud components are being separated into layers, seen here as thicker bands. The nebula is moving past Merope at approximately 7 miles per second (11 km/s). If it fails to be destroyed by the stellar radiation, it will pass the star several thousand years from now.

Hubble's Variable Nebula

The nebula NGC 2661 was discovered by William Herschel in 1783, but was given its current name after Edwin Hubble made a close study of it in 1916. He noted that the nebula was unusual in that it varied in its brightness over short periods of time. In astronomy, change typically occurs very slowly relative to our lifetimes on Earth but, in the case of this nebula, variations were noted over just three days.

The nebula is roughly the shape of a hollow conical shell, 0.2 light-years wide and 0.4 light-years tall. Its apex rests on R Monocerotis, a star 2,500 light-years from Earth and ten times the mass of the Sun. It is a naturally variable star itself, and its apparent magnitude varies between 10 and 12. Having formed only recently, it still has a thick disk of material surrounding it that prevents it from being viewed directly in visible light. However, its presence is given away by the nebula above the star reflecting the light toward Earth. The circumstellar dust spirals inward and is channeled into jets that emerge from the pole of the star. The dark jets cast shadows as they snake up the conical nebula and are seen as darker patches on the surface.

Since Edwin Hubble's first studies many observations of the nebula have been made, in an attempt to determine the dynamics of the system. It is expected that a symmetric lobe exists on the other side of R Monocerotis, consisting of dust ejected from the opposite pole. However, this has not been confirmed by observation.

Starlight from Merope in the Pleiades cluster is eroding this reflection nebula.

Hubble's Variable Nebula is a conical shell-shaped cloud of gas and dust, illuminated by R Monocerotis, the bright star at the bottom of the nebula.

DARK NEBULAE

The coldest of the nebula types, the clouds of dark nebulae are detectable only by the absence of light—a silhouette against a star field or bright nebula. Some contain stars born from collapsed regions of dust and hidden from the sight of telescopes such as Hubble.

With the vast numbers of stars in the galaxy, voids of starless space would not be expected. Despite this, early astronomers found regions where stars were absent or at least extremely scarce. William Herschel described one such region as "a hole in the heavens." It took the American astronomer Edward Barnard to determine that these apparent voids were, in fact, huge black dust clouds that blocked the light from stars farther along our line of sight. These clouds are known as dark nebulae, or Barnard objects. Barnard cataloged a large number of such objects in the period around 1900.

Dark nebulae are also referred to as absorption nebulae. Unlike emission and reflection nebulae, they are typically too dense to allow light to pass through them. Starlight that passes through the less dense edges of the cloud can be observed, however. Astronomers are able to identify the composition of the cloud by noting which wavelengths of light are absorbed. Carbon and silicon are common elements, bound up in grains of graphite and silica. Despite appearing dark in visible light, many dark nebulae shine strongly in infrared, indicating a heat source. Being the densest type of nebula, these clouds are often undergoing internal collapse to form stars.

The smallest types of dark nebulae are Bok globules, named for the astronomer who first observed them, Bart Bok. They are the last stage in the internal collapse of a dust cloud before the birth of a new star. Bok globules are about 1 light-year in diameter and will quickly collapse to form a small number of stars.

Horsehead Nebula

The Horsehead Nebula is one of the most photographed objects in the night sky. When a poll was conducted to decide the next object that should be imaged by the Hubble Space Telescope, this nebula won by a wide margin. It is instantly recognizable, because it is one of the few nebulae that looks like an everyday Earth-based object—in this case a horse's head similar to that of the chess piece known as a knight. The resemblance was discovered in 1888 when the nebula was first photographed. It appeared in Barnard's catalog of dark nebulae and received the designation Barnard 33.

The Horsehead Nebula stands out because it is silhouetted against the bright emission nebula IC 434. It is an extension of a much larger cloud of gas and dust that is itself part of the massive Orion Nebula. It lies south of the star Zeta Orionis, also known as Alnitak, the left-hand star in Orion's Belt. The glare from the nearby star makes observations of the Horsehead Nebula difficult even with a telescope.

Some light does pass through the dark nebula, especially in the region of the horse's mane. Small red spots are visible in the murk. They are protostars, forming from the collapse of localized regions in the dust cloud. At the top of the cloud a larger star is just visible as the dust is being eroded from around it. Above this image is the star Sigma Orionis, which provides the energy that causes the background IC 434 nebula to glow. Its light is eroding the cloud itself, and radiation pressure is creating the sharp edge on the top of the head.

Close-up of the "head" of the Horsehead Nebula.

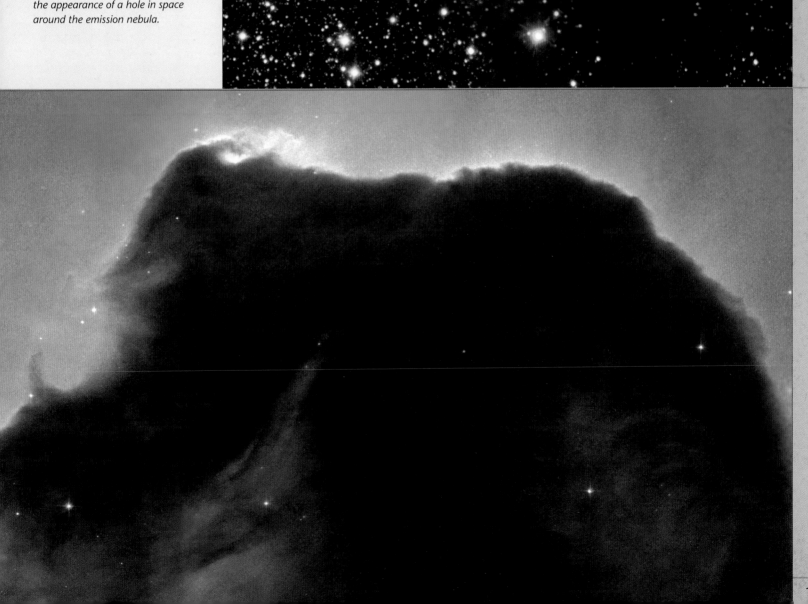

Infrared view of the region around IRAS 16362-4845, an emission nebula situated in a dark cloud in a Milky Way region called the RCW 108 Complex. While the infrared light from the nebula is strong enough to shine through the dust clouds of the dark nebula, the light from other background stars is not strong enough—giving the appearance of a hole in space around the emission nebula.

Thackeray's Globules

The emission nebula IC 2944 is 5,900 light-years away in the constellation Centaurus. It appears speckled with black marks overlying the pink glow of the hydrogen. These speckles are Bok globules, dark and dense clouds of dust that block any light from passing through them. South African astronomer A. D. Thackeray found these prime examples of Bok globules in 1950—three years after Bart Bok first proposed that small dark nebulae were forming stars within themselves. They are easy to see because they contrast well with the glowing nebula.

The Hubble Space Telescope is able to zoom in on the speckles and reveal their details. The image shows two large separate Bok globules that are aligned along the sight line of the telescope, which would make them appear to be one cloud if Hubble was less sensitive. The large clouds are highly irregular and have diameters of up to 1.4 light-years long. Some of the smaller globules have shrunk to near-perfect circles from our perspective. There is enough matter in these dark clouds to form 15 stars each with the same mass as our Sun. However, their harsh environment will probably cause them to dissipate before they can collapse enough to form any stars.

The larger globules show signs of being torn up by the radiation pressure from local stars. Internally, the clouds are turbulent, suggesting that they will break up in the short-term future. If they were better sheltered, these globules would be expected to produce a cluster of small stars, probably with less mass than the Sun.

Barnard 68

One of the closest dark nebulae in Barnard's catalog, Barnard 68 is only 410 light-years from Earth in the constellation Ophiuchus. It appears as a silhouette blocking the light from thousands of stars behind. Its size is comparable to the Oort Cloud in our solar system, with a diameter of roughly 0.4 light-years. The quantity of dust and gas in the cloud has twice as much mass as our Sun. Its temperature is a cold 16 K—only slightly warmer than deep space.

Around the edges of the cloud, stars appear fainter because a portion of their light output is blocked by the dust. The center appears black unless viewed in infrared, when stars hidden behind the cloud are revealed. A survey of the relative brightness of these thousands of background stars that interact with the dust has allowed astronomers to build up a model of the density distribution within the cloud. The level of light extinction for each star indicates the thickness of the dust in a specific region of the nebula. The results indicate that the cloud is currently stable. The internal pressure of the gas pushing out is balanced by the force of gravity pulling in. This situation will not continue indefinitely—eventually the cloud will collapse.

Closer examination of the cloud with radio telescopes has shown that it is in motion, with regions flexing in and out. Stars are known to vibrate in this way over periods of minutes, but Barnard 68 has a vibrational period of 250,000 years. Astronomers are unsure of the exact cause, but a shockwave from a nearby supernova is a likely culprit.

Dark clouds, known as Thackeray's globules, are silhouetted against nearby bright stars in the busy star-forming region IC 2944. This image was taken with the Hubble Space Telescope.

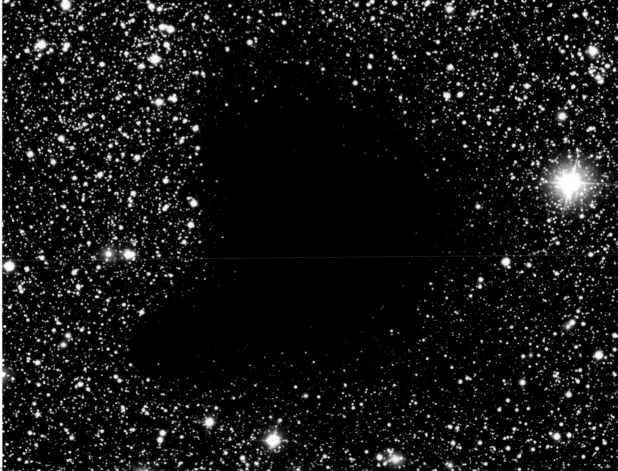

Combined visible and near infrared images of the dark nebula Barnard 68.

Elephant's Trunk Nebula

In the constellation Cepheus the large emission nebula IC 1396 glows in the characteristic pink hydrogen color. At a distance from Earth of 2,450 light-years, it covers an area on the sky roughly equivalent to ten full Moons, Silhouetted against it is the Elephant's Trunk Nebula, a long thin region of dust. IC 1396 is difficult to see even with a telescope, because it is not very bright. The dark nebula has been compressed into its sinuous shape by the pressure of the ionized gas that surrounds it. It will dissipate eventually, leaving behind a clutch of newborn stars.

The picture opposite shows the combination of images from two instruments on the Spitzer Space Telescope. Viewing the nebula in the near infrared (Infrared Array Camera, or IRAC) and far infrared (Multiband Imaging Photometer for Spitzer, or MIPS) provides much more information than visible light. The internal structure of the gas cloud appears rumpled, composed of tenuous filaments in the thinnest areas. Many more stars can be seen than in any visible-light image, since there are several hidden in the dust. Some are not yet bright enough to emit light in the visible spectrum.

The head of the trunk is a denser area under pressure from a single large star outside the cloud to the left of the image. Inside it are two stars that have pushed back their surroundings, blowing a vacated bubble in the nebula. The stars are of similar brightness in this infrared image, but in visible light one appears much less bright, suggesting that it still has a protoplanetary disk. Over time the disk will erode and possibly form a planetary system like our own.

PROTOSTARS AND PLANETARY SYSTEMS

The majority of processes in astrophysics occur over timescales that are huge when compared with the life span of a human, or even of humanity overall. The formation of the Sun and its attendant planets is one such process that has been investigated and debated for years.

Until recently our only information on the formation of planets and stars was the current state of the solar system, but with modern telescopes we are now able to look to other stars for knowledge. Although we cannot see the changes that occur during star formation, we can see various snapshots from different stages in other areas of the galaxy. Nebulae are often rich star-forming regions and are the focus of surveys by the space telescopes. They hold many protostellar systems that can provide clues to the birth of the Sun, the formation of Earth, and the origin of the outer belts.

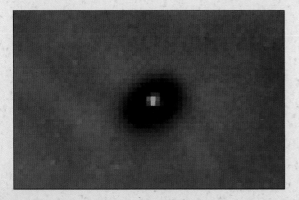

A view from the Hubble Space Telescope of a very young star (possibly a protoplanetary disk from which planets will form), seen against the background of the Orion Nebula.

Composite image in the infrared of the Elephant's Trunk Nebula from the Spitzer Space Telescope.

Intense radiation from a bright nearby star (Theta 1 Orionis C) is causing these protostars in the Orion Nebula to shed cocoons of dust. Their circumstellar disks can be seen as oval rings. Within these disks are the seeds of planets, but they may be destroyed by the relentless radiation from the nearby star.

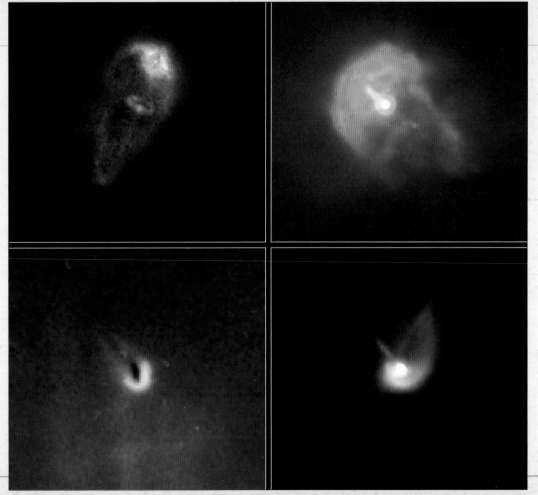

Star Formation

Stars begin forming when dense dust clouds lose their battle against their own gravity, and collapse. This can be triggered by a local event such as a supernova blast wave that shocks the material into a denser state. Over longer periods of time, compression waves that correspond to the spiral arms of the galaxy may pass through the region and tip the cloud into collapse. Cooling of the cloud by the radiation of heat as infrared light can reduce the internal pressure that opposes gravity. Whatever the cause of the initial collapse, once it begins it is self-perpetuating. As the center of the cloud becomes denser, the gravitational pull on the material increases. The collapse occurs from the inside out—particles in the dense middle will move rapidly to the center of gravity, whereas particles in the sparser edges of the cloud are slower. As the cloud decreases in size, other factors become more important in determining the dynamics of the system.

When a rotating dust cloud shrinks, its rotation speeds up. That is because without significant energy losses, the angular momentum of the cloud is conserved. Negligible initial rotation speeds become significant after a typically sized star-forming globule collapses to the size of our Sun. The inertia of a rapidly rotating object would make forming a star very difficult, if not impossible.

Several processes exist that reduce the angular momentum. Friction between particles in the dust cloud turn the kinetic energy into heat. Instead of compressing in a spherically symmetrical way, the cloud flattens and forms a disk at right angles to the axis of rotation. Also, many clouds are known to have magnetic fields that grow in strength as the cloud shrinks, producing a magnetic pressure that acts against the rotation and collapse. Astronomers are still trying to understand these processes and their influences on the resultant stellar system.

Appearing as a bulge, the central region of the protostar accretes more and more mass from the surrounding disk. It gets extremely hot as a result of the friction of the particles. All the gravitational energy is being converted into heat. Eventually it will become hot enough to ignite nuclear fusion, a process that requires extremely high temperatures and intense pressure. Even before that occurs, the protostar gets hot enough to emit stellar winds. These winds cannot flow out through the dense disk and are therefore projected out through the poles as symmetrical jets of ionized gas.

Planetary System Formation

Exactly how planetary systems are formed is not yet known, because we do not have enough detail on planetary systems other than our own. Instead, our knowledge of celestial mechanics and the geography of the solar system have provided a possible method, which is as follows.

As the circumstellar disk swirls around the central protostar, it gets denser. More gas and dust particles are channeled into a smaller cross-section. Atoms and molecules begin to accrete onto fine dust grains. Instabilities in the disk and uneven heating by the protostar produce eddies. They create areas of increased density, giving more opportunity for the growth of the dust grains in specific regions. Light materials such as gas and ice are less likely to stick to dust grains in the regions close to the protostar, since the temperature and stellar wind pressure are much higher. Heavy atoms such as those in metal and rock are preferred. This differentiation results in rocky planets occurring near the star, and gas giants and ice planets farther away.

Eventually some dust grains become the size of pebbles, then boulders. These objects are known as planetesimals, the seeds of future planets. Their gravity becomes important and enables them to attract material and grow faster. They sweep out the areas around their orbits, much like some shepherd moons in Saturn's rings. Soon they begin to interact and collide, smashing apart or sticking together depending on the circumstances. Eventually there will be a small number of surviving planets in stable orbits. Remaining planetesimals collide with the planets in an era known as the heavy bombardment period. The craters from this early stage can still be seen on the oldest surfaces in our solar system. The stellar wind pushes the majority of the leftover gas out of the inner part of the system.

False-color image of the protoplanetary disk around Beta Pictoris, showing the density of the dust. The warping of the structure may be because of the gravitational influence of a planet. The star itself is masked.

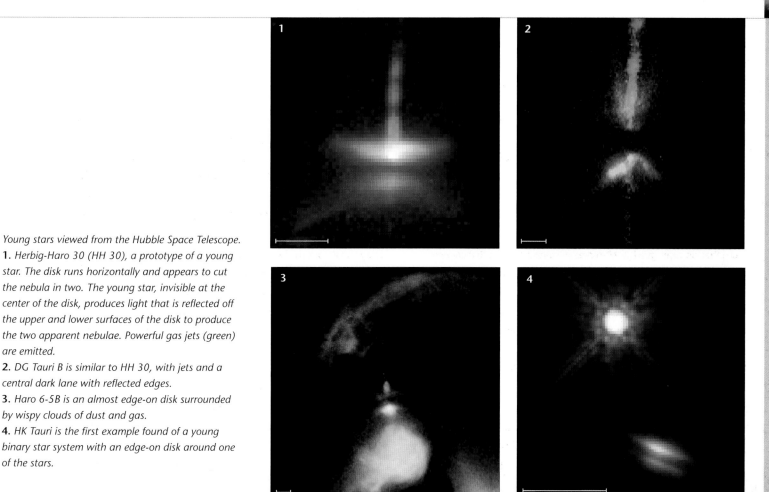

Young stars viewed from the Hubble Space Telescope.
1. *Herbig-Haro 30 (HH 30), a prototype of a young star. The disk runs horizontally and appears to cut the nebula in two. The young star, invisible at the center of the disk, produces light that is reflected off the upper and lower surfaces of the disk to produce the two apparent nebulae. Powerful gas jets (green) are emitted.*
2. *DG Tauri B is similar to HH 30, with jets and a central dark lane with reflected edges.*
3. *Haro 6-5B is an almost edge-on disk surrounded by wispy clouds of dust and gas.*
4. *HK Tauri is the first example found of a young binary star system with an edge-on disk around one of the stars.*

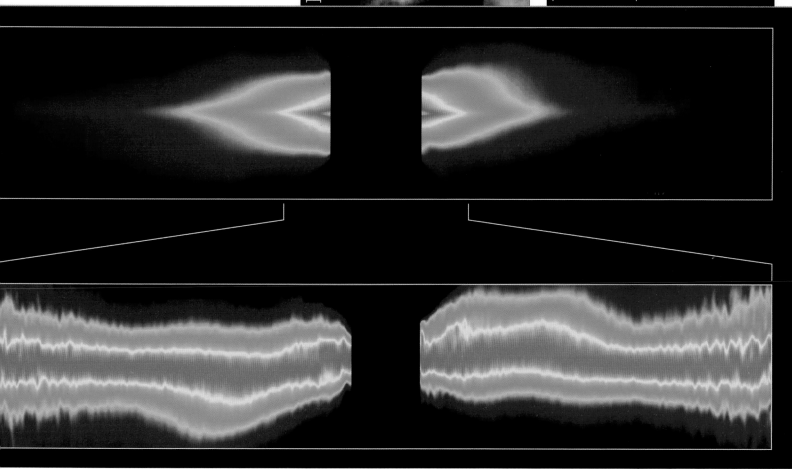

Fomalhaut

The 18th brightest star in the sky is in the southern constellation Piscis Australis. It has the name Fomalhaut, which comes from the Arabic phrase meaning the mouth of the southern fish. It is 2.3 times more massive than our Sun and 15 times more luminous. It is a young star, believed to have formed 200 million years ago. Its size gives it a life expectancy of only another billion years before it turns into a red giant.

In 1983 the Infrared Astronomical Satellite (IRAS) measured an unusually large amount of infrared light coming from the vicinity of Fomalhaut. The light was coming from its circumstellar dust ring, but the telescope was not sensitive enough to view the ring itself. The Hubble Space Telescope had the resolving power needed to image the dust. The Advanced Camera for Surveys instrument on Hubble contains a corona-graph. It is essentially a mask placed in the line of sight to block excessive radiation from a bright source and keep it from overwhelming the light from a fainter nearby object. In the image opposite (top right), the central dot surrounded by darkness marks the position of the star behind the mask. Some light from the star still leaks onto the picture, causing the radial spikes around Fomalhaut. They are not real patterns. The dust ring is clear, albeit slightly distorted by the imaging artifacts.

The dust ring is 2.3 billion miles (3.7 billion km) wide. Its inner edge is 130 times farther from Fomalhaut than Earth is from the Sun. Although our solar system has had longer to evolve, astronomers believe that the dust ring is similar in nature to the Kuiper Belt.

Fomalhaut is not in the center of the dust ring. It is offset from the central point by more than 1.4 million miles (2.25 billion km), indicating that some other body is having a gravitational influence on the system. Hubble is not sensitive enough to spot such a body, but a planet the size of a gas giant orbiting the star between 50 and 70 Earth–Sun distances away could be the cause. Further evidence for the existence of planets around Fomalhaut comes from the sharply cropped inner edge of the dust ring. Like Neptune interacting with the Kuiper Belt, a planet is capable of "pruning" a dust ring to form a precise shape.

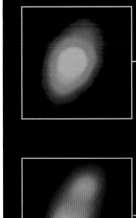

Below: *Infrared light images of the ring around Fomalhaut from Spitzer. Two different wavelengths of infrared light are combined in a single image (right).*

Extrasolar Planets

For life to exist elsewhere in the universe, planetary systems around other stars are necessary. Detecting objects as small as planets over interstellar distances is a task that was beyond the capabilities of our telescopes until the last two decades. Even now, only a handful of extrasolar planets have been directly imaged. Indirect techniques have been more successful and there are currently 530 known exoplanets, with the discovery of new ones almost daily.

The first extrasolar planet to be discovered was found around a pulsar (rapidly rotating neutron star) in the globular cluster M4. The regular timing of its radio pulses was varying because of the gravita-tional influence of a companion. Further analysis of the timing anomalies has led to the prediction that there are more than three planets roughly the size of Earth in the system.

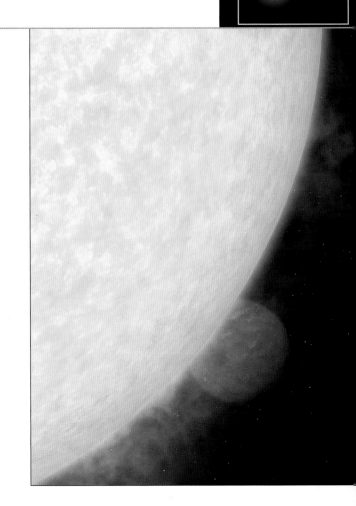

Right: *Visible-light image of the ring around Fomalhaut from Hubble.*

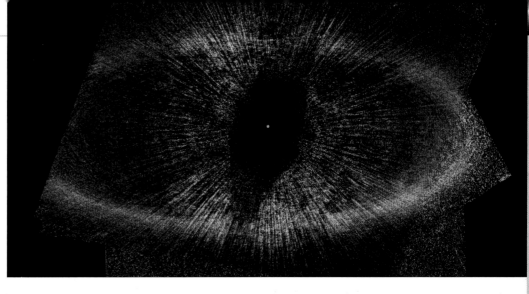

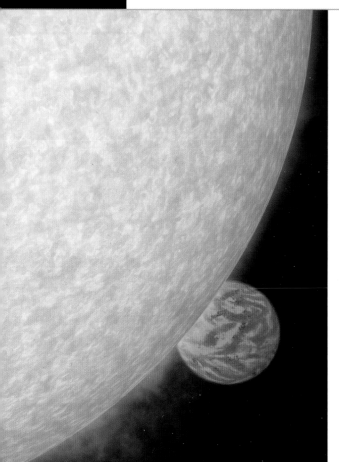

The infrared images (left) from the Spitzer Space Telescope show a different orientation than in the visible-light image, and they reveal the heat emission from the dust around Fomalhaut. The radiation from the star was blocked in order to examine the ring with the best possible contrast. The image taken at a wavelength of 70 microns (bottom left) shows the full extent of the disk—nearly five times wider than the solar system. The bottom lobe is one-third brighter. This may be due to the influence of the hypothetical gas giant planet or to an overabundance of dust in the area from a collision between planetesimals. The 24-micron image (top left) shows warmer dust closer to the center and the star. It is similar to the dust present in the inner solar system but much denser. If there are more planets orbiting Fomalhaut, they will be sweeping up the dust and probably undergoing bombardment from cometary objects falling in from the outer dust disk. The image center left is the combination of the 24- and 70-micron images.

An artist's concept of an extrasolar planet close to a star. At more than 1000 K, the planet's surface would be brightest in the infrared spectrum (right). In visible light (left), it is nearly invisible.

The most common technique used to find extrasolar planets is known as the radial velocity method. It relies on sensitive measurements of a star's light spectrum over time. In a system with a planet orbiting the star, the center of mass is not exactly at the same point as the center of the star. Both objects must orbit the center of mass, which results in the star wobbling slightly. (For example, because of Jupiter's mass, our Sun wobbles at a speed of 43 feet per second, [13 m/s].) This movement can be found by examining the Doppler Effect of the star's motion on its emitted light. When the star is moving toward us, the frequency of its light will increase; as it moves away, the frequency decreases. Speeds of as little as 10 feet per second (3 m/s) can be detected by instruments on Earth, but the system has to be nearby in order to achieve that level of accuracy.

A small number of planetary systems have their orbital planes along our line of sight. As the planets pass in front of the star, they will block a portion of the light. The star will appear to dim by a small fraction regularly. Observers looking at our Sun would detect a luminosity drop of 1 percent owing to an eclipse by Jupiter. Unlike the radial velocity method, this technique provides information on the planet's size rather than its mass. The two measurements can be combined to give the planet's density. More information can be gleaned from any change in the nature of the light as the planet moves in front of the star. Some of the light will be passing through the planet's atmosphere, allowing spectroscopic analysis of its composition. The picture opposite shows a gas giant planet that orbits a star 150 light-years away. It is lighter than Jupiter but larger, and its atmosphere has been found to contain hydrogen, oxygen, carbon, and sodium. These gases are being stripped from the gas giant by stellar radiation, since it orbits only 4.3 million miles (7 million km) from the star's surface.

Most of the extrasolar planets that have so far been discovered are large gas giants, orbiting closer to their star than Earth does to our Sun. This is not necessarily typical for planetary systems, because the techniques used to discover planets work best when the planet is large and near its star. Earth-sized planets are just beyond the range of our capabilities, although this will change with theintroduction of a new generation of instruments. The "hot Jupiters," as they are known, pose problems for the current theory of planetary system formation, which does not predict their existence.

An artist's impression of a "hot Jupiter" planet (with its own small satellite) orbiting the star HD 209458.

STARS

The vast majority of the visible matter in our galaxy exists in the form of stars. They are vast spheres of gas supported against gravitational collapse by the energy released from nuclear fusion.

The nuclear fusion that powers starlight imposes specific limits on the mass of a star. Stars weighing less than 8 percent of the mass of the Sun will not produce the extreme temperatures and pressures required for fusion to begin. The easiest fusion reaction turns four hydrogen atoms into a single helium atom, but it occurs only in temperatures above 3 million K.

Above: *A star field in the constellation Centaurus. Most of the stars in the image are near the center of our galaxy, but the blue curved streak is a much closer object—an asteroid orbiting the Sun only light-minutes away.*

Right: *This color composite shows the complex of bright nebulae and hot stars in the constellation Chamaeleon I, close to the southern celestial pole.*

At the other end of the scale, stars are not found with more than 150 times the mass of the Sun. This is probably because heavy stars are very hot and luminous. The radiation from a star above that limit would tear off its own outer layers and eject them into space. Stars close to that limit are highly unstable, resembling a continuous explosion. The mass of a star determines its life span. The most massive stars may only last a few million years before collapsing. Most stars are small enough to have life spans greater than the current age of the universe—13.7 billion years. Our own Sun is larger than average and has an expected life span of 10 billion years.

Estimating the total number of stars in our galaxy is difficult because most are hidden behind intervening dust clouds. The average mass of a star is not well known because stars with very low mass are difficult to detect. However, recent measurements suggest that there are more of them than expected. Estimates of the number of stars within the Milky Way range from 100 to 400 billion.

A timelapse photograph of stars over the Gemini Observatory at Mauna Kea, Hawaii.

Stellar Classification

Stars are typically characterized by three values: luminosity, mass, and temperature. A common astrophysical tool is the Hertzsprung–Russell (H–R) diagram. It plots stars on a graph with temperature across the x-axis and luminosity along the y-axis. When a large population of stars is plotted on this graph, a sinuous line stands out. This is the main sequence, which extends from low-temperature, low-luminosity stars (red dwarfs) to the large blue-white stars in the opposite corner. The main sequence is where stars spend the majority of their lives.

A system for classifying stars based on their color was developed at the beginning of the 20th century. Each received a letter of the alphabet according to the features in its spectra. It was subsequently discovered that the color of a star is directly related to its surface temperature, and the classification was reordered to take this fact into account. However, the alphabetic order became jumbled up. A mnemonic is used to remember the current order, OBAFGKM: Oh Be A Fine Girl (or Guy), Kiss Me. The Sun is a G-type star and sits in the center of the H–R diagram.

The average statistics for main-sequence stars of each type are listed in the table below.

Class	Temperature	Color	Luminosity	Radius	Mass
O	45,000 K	Blue	1,000,000	15	60
B	15,000 K	Blue-white	10,000	4	6
A	8000 K	White	35	2	2
F	6500 K	Yellow-white	3	1.4	1.4
G	5500 K	Yellow	0.8	0.9	0.9
K	4000 K	Orange	0.3	0.7	0.7
M	3000 K	Orange-red	0.03	0.4	0.2

The stars on an H–R diagram are not randomly scattered. Most of them, such as the Sun and the Dog Star, Sirius A, lie on the main sequence, a line that runs from bright, hot, blue stars at the top left to cool, faint, red stars at the bottom right.

Antares is one of the largest known stars, a red giant 500 times bigger than the Sun.

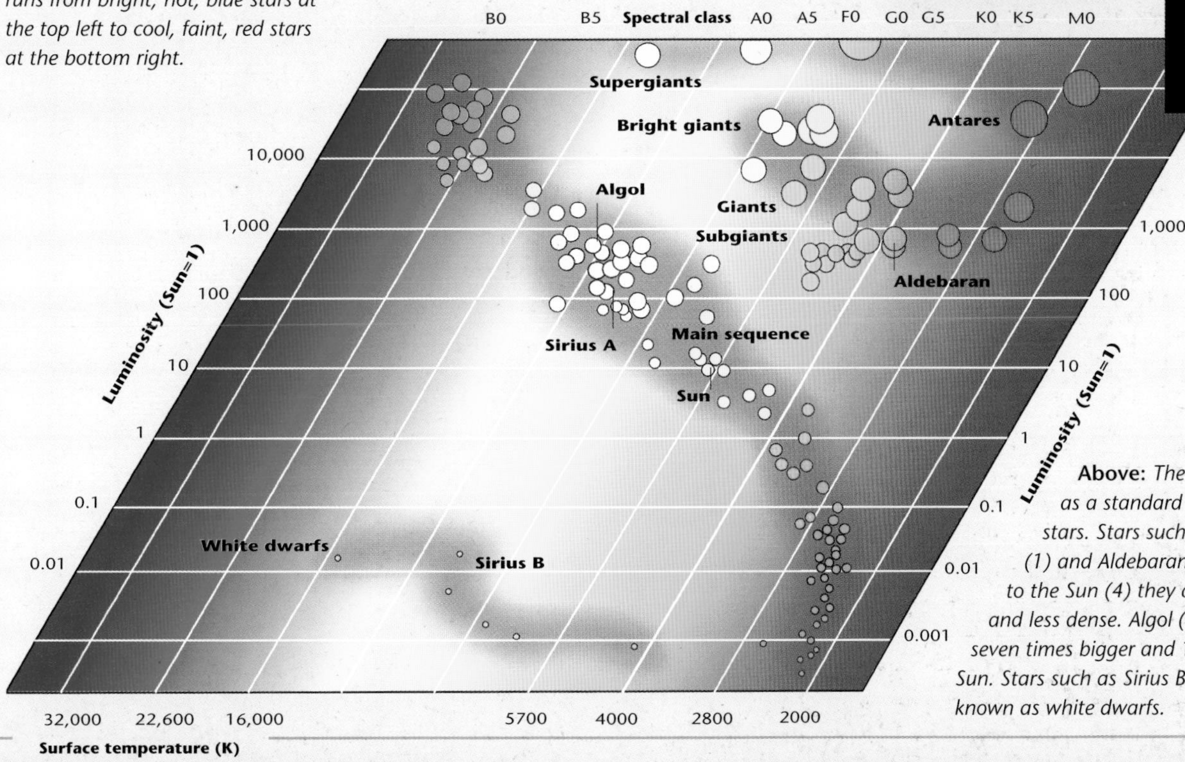

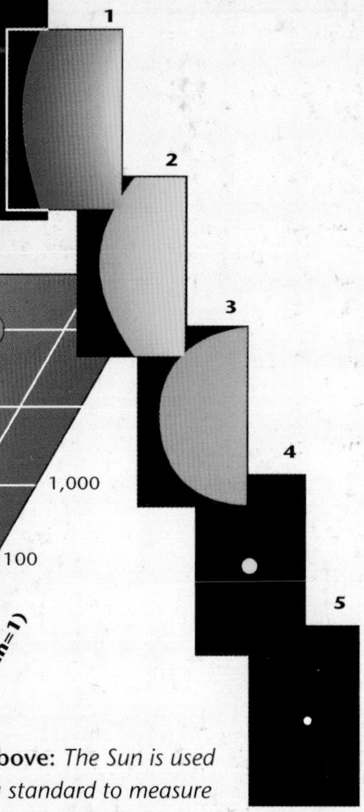

Above: The Sun is used as a standard to measure stars. Stars such as Antares (1) and Aldebaran (2) are red giants. Compared to the Sun (4) they are very large but fairly cool and less dense. Algol (3) is a main-sequence star seven times bigger and 100 times brighter than the Sun. Stars such as Sirius B (5), are small, hot, faint stars, known as white dwarfs.

H–R diagram labels:

Spectral class: B0 B5 A0 A5 F0 G0 G5 K0 K5 M0

Supergiants
Bright giants
Antares
Algol
Giants
Subgiants
Aldebaran
Sirius A
Main sequence
Sun
White dwarfs
Sirius B

Luminosity (Sun=1): 10,000 1,000 100 10 1 0.1 0.01

Luminosity (Sun=1): 1,000 100 10 1 0.1 0.01 0.001

Surface temperature (K): 32,000 22,600 16,000 5700 4000 2800 2000

Red Dwarfs

Members of the most common class of star— red dwarfs—are between 8 and 50 percent of the mass of the Sun. They are relatively cool stars with a surface temperature of below 4000 K. Their size and lower energy output makes them faint and therefore almost unknown outside the region around the Sun. The rate at which they use up their nuclear fuel is low, meaning that their expected lifetimes are much longer than the current age of the universe. We do not know what happens to a red dwarf that has come to the end of its hydrogen-burning life, because it has never happened yet.

Red dwarfs contain large convection zones in which plasma rises to the star's surface and falls back again. Such movement of conductive material generates a magnetic field. The magnetic field of our Sun is responsible for solar flares and sunspots, and analogous features have been observed on nearby red dwarfs. Some are known as flare stars, since their brightness varies over periods as short as minutes. Red dwarfs are not intrinsically bright in the visual spectrum, although a solar flare can increase its luminosity by a significant amount as well as releasing radiation in higher energy wavebands such as X-rays.

The closest star to our Sun, Proxima Centauri, is a flaring red dwarf. It is too dim to be seen by the naked eye, despite being only 4.2 light- years away. It is the outlying member of the Alpha Centauri triple star system, orbiting the other two stars once every 500,000 years or so.

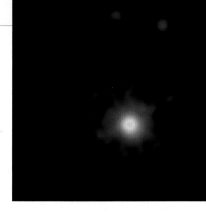

Proxima Centauri stands out in this X-ray image.

GIANTS

Although rare in terms of numbers, their significant luminosity makes giant stars common in astronomical studies. They are also much more dynamic than their smaller cousins, some lasting no more than a million years before undergoing a pyrotechnic death.

Giant stars occupy the area on the Hertzsprung–Russell diagram above and to the right of the main sequence, red giants on the right and blue giants on the left. They are typically either blue-white (hot surface) or red (cooler surface). Giants in the middle ground are very rare. All giant stars are larger, heavier, and more luminous than our Sun. The widest stars are nearly 2,000 times larger in diameter than the Sun, and the heaviest contain 150 solar masses of material. A star's lifetime is tied to its mass—the largest stars last only a couple of million years.

The large mass of giants produce higher temperatures and pressures than would be found within lighter stars. The temperature in their core is hot enough to begin fusion reactions with heavier elements. Three helium nuclei produced by the fusion of hydrogen can combine to form a nucleus of carbon, but only if the density of helium is high and the temperature is above 100 million K. This fusion reaction is called the triple alpha process. Fusion reactions that create even heavier elements require further increases in temperature. Because all the fuel for a particular reaction has been consumed and turned into heavy elements, the core of the star will collapse under gravity. This heats up the core and may initiate a new set of higher-temperature fusion reactions. Most giant stars are variable in their luminosity because these fusion processes turn off and on in their core. Eventually the chain of heavier elements ends at iron—there is no fusion reaction with iron that will release energy. A core of iron will soon result in a star imploding under gravity and then exploding as a supernova.

The blue supergiant Sher 25 (circled) is captured here by the Hubble Space Telescope in the stellar cluster NGC 3603.

Brown Dwarfs

Stellar objects that have less than 8 percent of the mass of the Sun are known as brown dwarfs. The temperatures in their cores are too low to sustain hydrogen fusion. For this reason, they are often called "failed stars." They have surface temperatures of below 2500 K, making them visible only by telescopes that image in the infrared part of the spectrum. Because of this, they are usually discovered in binary systems, owing to their gravitational effect on their larger companion star.

The lack of nuclear fusion reduces the pressure within a brown dwarf compared to a normal star, making it smaller and denser. They are all roughly the size of Jupiter, despite the fact that their mass varies from 10 to 80 times greater than that of the gas giant planet. The heat energy in the brown dwarf is generated by the mass contracting and settling toward the center. Over time, without a consistent energy-generating process, the star will cool.

Apart from their mass, brown dwarfs can be characterized by the presence of lithium. At a temperature just below the threshold for hydrogen fusion, lithium nuclei can be broken up into two helium nuclei after a collision with a proton (hydrogen nucleus). Small stars are fully convective—material within the star will pass through all the internal regions from the core to the surface. In a dwarf star that is hotter than the typical values for a brown dwarf, this reaction will soon eliminate all lithium. Stars larger than the Sun can also retain lithium because their internal material is not fully mixed.

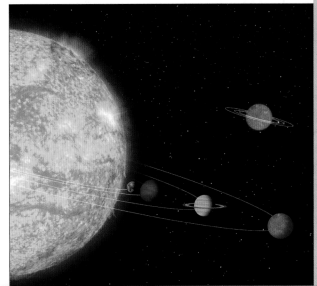

Artist's impression of a brown dwarf orbiting 55 Cancri, a Sunlike star. The brown dwarf itself is portrayed as having a protoplanetary disk, capable of forming a few small planets.

Eta Carinae

At the center of the Carina Nebula, within the Keyhole Nebula, lies the most massive visible star known. Eta Carinae is on the theoretical upper limit for stellar mass. As a result, its radiation pressure renders it highly unstable, and it belongs to the rare class of star called luminous blue variables. Its brightness alters in bursts, not periodically. Despite being 8,000 light-years away, it became the second-brightest star in the sky behind Sirius (8.7 light-years away) in the 1840s. The outburst was of such magnitude that it would have destroyed a smaller star. It is currently at a magnitude of about 4.4 and is visible to the naked eye in a dark sky. Several unpredictable outbursts have followed, and it is expected to fully explode as a large supernova at any point in the next few hundred thousand years. Luminous blue variable stars live fast and die young, lasting only about a million years.

The Hubble Space Telescope picture opposite shows the explosive outflow of gas. The white circle in the center is the star itself. Two lobes of gas—ejected during the outburst in 1843—are visible. Each lobe contains as much mass as our Sun. They are known as the Homunculus Nebulae. There is also a thin equatorial disk, emitted from between the lobes. The gases are traveling outward at a speed in excess of 1.5 million miles per hour (2.5 million km/h). The detail in the gas lobes was hard for the telescope to achieve because the central star is a hundred thousand times brighter than the nebulae.

Eta Carinae and the Homunculus Nebulae.

Wolf–Rayet Stars

Wolf–Rayet stars are hot blue-white giants nearing the ends of their lives. They are rare objects, named for their discoverers, who noted that certain stars had unusually strong emissions from the elements nitrogen, carbon, and oxygen. These are the products and ingredients of high-temperature fusion reactions that occur within massive stars. Their signatures appearing in stellar spectra so strongly suggested that these elements had moved out from the core to the surface.

Wolf–Rayet stars are classified by the dominant element in their spectra: WN for nitrogen-heavy, WC for carbon-heavy, and WO for the rarer oxygen-heavy. All are more than 16 times heavier than the Sun and have surface temperatures ranging from 25,000 to 100,000 K. Their stellar winds are so strong that they lose entire layers of material. Some can lose as much as the mass of the Sun in only 10,000 years.

Over 200 Wolf–Rayet stars have been cataloged in the Milky Way, with as many as ten times that number expected to exist throughout our galaxy. The strong spectral signature of Wolf–Rayet stars makes them easy to identify even in other galaxies. They are commonly found in galaxies known to be undergoing a burst of star formation.

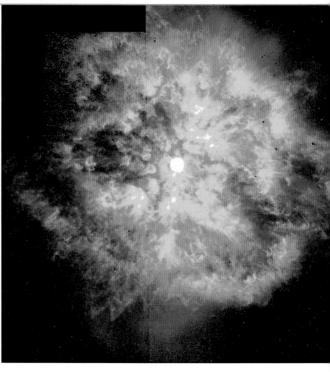

The star WR124 resembles a permanent explosion. Superhot plasma is being ejected into space at more than 93,000 miles per hour (150,000 km/h). This process appears to have begun recently, since the expanding gas cloud is no more than 10,000 years old.

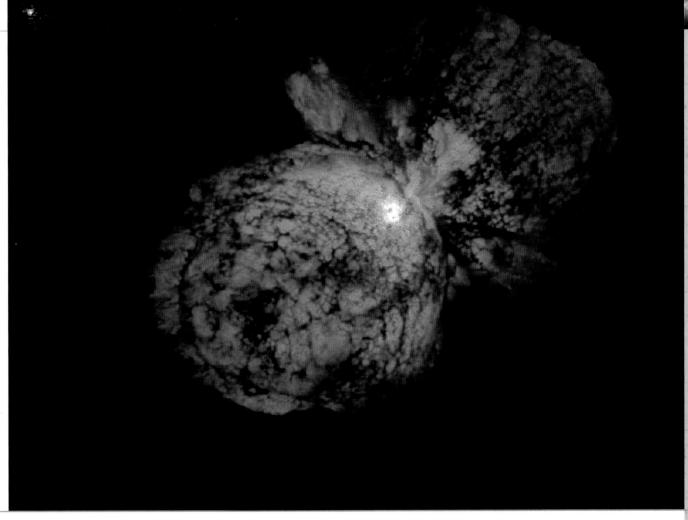

Red Giants

When our Sun (or any star of similar mass) comes to the end of its life on the main sequence, it evolves into a red giant. Red giants are low-temperature, high-luminosity objects that inhabit the top right- hand corner of the Hertzsprung–Russell diagram. Their cores have run out of hydrogen fuel and have begun fusing helium instead. Hydrogen fusion continues in a shell around the core, pushing the outer layers of the star out. As it expands, the surface of the star gets cooler as less radiation passes through it per unit area, although the overall luminosity increases. The core gets hotter and is able to fuse ever larger elements, detectable in the stellar spectra. The presence of technetium, a radioactive element, shows that the fusion process is ongoing. Technetium has a half-life of 200,000 years, short enough to rule out its presence when the star first formed.

One of the largest red giants is the star Betelgeuse, the highly visible left shoulder of Orion. Its radius is variable, driven by the increasingly unstable fusion in the shell outside the core. At its largest, Betelgeuse has a diameter of nearly 1 billion miles (1.6 billion km). Its irregular outer atmosphere extends much farther— if it were placed at the center of our solar system, it would extend past the orbit of Jupiter. Despite its vast size, Betelgeuse has an estimated total mass of only 20 times that of the Sun. The average density of the star is therefore less than that of Earth's atmosphere at sea level.

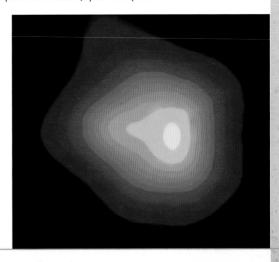

A radio image of Betelgeuse.

WHITE DWARFS

White dwarfs represent the final stage in the process of evolution for stars not large enough to be considered giant. They are the supercompact ashes of burned-out stellar cores.

After a dying star has reached its peak as a red giant, its outer layers are lost and they form a planetary nebula, leaving behind the stellar core. This is known as a white dwarf, and it consists of the ashes of fusion—mainly carbon and oxygen. It retains the intense heat of the core, with a surface temperature that is often more than 100,000 K, but it is not hot enough to use carbon as a fuel for further fusion reactions.

A white dwarf has a maximum mass. Without an internal source of power to counteract gravity, it is crushed to an incredibly high density. The only force holding it up is the great pressure created by the compressed electrons (known as electron degeneracy pressure). A teaspoon of this material would weigh about 5 tons. Even this pressure can be overcome by gravity as long as there is enough mass. A white dwarf with more than 1.4 times the mass of the Sun would rapidly implode and create a supernova. A lot of the mass of a star is dissipated as gas and dust in the red-giant stage, so stars up to eight times the mass of the Sun will usually leave behind stable white dwarfs.

The white dwarf will slowly cool over time. There is evidence that the carbon atoms crystallize as the white dwarf cools. This would produce a diamond approximately the size of the Earth but 300,000 times denser. Eventually the white dwarf will cool to the point where it no longer shines, becoming a dark cinder. Astronomers believe that the cooling rates are such that no white dwarf has yet completely cooled down in the 13.7 billion years since the big bang.

Above: *The planetary nebula NGC 6369 surrounds a new white dwarf.*

Right: *The central star of the planetary nebula NGC 2440 is one of the hottest known. The nebula is rich in dust clouds, some of which form long dark streaks pointing away from the central star.*

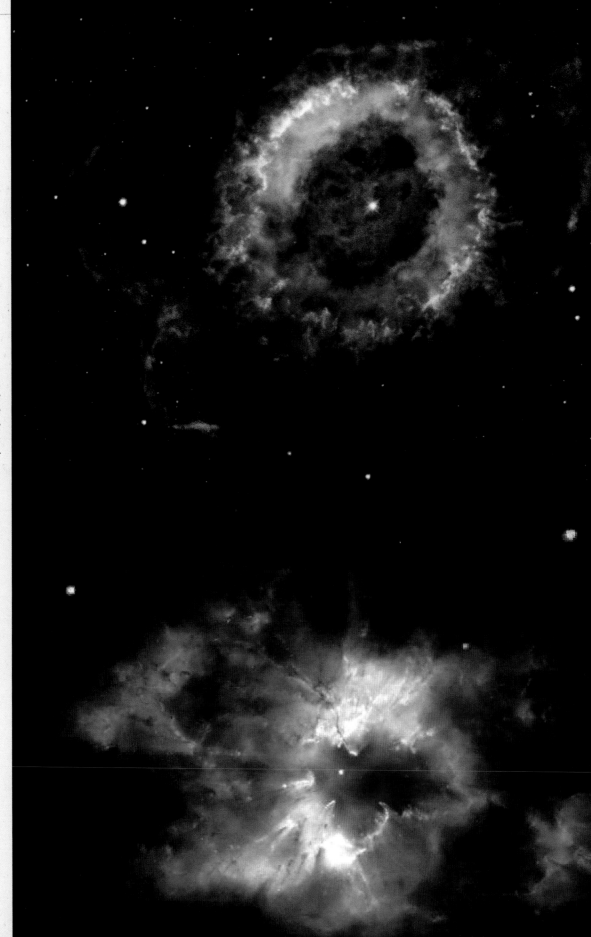

Right:
Astronomers have found a white dwarf among the 20,000 or so stars in the globular cluster NGC 1818.

Sirius A and B

The brightest star system in the night sky is Sirius, a binary star. Sirius A is a bright yellow-white star 2.5 times the mass of the Sun. In 1841 German astronomer Friedrich Bessel noted that the star wobbled as it crossed the sky, and predicted the existence of an unseen companion. Sirius B was viewed directly for the first time in 1862. Since Sirius was known as the Dog Star, Sirius B received the nickname "the Pup."

The two stars orbit at a distance of 1.9 billion miles (3 billion km) from each other. It takes Sirius B about 50 years to complete an orbit around Sirius A. The system is only 8.6 light-years from the Sun. Close examination by the Hubble Space Telescope has allowed astronomers to make an accurate measurement of the size and mass of Sirius B. Because of its great density, Sirius B has an exceptionally strong gravitational field at its surface—350,000 times stronger than that on Earth. Light emitted from its surface has its wavelength lengthened slightly by gravity's pull, and the extent to which this occurs translates directly into a measurement of its mass. Sirius B is only 7,500 miles (12,000 km) across, slightly smaller than Earth, but it contains as much mass as the Sun. This marks it as a white dwarf star. The precision of Hubble was needed to measure it, because Sirius A outshines Sirius B by a factor of 10,000.

Sirius B was probably originally a star four times the mass of the Sun, heavier than Sirius A. Heavier stars evolve faster, so although they were born together, Sirius B has evolved through the red-giant stage and become a white dwarf well before its companion.

An artist's impression of Sirius A and its tiny companion, Sirius B (bright blue).

BINARY STARS

A binary system comprises two stars that orbit the point of their combined center of mass (known as the barycenter). It is believed that the majority of stars in our galaxy are in binary systems. The interaction between such close companions provides us with useful information on stellar evolution.

The existence of binary star systems first entered astronomers' thoughts at the end of the 16th century. Later, William Herschel, also known for the discovery of Uranus, made an extensive study of double stars. He expected to see stars that appeared close together because they shone from almost exactly the same direction but were otherwise unrelated. The components of such systems typically had measurably different magnitudes, leading astronomers to think that the fainter of the pair was much farther away. Herschel planned to observe a binary, note the separation, and observe it again six months later. The second observation would record a different separation, since the closest star would have moved farther relative to the background stars than the more distant binary component. This phenomenon is known as parallax and was used to measure the distances to nearby stars. When Herschel analyzed his measurements, he came to realize that the variation was sometimes caused not by parallax but by the stars' own orbital motion about each other. This was the first evidence of orbital motion outside of the solar system.

Herschel studied the Castor system in the constellation Gemini. Through a telescope he could see that it was a double star and he was able to determine that the two stars were gravitationally bound to each other. A binary star system in which the two components are far enough apart to be viewed separately through a telescope is known as an optical binary.

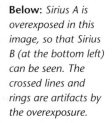

Below: *Sirius A is overexposed in this image, so that Sirius B (at the bottom left) can be seen. The crossed lines and rings are artifacts by the overexposure.*

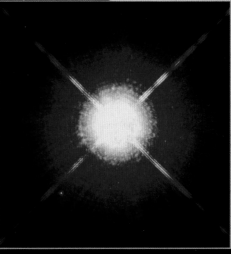

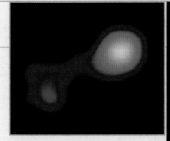

Above: *A Chandra X-ray image of the Mira star system.*

Below: *Hubble views the tiny companion of the star HD 16160 (seen at lower left). The companion star (upper right) is a brown dwarf, one of the smallest stars known. The white bar across the large star is an image artifact caused by the over-exposure needed to view the fainter star.*

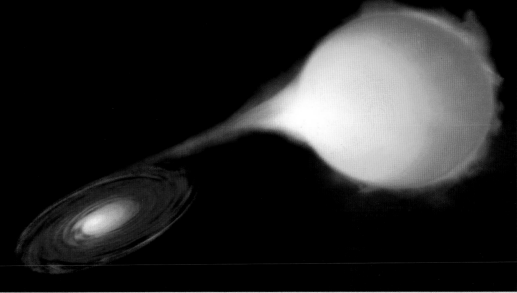

Above: *Artist's concept of the Mira star system: Mira A (right) and Mira B (left). Mira B's gravitational tug makes a gaseous bridge between the stars, and gas from Mira A's stellar wind and the bridge accumulates in an accretion disk around Mira B.*

Spectroscopic Binaries

The two stars Castor A and Castor B orbit once every 420 years. Unknown to Herschel, both components are, in fact, binary stars themselves. The binaries were impossible to resolve with telescopes because they are much too close together to be separated visually. Such binary systems are called spectroscopic binaries, since they are identified through the effect of their orbital motion on their light spectra.

Spectrographic analysis of starlight can reveal two important facts about the star itself. A star produces light at a wide range of frequencies. Elements in the star shine at characteristic frequencies, resulting in a peak in the spectrum. Elements can also absorb light at specific frequencies, resulting in a blank section in the spectrum. The spectra can therefore provide an ingredients list for

stars. Secondly, the characteristic patterns of peaks and troughs can be shifted up or down the spectrum owing to the motion of the star toward or away from us. This is the Doppler effect working on light waves, and it provides a sensitive test for orbital motion. If the spectrum of a star is alternately shifted up and down, it is moving back and forth—indicating that it is in orbit around another object. Both Castor A and Castor B were found to be spectroscopic binaries with very short orbital periods of a few Earth days. Further observations have found another component of the system, Castor C. It orbits the A and B components at a distance of roughly 93 billion miles (150 billion km).

The white dwarf Sirius B, viewed in X-rays. Sirius A is 10,000 times brighter in the optical part of the spectrum, but can only just be seen here as a spot above the dwarf. The six radial lines are imaging artifacts.

Eclipsing and Astrometric Binaries

Castor C is another binary system, with two faint red dwarf stars closely orbiting each other. The plane of their orbit is aligned with our line of sight, which means that the brightness of the system varies periodically as one star occults the other. This is known as an eclipsing binary. The rate and magnitude of the change in emitted light from such a system provides a lot of information about the sizes of the stars involved. The spacing of dips in the light curve gives the orbital period of the stars, which can be as low as minutes in extreme cases. This technique has been expanded to hunt for large planets that may be orbiting other stars.

Castor is therefore not just a binary system—it is a sextuple star system made up of three separate binaries. Such a system is rare—the majority of stars are doubles, singles (like the Sun), or triples (Alpha Centauri).

Some stars appear to orbit objects that cannot be observed through visual methods such as telescopes or by using spectro-graphic techniques. The periodic wobble of the bright component is often the only clue to the existence of a dark companion. Such systems are called astrometric binaries. Often the dark companion has been detected subsequently through better observational

techniques. Sirius was found to move in a sinusoidal (sine-curve shaped) path across the sky by Friedrich Bessel in 1841, leading him to propose the existence of a companion. Later observations with better telescopes were able to view the companion star—a white dwarf—directly.

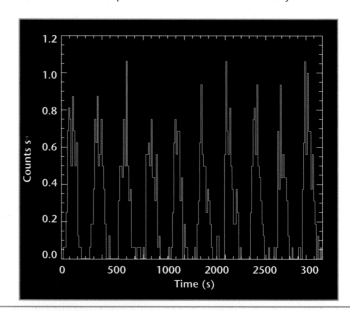

Right: *An X-ray intensity graph of a double white dwarf binary system. The stars are only 50,000 miles (80,000 km) apart.*

Above: *The illustration shows two white dwarf stars orbiting each other and destined to merge. As they do so, they flood space with gravitational waves. A ripple from these waves would cause Earth, the Moon, and all matter to wobble, subtly altering the distance between them. Although such waves were predicted by Einstein, they have never been detected.*

Binary Star Formation

The formation of systems with two or more gravitationally bound stars is a subject typically investigated through computer simulations. For a pair of previously separate stars to become a binary system, they need to come close—but not close enough to collide—and then lose enough momentum to avoid flying off in different directions. This is most likely to occur in dense star clusters where other nearby stars have a gravitational impact on the pair. However, studies show that binaries are more likely to be created when stars are formed than when pairs of mature stars come together. Gas clouds collapse under their own gravity to form stars, and an unevenly distributed gas cloud can easily form two stars that begin their lives very close to each other.

A star-forming gas cloud is a large dispersed object, probably moving as well as rotating. As it collapses, it spins faster and faster (much as a spinning ice skater brings the arms close to the body to speed up rotation). This is because, in the absence of intervening forces, the rotational momentum of a system is conserved—if mass is moved closer to the axis of rotation, the rotation speed increases. A single star born from such a cloud would need to spin very fast to preserve the rotational momentum of the original cloud. The inertia of the gas makes it harder for a single star to form—two stars orbiting a central point would be more feasible.

Close Binaries

Most binaries consist of separate stars, but there is also a class of systems known as close binaries. The stars can be close enough for stellar material to pass from one to the other. These systems are known either as semidetached binaries or contact binaries. If one star's outer layers are being pulled off by the gravity of its companion, the system is a semidetached binary, but if both stars share material, it is said to be a contact binary system. In a semidetached binary the atmosphere of the larger star is pulled into a teardrop shape, with the sharp end trailing onto its companion, which is a smaller, denser object.

Gas flowing from one star onto another often forms an accretion disk, much like water going down a plughole. As the gas spirals around, it heats up through friction. If the accreting object is dense enough—for example, if it

is a neutron star or a black hole—the gas will be so hot that it emits X-rays. Novae occur when accreting gas causes a flare-up on the surface of the denser binary component, and the star shines very brightly for a short amount of time. A particular kind of supernova (known as a Type Ia supernova) occurs when a white dwarf's mass is increased through accretion to more than 1.44 times the mass of the Sun. The additional gravitational pressure is too much for the star to support, and it implodes violently.

Contact binaries are so close to each other that they begin to merge. Friction between the gases and magnetic fields of the two components slows the orbital rate and reduces the distance between them. Eventually, the stars will coalesce into one.

An artist's illustration of a classical nova binary system just before an explosion on the surface of the white dwarf. Gas is flowing from the large red companion star into a disk and then onto the white dwarf hidden inside the white area. As the gas flows closer to the white dwarf, it gets hotter, shown by the change in colors from yellow to white.

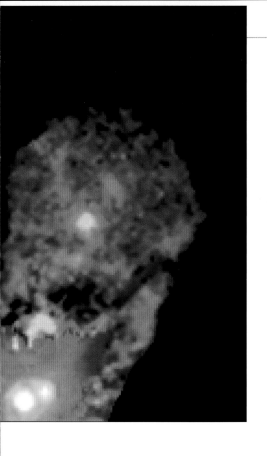
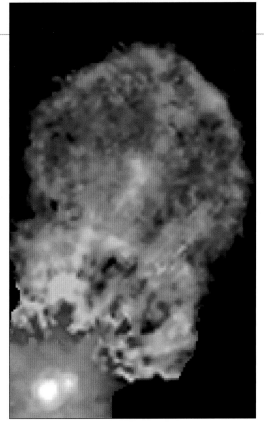
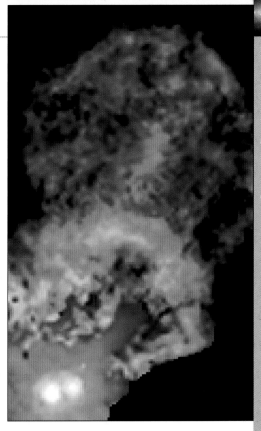

Hubble Space Telescope images show bubbles of glowing gas being blown from the young binary star system XZ Tauri. A dramatic change occurred between 1995 (left) and 1998 (center)—in 1995 the bubble's edge and interior were the same brightness, but by 1998 the edge was brighter (probably caused by the hot gas cooling off). The third picture (right) was taken in 2002.

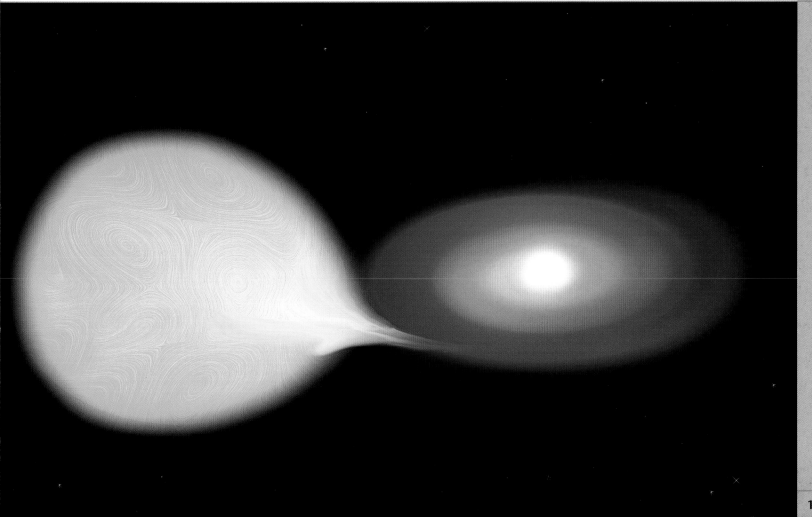

OPEN CLUSTERS

Open clusters are groups of stars that have recently formed from the same dust cloud. The component stars vary greatly in mass, but all are the same age and roughly the same distance from Earth, making them very useful in comparative studies of stellar properties.

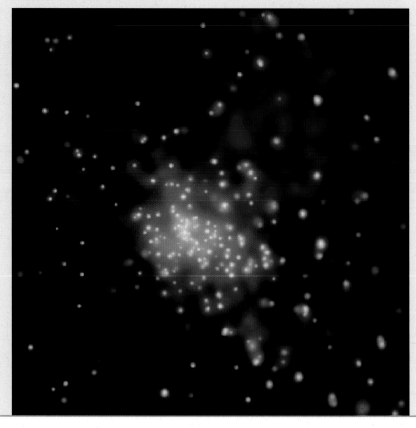

Right: *The dense cluster RCW 38 is five light-years across. The interstellar gas and the stars are hot enough to emit X-rays, as seen here by Chandra.*

Pleiades

The most visibly striking and famous open cluster is the Pleiades (M45) in the constellation Taurus. In Greek mythology the Pleiades were the seven daughters of the titan Atlas and the nymph Pleione. The names of both parents and their children are given to the nine brightest stars in the cluster, which are all just visible in good conditions with the naked eye. In total there are between 500 and 1,000 stars in the cluster, spread over an area 12 light-years in diameter.

The stars are all roughly 100 million years old. The named stars are large and bright. Alycone, the brightest, is 1,000 times more luminous than our Sun. A large number of the stars are low-mass brown dwarfs, each less than one-tenth the mass of the Sun. Some are white dwarfs that are unexpected in a young system such as an open cluster, because typically they take billions of years to form. It is be-

lieved that these white dwarfs resulted from mass transfer between stars, probably in binary systems. This process would remove matter from a massive star and accelerate its evolution.

The blue glow within the cluster is due to a nebula reflecting the light of the stars. This dust cloud is not believed to be the same one that gave birth to the stars, since their combined radiation pressure would have dispersed it well before now. Instead, it is thought that the Pleiades are currently in a region of high dust density. The Pleiades stars are currently moving across the sky in the same direction, but in about 250 million years time the stars will scatter, breaking up the cluster.

The open star cluster known as the Pleiades, or Seven Sisters, embedded in its reflection nebula.

Right: *Hubble views the cluster R136 embedded in the 30 Doradus Nebula. Cluster R136, at the center of the image, contains dozens of massive bright stars.*

Trapezium Cluster

Hanging from Orion's Belt, Orion's sword consists mainly of the emission nebulae M42 and M43. At the center of M42, at a distance of 1,500 light-years, lies one of the densest and youngest star clusters known. It is called the Trapezium Cluster, named for the shape marked out by the four brightest stars. They are several hundred thousand years old, and their violent output radiation is a major source of the energy that makes the surrounding nebula glow.

Images of the region by the Chandra X-Ray Telescope show that the stars are exceptionally hot, with temperatures flaring to more than 60 million K. Star formation is still occurring in M42 (the Orion Nebula), but protostellar dust clouds in the vicinity of the Trapezium stars are in danger of being scattered before a star can be born. Several can be seen in the large image opposite. With their trailing dust tails, they appear like comets nearing the Sun. The physical process is very similar—the radiation is eroding the outer layers from the protostar's globule. In its shadow some of the material remains, producing a teardrop shape. There are well over a hundred protostars in this region alone, with many more fully formed lower-mass stars that cannot be seen in this image. Hubble has been able to locate them by looking at the near-infrared part of the spectrum, which is able to pierce the thick nebula dust to reveal the stars hidden within. Fifty brown dwarfs were seen, suggesting that they are more numerous than astronomers initially thought. They are easier to detect in star-forming areas because they are young and retain much of the heat generated from the process of formation.

Right: *However, when seen in near infrared, not only do the main stars become apparent, but numerous smaller brown dwarfs can be seen.*

GLOBULAR CLUSTERS

Globular clusters are gravitationally bound conglomerations of stars. They contain between 10,000 and 1,000,000 stars in a spherical formation that is highly concentrated toward its center. These balls of stars can vary in size from as little as 30 light-years to as much as 300 light-years across, and the central regions can contain several thousand stars per cubic light-year.

Above: *Globular cluster M80, some 32,600 light-years away, is one of the densest star clusters in the Milky Way, comprising hundreds of thousands of stars held together by their mutual gravitational attraction.*

Right: *The Trapezium Cluster is a stellar nursery with numerous newborn brown dwarfs. They cannot be seen in visible light, in which dust masks the light.*

Right: *globular cluster 47 Tucanae, located 16,700 light-years away, appears almost as a single object when seen from ground-based telescopes. But observed by the Hubble Space Telescope, it is seen as an immense globular cluster of stars. Most are old stars, but a few are so-called blue stragglers. They may have formed by the slow merger of stars in a double-star system.*

The first globular clusters to be discovered were described as nebulae, because their density was too great to be resolved into individual stars with the telescopes available in the 1600s. We now know that Messier's catalog of nebulae contained 29 globular clusters.

The term globular cluster was invented in the 18th century by William Herschel, who determined that the fuzzy blobs were, in fact, made of stars. His surveys increased the number of known globular clusters to 70. Today we recognize 150 such clusters in the Milky Way and estimate that there could be another 50 currently hidden from our sight. Other galaxies are also known to contain globular clusters. Some contain as many as several thousands.

The double cluster NGC 1850 is located in one of our neighboring galaxies, the Large Magellanic Cloud. Both components are relatively young, with the smaller cluster to the right and below being the younger. NGC 1850 is surrounded by a filigree pattern of diffuse gas that scientists believe was created by the explosion of massive stars. No similar classes of clusters have been found in the Milky Way.

Omega Centauri

The largest globular cluster in the Milky Way was originally classified as a star in the southern constellation Centaurus. In 1677 Edmund Halley noted that the object was not a single star but many. In fact, Omega Centauri contains several millions of stars in an area 150 light-years wide. Its size is unusual, being ten times larger than the next largest globular clusters in the galaxy. Of the examples we know about, only the G1 cluster situated in the Andromeda Galaxy is more luminous. Omega Centauri is visible with the naked eye despite being 17,000 light-years distant.

Most telescopes are unable to see the centers of globular clusters as anything more than a fuzzy white area, but the Hubble Space Telescope is powerful enough to separate the stars from each other. In this image, which covers an area 13 light-years across, there are 50,000 stars. By comparison, the stellar density in the vicinity of the Sun is several thousand times less. With stars in such close proximity to each other, close encounters and even collisions are possible.

Although like most other globular clusters it is old, analysis of the light from Omega Centauri reveals that not all the stars formed at the same time. There is a spread of a few billion years, the average age being around 12 billion years. This fact, combined with Omega Centauri's unusual size, has led astronomers to speculate that the cluster is the core of an old dwarf galaxy that has collided with the Milky Way and lost most of its stars.

The center of Omega Centauri, as seen by the Hubble Space Telescope. Most of the stars in the image are faint yellow-white dwarf stars similar to our Sun. But there are also a few red giants (yellow-orange stars) and some faint blue stars that are midway between dwarf stage and red-giant stage.

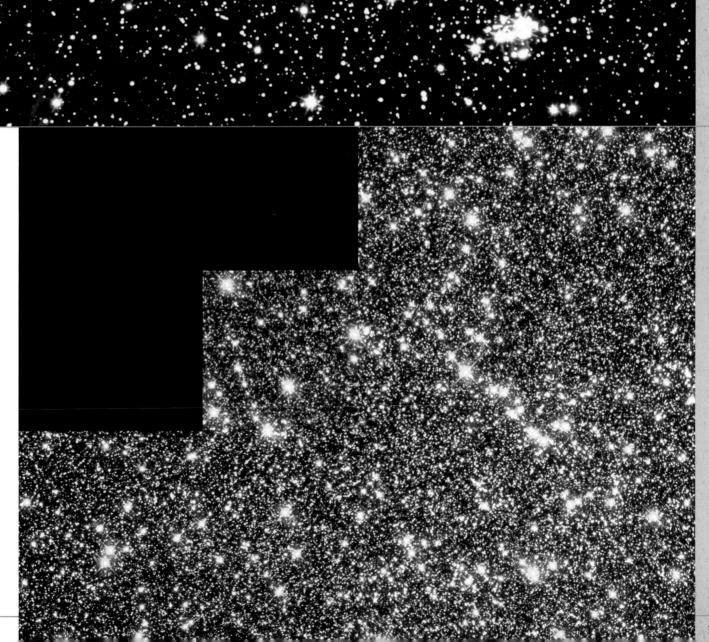

NGC 6397

NGC 6397 is the second closest globular cluster to Earth, lying at a distance of 7,200 light-years. It contains about 400,000 stars. Ten percent of the globular clusters in the galaxy, including NGC 6397, have undergone a process known as core collapse, which results in a much denser central region. The stellar population there is a million times more crowded than the space around the Sun. Collisions between stars occur every few million years. The resulting objects are known as blue stragglers. Their increased temperature, which is gained from the energy released in the collision, makes them appear hot, and the increase in hydrogen fuel effectively regenerates them into young stars. These bodies initially confused observers, who expected to find only ancient yellow and white stellar populations in globular clusters. Astronomers have also detected in this cluster the effects of near collisions. Cataclysmic variables are binary systems in which a normal star and a white dwarf have captured each other and are orbiting closely. Material from the normal star is torn off and accreted onto the surface of the white dwarf. This generates fluctuating bursts of ultraviolet radiation, which has been detected in this image by Hubble.

Faint low-temperature stars in NGC 6397 were used to determine the age of the Milky Way. The stars in the cluster are 13.4 billion years old (plus or minus 0.8 billion). They are not the first stars born in the galaxy, since they contain elements that could only have been created as a result of fusion in hot massive stars. The abundance of these elements (in this study beryllium was used) suggests that the very first stars formed 200 million years earlier. That makes the Milky Way 13.6 billion years old—in other words, it formed only a few hundred million years after the big bang.

The stars in NGC 6397 are in constant motion and so close together that collisions and near misses occur regularly.

PLANETARY NEBULAE

Possibly the most beautiful objects in astronomy are planetary nebulae. They have a wide array of shapes and structures that can be appreciated even through small telescopes. The most powerful telescopes, such as the Hubble Space Telescope, are able to distinguish the compositional and color differences. The end result resembles abstract art, generated by the laws of physics.

The ejected layers of stellar material lost by a red giant star as it changes into a white dwarf can produce stunning emission nebulae. These objects were given the name "planetary" nebulae by the 18th-century astronomer William Herschel. They do not have anything to do with planets, but Herschel perceived them as being similar to the planet Uranus, which he discovered. He later observed such an object with a bright central star, and considered the possibility that the nebula was made from material emitted by the star.

Planetary nebulae are too faint to be seen by the naked eye. Charles Messier discovered the first one in 1764. The Dumbbell Nebula is listed as M27 in his catalog, and there are three other planetary nebulae listed.

Early spectroscopic studies of planetary nebulae revealed that they emitted light strongly in a few specific wavelengths. The brightest of them did not appear to correspond to any known element at the time. To explain the observation, scientists proposed the existence of a new element and called it "nebulium." Similarly, the element helium was discovered by observing its spectra in the Sun before it was isolated as a gas on Earth. Further studies showed that nebulium was, in fact, a particular state of oxygen that existed only at high temperatures and in a very low-density environment. Additionally, the spectra of planetary nebulae show that they are all expanding away from their central star, typically at speeds of a few miles per second. Therefore, by extrapolation, these structures are no more than a few tens of thousands of years old.

Right: *NGC 2392's unique shape has given it the nicknames Clown Face Nebula, and Eskimo Nebula.*

Below: *M2-9 is a striking example of a bipolar planetary nebula. It is also known as the Twin Jet Nebula.*

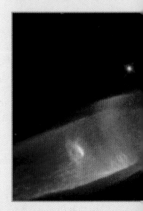

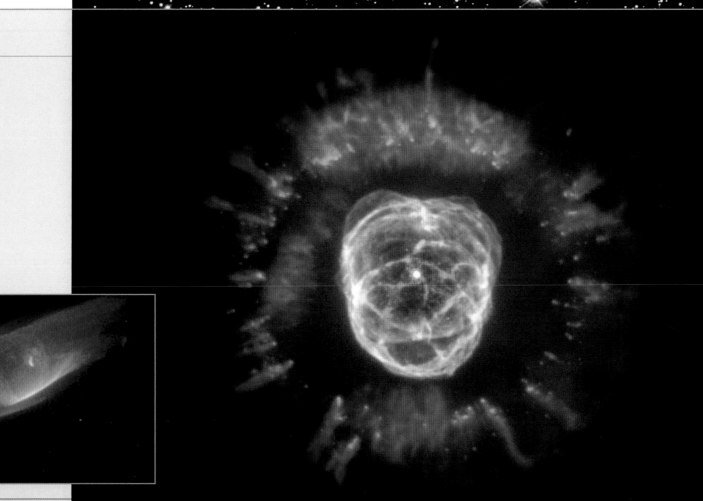

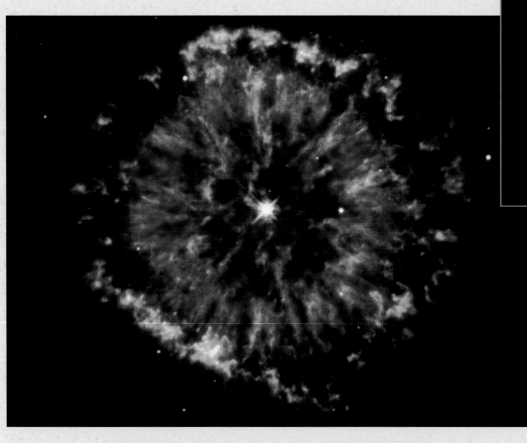

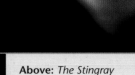

Above: *The Stingray Nebula (Hen-1357).*

Left: *NGC 6751 glows like a giant eye.*

Retina Nebula

In a universe in which the nature of gravitational forces produces so many spherical, circular, or even spiral structures, a rectangle is an unusual sight. This is what the Retina Nebula (IC 4406) appears to be when seen face-on. Viewed side-on, however, it is cylindrical. It is bipolar, the axis of symmetry being perpendicular to our line of sight. If we could view it end-on, it would probably appear very similar to the Ring Nebula. As with many planetary nebulae, it has a high degree of symmetry over a large area. The central circumstellar disk and the narrow waist structure seen in most bipolar nebulae are almost entirely absent in the Retina Nebula. The reasons for these particular variations are still subject to speculation.

The Retina Nebula lies in the southern constellation Lupus (the Wolf) at a distance of roughly 5,000 light-years. In this color-corrected image taken by the Hubble Space Telescope, ionized hydrogen gas is green, oxygen is blue, and nitrogen is red. The dark lines that cross the central region of the nebula are made from dust, perhaps a thousand times denser than the glowing material. Each line is about 15 billion miles (25 billion km) wide—four times the distance from the Sun to Pluto. They exist at the boundary between the glowing gas against which they are silhouetted and the unseen neutral gas beyond. These dark lines give the nebula its name, since they appear similar to the network of tiny blood vessels that cross the retina of a human eye.

In this edge-on view of the dying star at the center of planetary nebula IC 4406, the Hubble Space Telescope reveals a rainbow of (false) colors.

Most stars that have between 0.8 and five times the mass of the Sun will enter the red-giant phase and cast off their outer layers. However, because planetary nebulae are relatively short-lived structures, they are rare. At any one time our galaxy contains perhaps 10,000 of these objects. We can only see between 10 and 15 percent of them because their light does not penetrate through dark interstellar clouds. They are effectively emission nebulae, but on a smaller scale. The central star is a newly formed white dwarf that is capable of heating the cast-off dust to over 10,000 K. Observed examples are typically on the scale of one light-year wide. They are limited either by the extent of the dust cloud or by the fact that the stellar radiation has not ionized all the available material.

Planetary nebulae come in a wide range of shapes, from the simple to the intricate. Most are spherical, with perhaps a slight bipolar influence caused by the spin of the parent star. Stronger bipolar influence to the final shape may come from companion stars or even planetary systems, and occurs in roughly 15 percent of nebulae observed. They are narrow around the central region and have two lobes that spread from the poles. The glowing dust often obscures the central region, so it is not always known whether companion stars are present. In some cases thick debris rings around the star are thought to provide a funnel for the outflowing gas, keeping it from escaping in a spherically symmetric way. An alternative cause of strong bipolar shapes is the star's own magnetic field.

Right: *A false-color image of the Red Rectangle (HD 44179) from the Hubble Space Telescope. In this picture the white areas represent the brightest parts of the nebula, and red the faintest.*

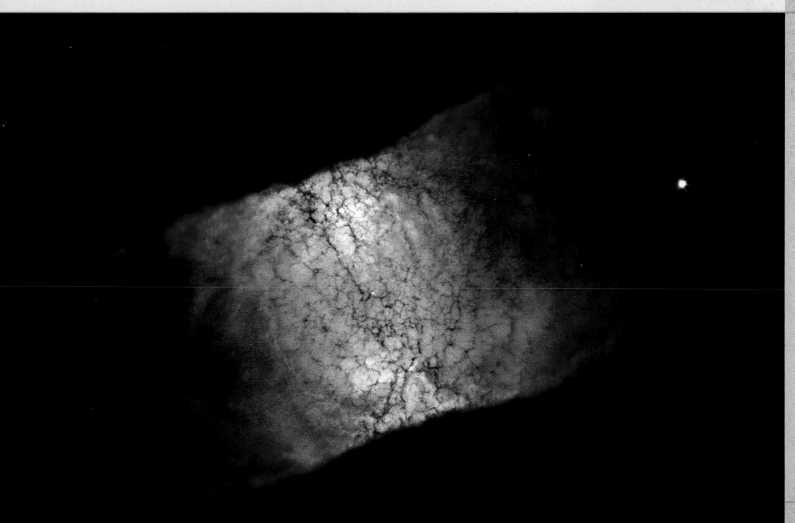

Eight-Burst Nebula

The planetary nebula catalogued as NGC 3132 is also known by the names Eight-Burst or the Southern Ring. It is 2,000 light-years away, half a light-year across, and has an oval shape as viewed from Earth. The outer edges of the heated dust and gas are traveling away from the central star at around 9 miles per second (15 km/s).

Two stars are visible in the center of this image. The smaller one is a newly formed white dwarf. It is the source of the nebula. Its companion star will have had an influence on the structure of the nebula, but the details of such an interaction are not understood. The star in the top right is not related to this system.

The colors in this Hubble Space Telescope image are chosen to reflect the temperature. Blue indicates the hottest gas, while red is cooler. A dark lane of dust can be seen curving across the central blue area, but the cause of this structure is unknown.

This image of NGC 3132 shows two stars near the center of the nebula—a bright white one and an adjacent fainter companion to its upper right.

Helix Nebula

One of the closest planetary nebulae to Earth is the Helix Nebula, thought to be between 400 and 800 light-years from Earth. Its proximity and face-on nature make it the largest planetary nebula seen on the sky, with an apparent diameter nearly that of the full Moon. It is too faint to be seen by the naked eye. Neither Messier nor Herschel, the early catalog compilers, was able to see it. It is a bright example of a planetary nebula, but because its light is spread over a large area, they probably looked right through it.

In the main image Hubble shows the temperature range through the nebula. The central, hottest areas are blue (to represent the light given off by ionized oxygen). Outside, the temperature drops below the threshold needed to ionize oxygen, and the colors show nitrogen and hydrogen emissions. Resolving a three-dimensional shape from a two-dimensional image in this way can be very difficult, but there is evidence to suggest that the Helix Nebula contains two dust disks perpendicular to each other.

The inner edge of the yellow ring has a large number of cometary knots. The heads of these tadpole-shaped dust clouds are larger than our solar system, and the trailing tails are about 93 billion miles (150 billion km) long. They are created by the collision of hot gases newly emitted by the star and old, colder gas possibly emitted in the red-giant stage. The collision of gases is not smooth—denser pockets of older gas will shield other areas and keep them from being pushed out. The resultant shapes resemble spokes radiating away from the central star.

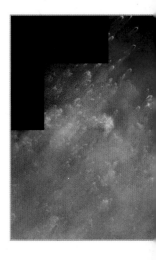

Right: *Composite image of the Helix Nebula.*
Inset: *Detail of cometary knots.*

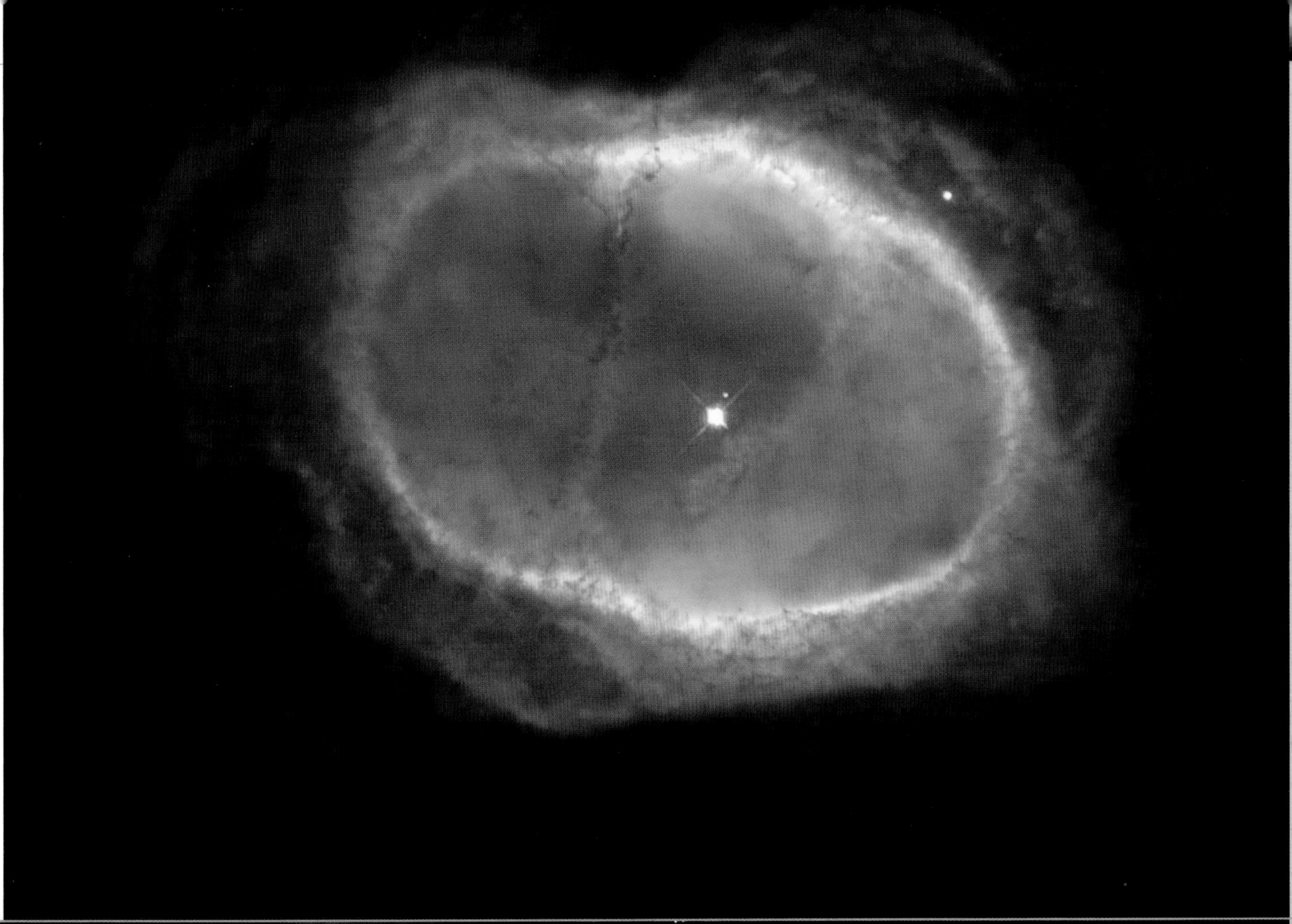

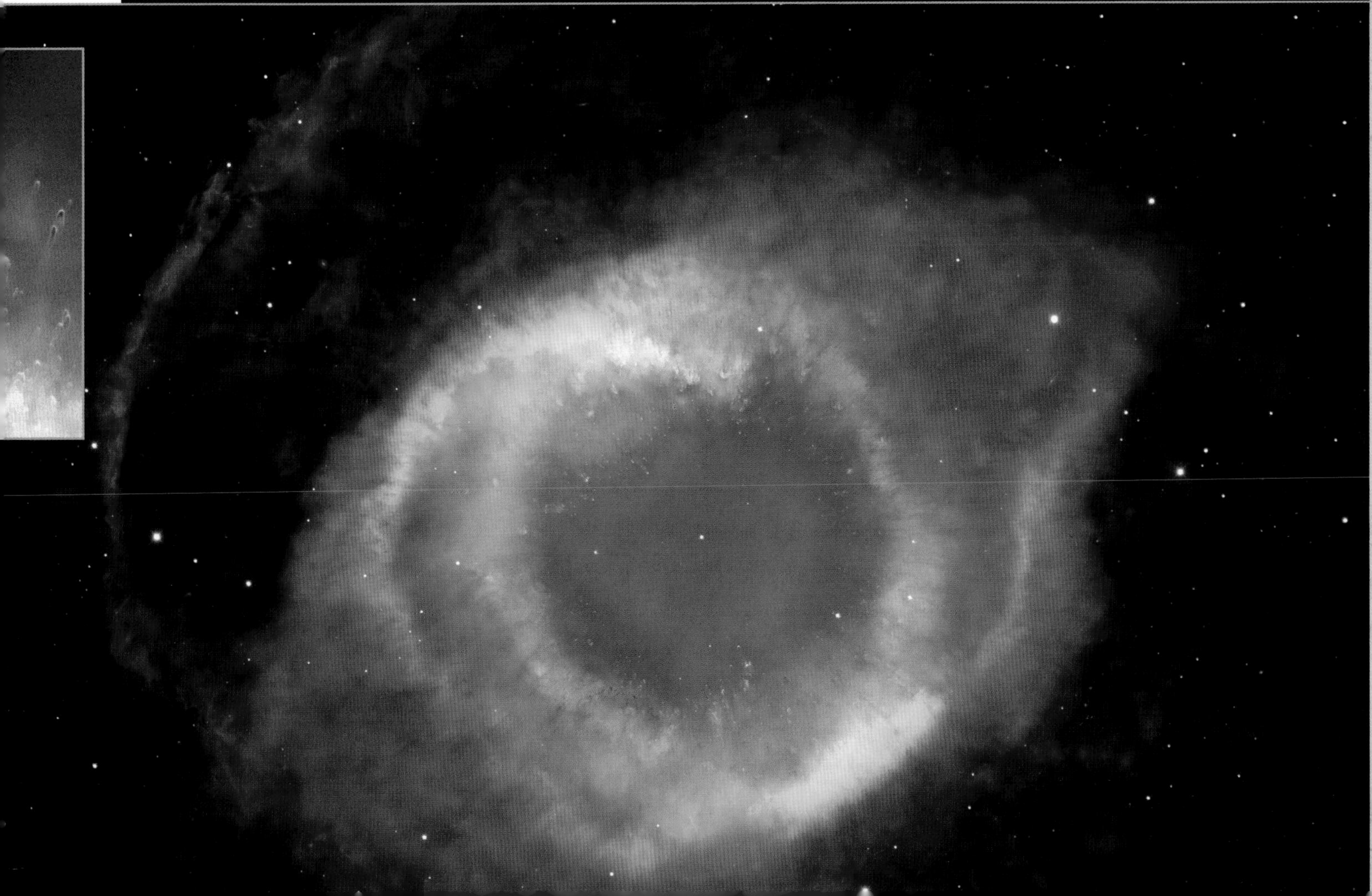

Cat's Eye Nebula

Discovered in 1786 by Herschel, this nebula is situated in the constellation Draco. It is almost precisely on the ecliptic pole. An observer at the Cat's Eye would see our solar system face-on. It is approximately 3,000 light-years away and 1,000 years old.

The Cat's Eye Nebula is one of the most visually complex planetary nebulae known. Its shape includes rings, arcs, and shock fronts. Astronomers believe that the shape must be the result of interactions between two close binary stars at the nebula's center, although this has not been confirmed. Mass transfer between the components of the binary could produce outflow jets that would influence the pattern of the planetary nebula.

The image opposite is a composite of visual spectrum data from the Hubble Space Telescope and X-ray detection by the Chandra X-Ray Observatory. Ionized hydrogen appears as red, and nitrogen is green in the Hubble picture. X-ray emission is shown in blue. The central star has a surface temperature of 60,000 K, but the blue regions indicate that the surrounding gas has been heated to several million Kelvin. This is caused by the shock created when the fast-moving gas slams into previously ejected material and heats up. Investigations by the Infrared Space Observatory (a precursor to the Spitzer Space Telescope) revealed traces of other elements, including sulfur, neon, and argon.

The Cat's Eye Nebula, combining data from both Hubble Space Telescope (visible-light) and Chandra (X-ray) observations.

Ring Nebula

The most famous and easily recognized planetary nebula is M57, the Ring Nebula. It was the second such object to be discovered (after the Dumbbell Nebula). Messier described it as faded and similar in size to Jupiter. It is visible to small telescopes as a faint greenish object shaped like a smoke ring, lying in the constellation Lyra. The green tinge is caused by the majority of the light coming from low-density ionized oxygen. Images from the largest telescopes have their color balances altered in order to reveal as much useful detail as possible.

In this picture, blue indicates emission from the hottest gases (helium), green is cooler (oxygen), and red cooler still (hydrogen). Based on estimates of the size and speed of the gas ring, the nebula was ejected from the central star between 6,000 and 8,000 years ago. The white dwarf responsible is clearly visible in the picture and remains very hot at 120,000 K.

The Ring Nebula is similar to the Helix Nebula, in that the boundary between the hottest gas (blue) and the cooler regions (yellow) contains many cometary knots. The ring is 1 light-year in diameter and 2,000 light-years from the Sun. The three-dimensional shape of the nebula is believed to be either a torus or a cylinder. In either case, our line of sight is close to the axis of symmetry, as if we were looking into a barrel. The slightly elliptical nature of the ring suggests it is not quite face-on.

The colors in this Hubble Space Telescope image of the Ring Nebula are approximately what the eye would see. The image was assembled from three black-and-white photos taken through different color filters.

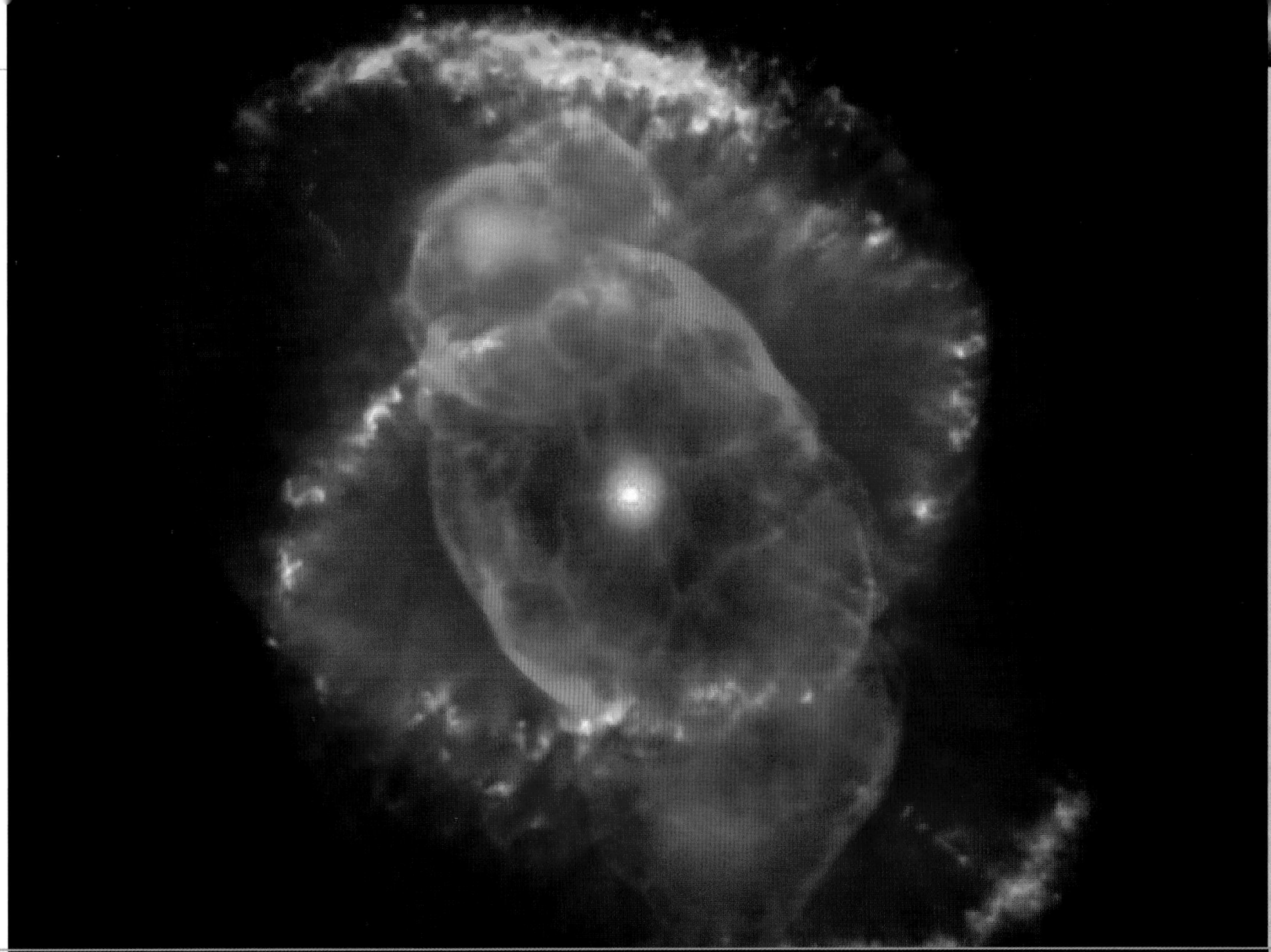

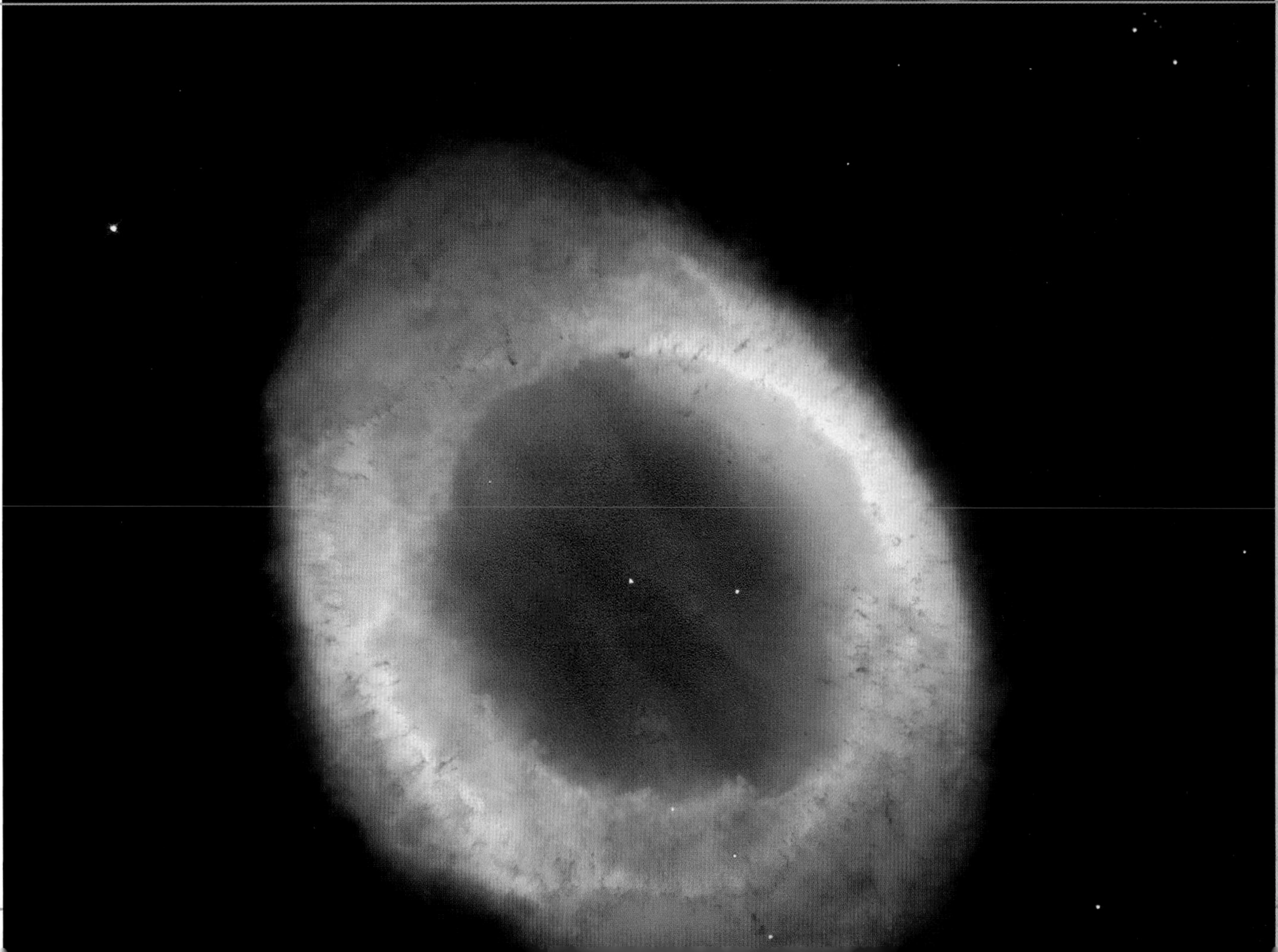

SUPERNOVA REMNANTS

The term *supernova* comes indirectly from the Latin *nova stella,* meaning new star. Despite the meaning, it is the event that signals the explosive death of a star. A nearby supernova can cause a previously unseen star suddenly to shine as brightly as any other in the night sky for a short period of time.

Supernovae occur in two different ways and are referred to as types. Analysis of the light spectrum determines the type of a particular event. Historically, the types were split up into Types I (with subtypes Ia, Ib, and Ic) and II. Continued research has found that Ib and Ic are variants of Type II supernovae, with Ia being substantially different. The terms have remained, however—an example of old classification systems conflicting with newer scientific revelations.

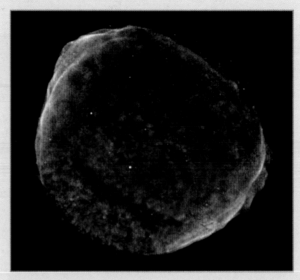

Above: *A remnant of a Type Ia supernova that was visible on Earth in 1006. Chandra's X-ray image shows the formation of glowing blue shockwaves as the material rushes out.*

Right: *The Veil Nebula in Cygnus. This fragment is also known as the Cygnus Loop and came from a supernova that would have exploded more than 5,000 years ago.*

Type Ia Supernova Remnants

Type Ia supernovae are caused by the accretion of matter onto a white dwarf, typically from a large binary companion star. A white dwarf is almost entirely made of oxygen and carbon, with no fusion occurring. It is stable as long as conditions within it remain below the threshold required to fuse carbon. A binary companion star that is close enough will have its outer layers torn off by the white dwarf, adding to its own mass. As the mass goes up, the density and temperature in the white dwarf's core increase. At the Chandrasekhar limit of 1.44 times the mass of the Sun, the conditions are right for carbon–oxygen fusion. This is a runaway process that consumes the entire star almost instantly. The resultant explosion releases a staggering amount of energy—as much as 10^{44} joules (the total amount of energy produced by our Sun in its entire 10-billion-year life-time). For a few days the supernova will outshine all the stars in an average galaxy combined. The resultant elements (mostly iron) produced by the fusion reactions are scattered at initial speeds of 6,000 miles per second (10,000 km/s). The expanding cloud appears similar to a planetary nebula, but in this case the progenitor star is entirely destroyed. Also, the physical scales are vastly different; where supernova remnants can be up to 100 light-years in diameter, planetary nebulae are typically only one light-year across.

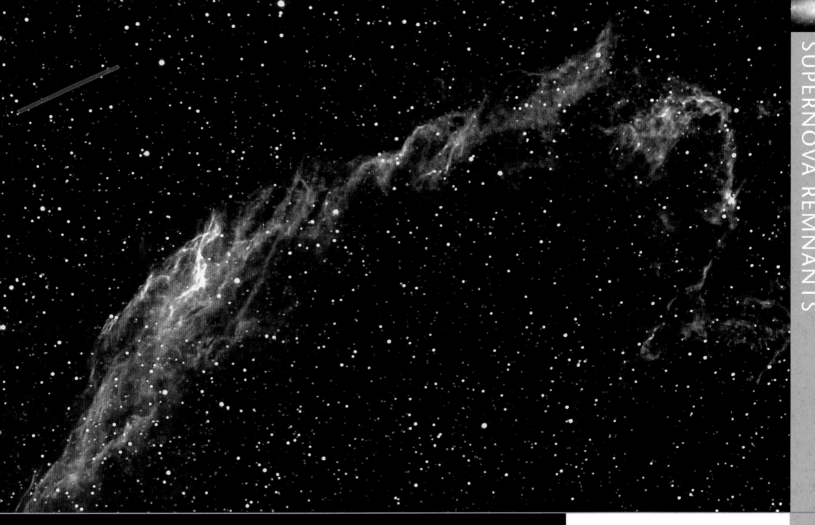

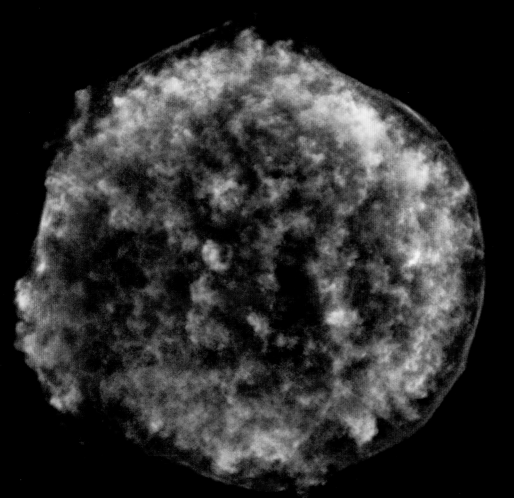

Turbulent debris created by a supernova explosion first observed by the Danish astronomer Tycho Brahe in 1572 and named for him as Tycho's Supernova. This X-ray image of the remnant shows an expanding bubble of million-degree debris (green and red) inside a shell of extremely high-energy electrons (blue).

Type II Supernova Remnants

Type II supernovae (and types Ib and Ic) result from the death of a giant star. Toward the end of its life, a giant star will have run out of hydrogen in its core. The loss of energy generation leads to contraction of the core, raising the temperature to the point at which helium fusion can begin. When the helium runs out, carbon takes over, followed by oxygen, silicon, and finally iron. Iron does not generate energy in fusion reactions and therefore collects in the core of the star. The core is heavy enough (more than 1.4 solar masses) to collapse rapidly. The normal pressures that hold up white dwarfs are overwhelmed as protons and electrons fuse together to form more neutrons. The core reaches the density of an atomic nucleus—around 500 million tons per cubic centimeter. At that point the internal pressure is greater than the compression effect of gravity, and the core rebounds like a spring. The resultant shockwave passes through the outer layers of the star, heating and propelling the material outward. Elements heavier than iron are formed in the extreme conditions at this stage.

For each proton and electron pair turning into a neutron, another particle called a neutrino is produced. Neutrinos are unusual in that they move at very close to the speed of light and very rarely interact with ordinary matter. Many thousands of these particles pass through our bodies every second, yet we would have to live for many thousands of years before we had an even chance of one actually hitting an atom inside us. Of the energy produced by a supernova, 99 percent is carried away in the form of vast numbers of neutrinos that can be detected by instruments on Earth. A sudden peak in the detection of neutrinos indicates a supernova event.

Cassiopeia A

Supernovae are rare events. Within the Milky Way, it is estimated that they occur on average once a century. The most recent nearby supernova occurred 10,000 light-years away in the constellation Cassiopeia. It was a Type II supernova and has left behind an expanding gaseous remnant and a central neutron star. Measurements of the remnant from the explosion suggest that it is about 300 years old as we see it. Unlike other supernova events—such as those seen by the astronomers Kepler and Tycho—there are no records of a "new star" appearing at that time. This could be due to the progenitor star casting off so much matter in its last years that it obscured the visible light flash that would have occurred.

The shell of matter seen here at various wavelengths is moving outward at 2,500 miles per second (4,000 km/s), and its average temperature is 30,000 K. The chemical composition of the cloud is of great interest because the distribution of elements in the ejected cloud can point to the internal structure of the star and how it exploded. Iron-rich clouds would come from the core material that has not remained to form a neutron star. The temperature of this matter would have been around four to five billon Kelvin. Silicon- and sulfur-rich clouds would have come from the layers outside the iron core, at a temperature of three billion Kelvin. The iron-rich clouds were flung farther out than the silicon and sulfur by the supernova, suggesting that the star has turned itself inside out.

Right: *The combined efforts of Hubble (visible light, yellow), Chandra (X-rays, blue and green), and Spitzer (infrared radiation, red) produced this composite image. The central neutron star appears as a small turquoise dot.*

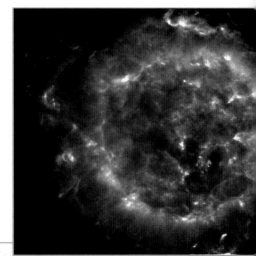

Right: *Radio image of Cassiopeia A from the Very Large Telescope (VLT) in New Mexico.*

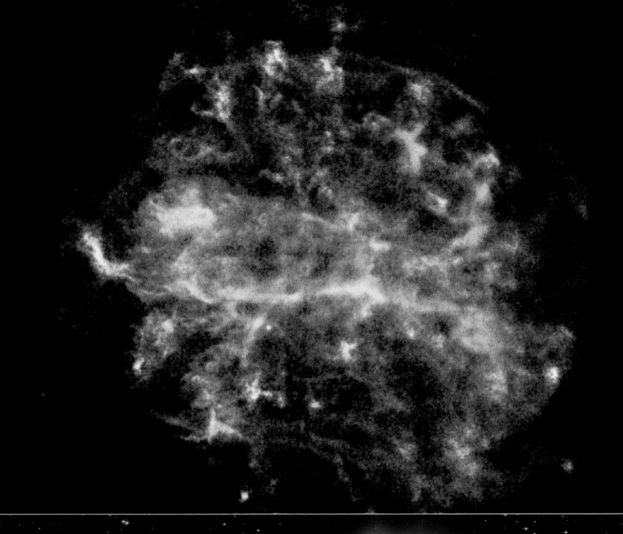

The supernova remnant known as G292.0+1.8 is about 1,600 years old. Chandra's image of the remnant shows the X-ray emissions from various elements, including oxygen, neon, silicon, and sulfur. This is one of only a few oxygen-rich supernovae known in our galaxy. There is a pulsar at the center of the expanding shell of gas.

Crab Nebula

The Crab Nebula is the result of a Type II supernova 6,500 light-years away in the constellation Taurus. The light from the explosion reached Earth in 1054. Records made by Chinese astronomers at the time mention a new star that was bright enough to be visible during the day for 23 days. They refer to the supernova as a "guest star." It took just two years for the "guest" to fade to the point where it was no longer visible at night. There are also records of the supernova kept by Native North Americans—preserved as an astronomical diagram painted on the wall of a canyon in New Mexico and in the form of starburst-patterned pottery—from about the correct period.

The remnant itself was discovered by the amateur astronomer John Bevis in 1731. Independently, Charles Messier discovered it when searching for Halley's Comet in 1758. The Crab Nebula became the first entry in Messier's catalog of nebulous objects. The name came from a drawing of the detail within the nebula by Lord Rosse in 1884 that had crablike elements to it. Unfortunately, the drawing bore little resemblance to the nebula as seen through improved telescopes later on. Despite this, the name stuck.

Viewed in the present day by the Hubble Space Telescope, the cloud of filaments is 10 light-years in diameter and is expanding at about 1,110 miles per second (1,800 km/s). The central bluish glow is known as synchrotron radiation, emitted by electrons moving at close to the speed of light in a strong magnetic field. The power source for this is the rapidly rotating neutron star at the nebula's core: the Crab Pulsar.

Hubble image of the Crab Nebula, an expanding remnant of a star's supernova explosion.

Neutron Stars and Pulsars

As well as a nebula, Type II supernovae leave behind a compact remnant. For stars between 15 and 30 times the mass of the Sun, that object will be a neutron star. Neutron stars are highly exotic objects forged from extreme conditions. The iron core of the progenitor star has remained intact after the explosion, but cannot be called simply "iron" any longer because the intense gravitational pressures crush the atomic nuclei into a complex particle soup, sometimes called neutronium.

Neutron stars contain between 1.4 and 5 solar masses of material squashed into a sphere just 12 miles (20 km) in diameter. The surface gravity of such an object is between 10^{11} and 10^{12} times greater than that on Earth. The interior is fluid and superconducting, producing a magnetic field up to 10^{15} Gauss. A Gauss is a unit of magnetic field strength. Earth's own magnetic field is only 0.6 Gauss, and a refrigerator magnet's field may be 100 Gauss. Earth-based experiments cannot reach anywhere near these conditions, so our knowledge of the star's interior consists mostly of a list of unconfirmed theories.

All stars spin, and a massive star core that has reduced in size dramatically will be spun up dramatically as well. The strong magnetic fields on a neutron star send out radio pulses as it rotates, in a similar way to a lighthouse sending out flashes of light. Thus these objects are also known as pulsars. First discovered in 1967 by Jocelyn Bell, these radio pulses repeated so rapidly and regularly that they were initially thought to be from alien civilizations signaling to Earth. The Crab Pulsar pictured here rotates 30 times a second—impossible for a large star, but not for an object as small as a neutron star.

This composite image of the Crab Pulsar by Hubble (red) and Chandra (blue) shows a ring at the nebula's heart. The central pulsar is surrounded by tilted rings of high-energy particles.

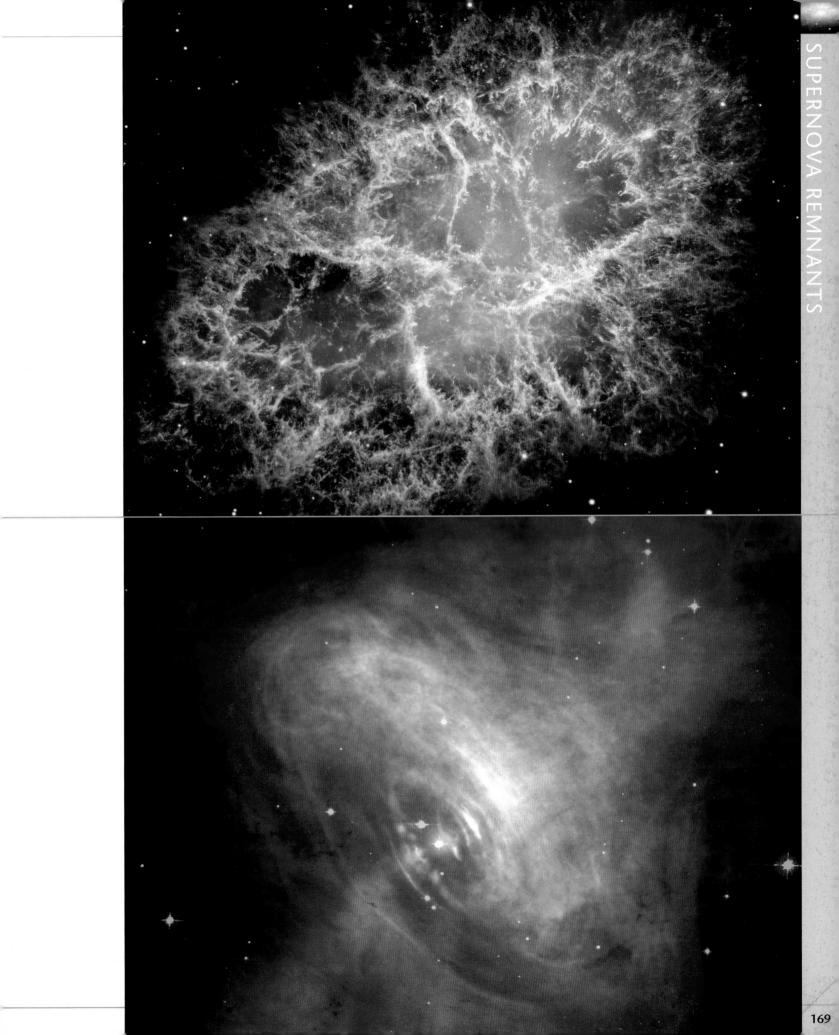

Black Holes and Supernovae

Black holes were first proposed as a theoretical construct in 1783 by John Mitchell, a geologist. He suggested that a "dark star" massive enough to have an escape velocity at its surface equal to that of the speed of light would be invisible. The idea was not considered much more during the 19th century, since it was believed that light was a wave unaffected by gravity. In 1915 Einstein's general theory of relativity showed that light would be affected by gravity. Shortly afterward, the German mathematician Karl Schwarzschild derived the equations describing a radius for a given mass within which light could not escape. For an object with the mass of the Sun, the Schwarzschild radius is about 1.9 miles (3 km). If all the mass of an object is within the Schwarzschild radius (also known as an event horizon), it is a black hole.

Type II supernovae resulting from stars containing over 30 solar masses are thought to produce compact remnants that are too dense to remain as neutron stars. They therefore become black holes. The matter is crushed to a single point known as a singularity. Nothing escapes from within the event horizon, but the black hole can be detected by the gravitational effect on nearby matter. Swirling like an entire sea of water down a small plughole, surrounding matter heats up through friction. The temperature is enough to produce high-energy X-rays.

Most black holes are the mass of a large star, but the centers of galaxies contain supermassive black holes that can be a billion times the mass of the Sun. They can be detected by the powerful gravitational effect they have on nearby stars. They orbit far too closely for the central object to be anything other than a black hole, since nothing else can contain so much matter in so little space.

Sagittarius A. The arrow points to the supermassive black hole at the center of the Milky Way. It has a mass 2.6 million times that of the Sun and is a powerful source of radio emission.

BLACK HOLES
Other Black Holes

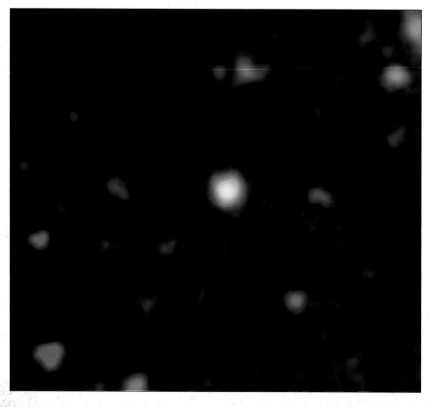

Left: *The yellow dot in the center of this image indicates a quasar, powered by a supermassive black hole in the heart of a galaxy. The blue and green dots show infrared emissions from less active galaxies.*

Right: *Elliptical galaxy NGC 4697 contains a number of pointlike X-ray sources. Each one marks the location of a black hole or neutron stars in binary systems.*

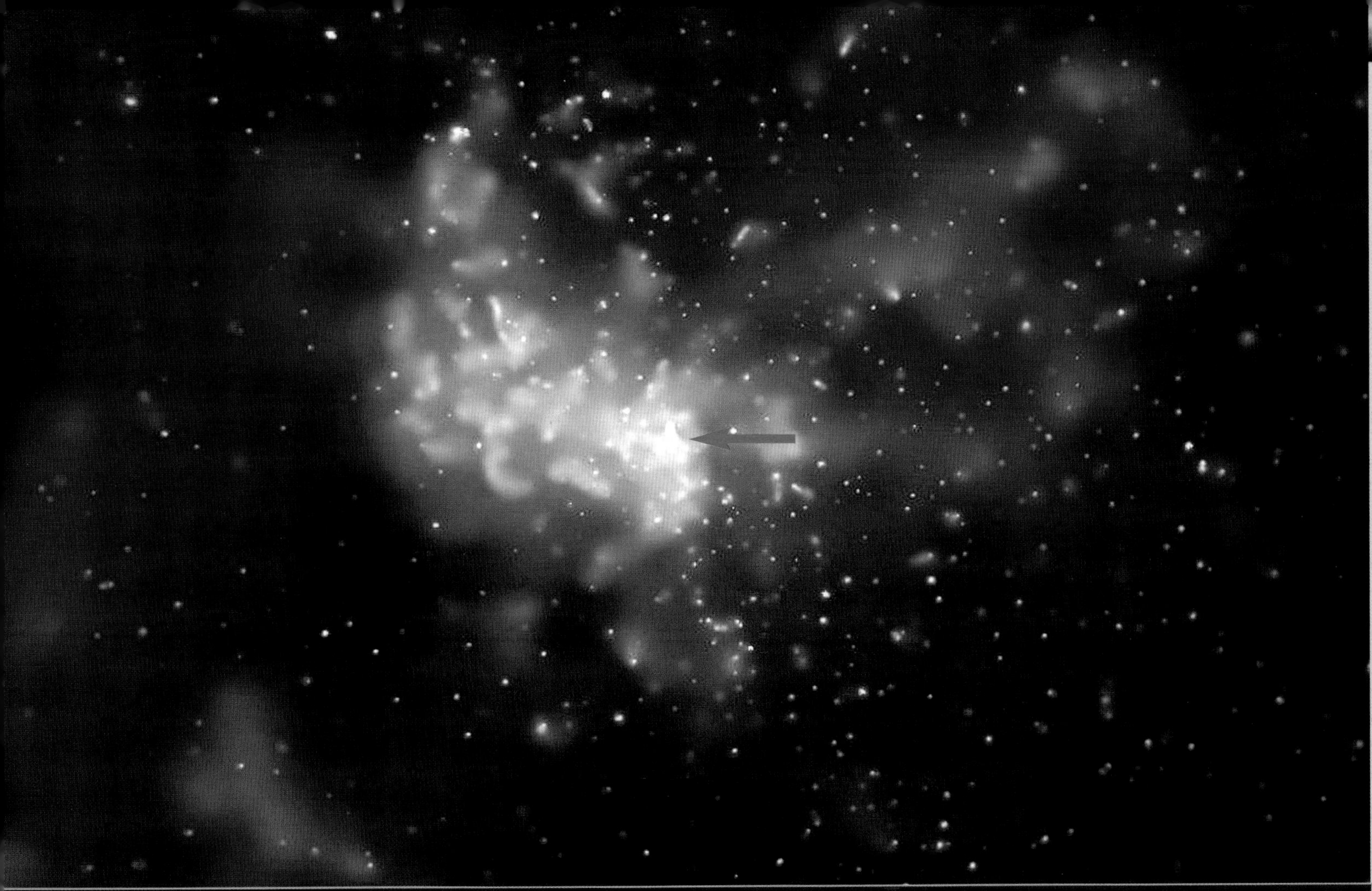

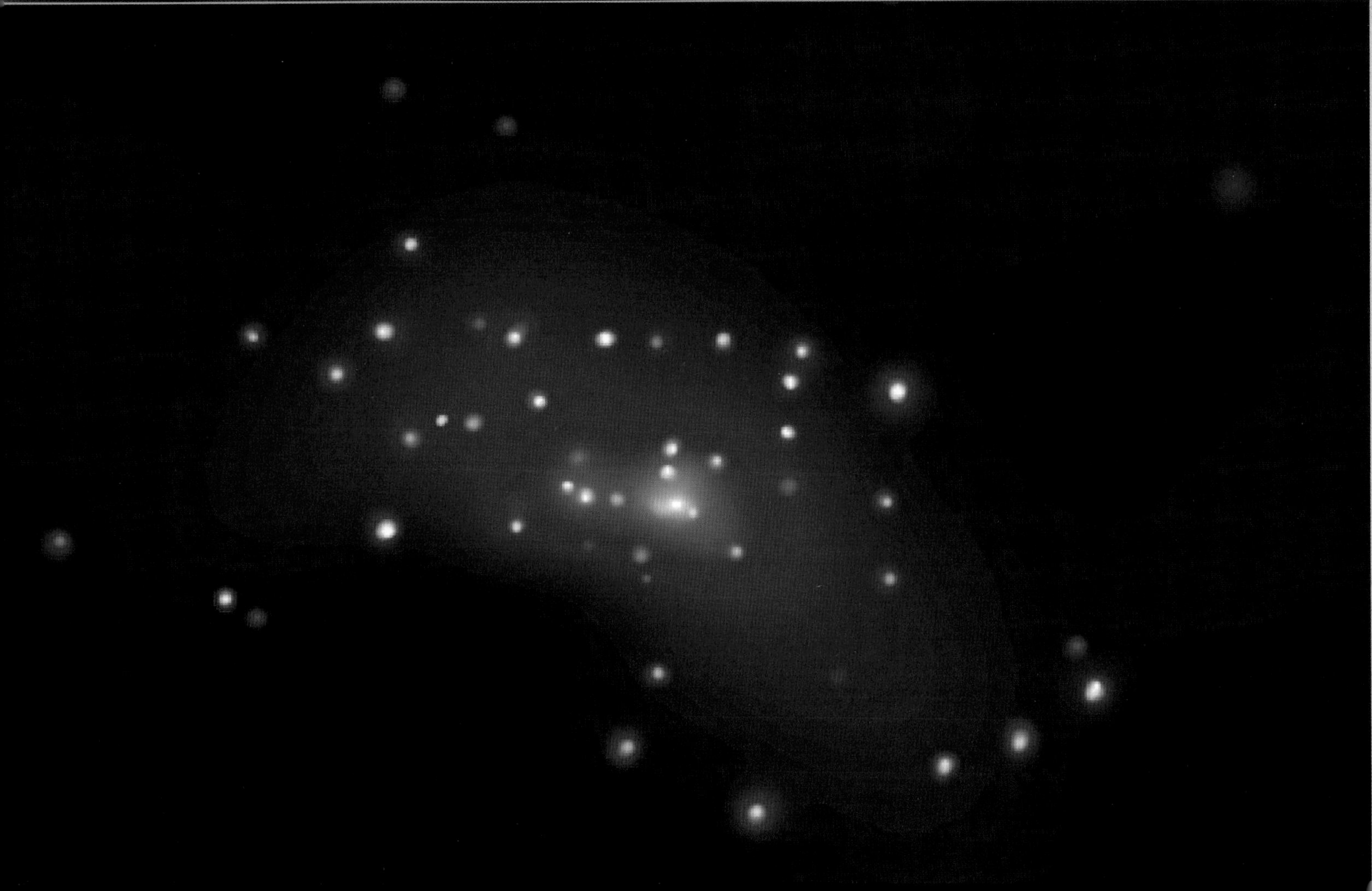

GALAXIES AND BEYOND

Dwarf galaxy NGC 1569.

INSIDE THE UNIVERSE

Galaxies are large collections of stars. Small dense clumps of stars are known as star clusters. If clusters contain more than about ten million stars, they are usually said to be small galaxies, although this distinction is not always exact. The stars are all orbiting around the center of mass of the galaxy, under the influence of the gravity of all the other stars. Among the stars are gas and dust that have not yet formed stars or are the remnants of stellar deaths.

The distribution of stars in the galaxy give it its shape. There are four main shapes: elliptical, spiral, barred spiral, and irregular. The Milky Way is an average size, around 100,000 light-years across. The small dwarf galaxies, which tend to be irregular or elliptical, are only about 10,000 light-years in diameter. The largest galaxies, found at the centers of clusters, are millions of light-years across.

Most galaxies lead boring lives, but some are more exciting. Although on the whole galaxies are well separated, they can get too close. Such interactions can cause the galaxies to collide and merge, resulting in the birth of new stars and, eventually, a new, larger galaxy. Other galaxies harbor a supermassive black hole in the center capable of influencing the environment surrounding these so-called active galaxies.

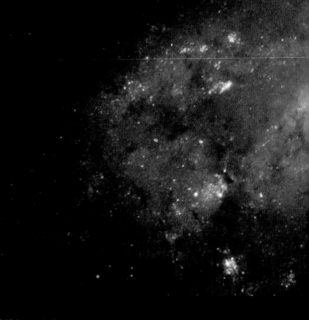

NGC 3949 is a spiral galaxy about 50 million light-years from Earth. It is similar in shape and structure to our own galaxy. Like the Milky Way, this galaxy has a blue disk of young stars peppered with bright pink star-forming regions.

Some nearby galaxies that have been known for a long time have individual descriptive names, such as the Whirlpool Galaxy. Many have other identifiers arising from their listing in astronomical catalogs. The two most common catalogs are the Messier catalog and the New General Catalogue (NGC). Charles Messier was a French comet hunter in the late 18th century who cataloged 110 nebulae. Some of them turned out to be gaseous nebulae in our Milky Way, and others are relatively nearby galaxies. Objects in the Messier catalog have a number with the prefix M. The New General Catalogue is more comprehensive, with 7,840 objects. Not all are galaxies, and there is some overlap—every Messier object also has an NGC number. New catalogs are produced all the time, from almost each new space telescope or ground-based observation campaign, so a single object can have many different names.

The Milky Way is not alone in space. It is part of the "Local Group" of about 40 galaxies. Most are small, but there are two other large spiral galaxies: Andromeda and Triangulum. A large number of galaxies in the Universe occur in groups. Sometimes hundreds of galaxies exist within a small region of space. These are called clusters of galaxies and are the most massive objects in the universe that are held together by their own gravity. These clusters and groups define the large-scale structure of the universe.

LOCAL GROUP

Our galaxy, the Milky Way, is a member of the Local Group, which contains at least 40 galaxies of various sizes. There are two other large galaxies (Andromeda and Triangulum) nearby. Along with the Milky Way, they dominate the group. There are many smaller satellite galaxies that orbit these three. The Andromeda Galaxy is the largest member of the Local Group, being half as large again as the Milky Way. It is the most distant object that can be seen with the naked eye.

The most prominent members of the Local Group in the night sky are the Large and Small Magellanic Clouds (LMC and SMC). When European explorers began to sail around the world, they used the stars to navigate by. The northern constellations were already familiar to them, and when they ventured south they devised new ones. Ferdinand Magellan was the first European to circumnavigate the globe between 1519 and 1522. He was the first person to bring to the attention of the Western world the two fuzzy patches seen in the southern sky, and the Large and Small Magellanic Clouds were named for him.

The closest member of the Local Group (Sagittarius Dwarf Elliptical Galaxy) was only discovered in 1994 because it is in the process of colliding with the Milky Way and so is seen through the disk of our galaxy. It is only 65,000 light-years away. Several new satellite galaxies have been discovered in recent years.

Large Magellanic Cloud

Of the two Magellanic Clouds, the Large Magellanic Cloud (LMC) is the brighter. It is closer to the Milky Way and is the fourth largest member of the Local Group. The galaxy looks similar to the bar of a barred spiral. In fact, in the past it may have been a small barred spiral until the Milky Way started pulling it apart. Its most prominent feature is the Tarantula Nebula, a vast stellar nursery that can be seen with the naked eye. It is about 1,000 light-years across. If it were at the distance of the Orion Nebula—one of the largest star formations in our region of the Milky Way—it would be the size of 60 full Moons.

It was in the outskirts of the Tarantula Nebula that the first close supernova in the age of modern astronomy was seen to explode. In 1987 the first supernova since Kepler's "star" of 1604 was observed. The proximity of the Magellanic Clouds made it possible for scientists to find an image of the original star that exploded in pictures taken before 1987. Sanduleak –69°202 was a blue supergiant star, a type that astronomers had thought did not go supernova. The Hubble Space Telescope has provided images of the remnant, revealing three rings surrounding what is left of the star. This was surprising: astronomers had expected to find an hourglass-shaped nebula. As a result, they had to provide new theories for the rings.

Although the Large Magellanic Cloud is close enough to the Milky Way for ground-based telescopes to resolve individual stars, the Hubble Space Telescope can see many more. It can show the shapes of some of the planetary nebulae that are present.

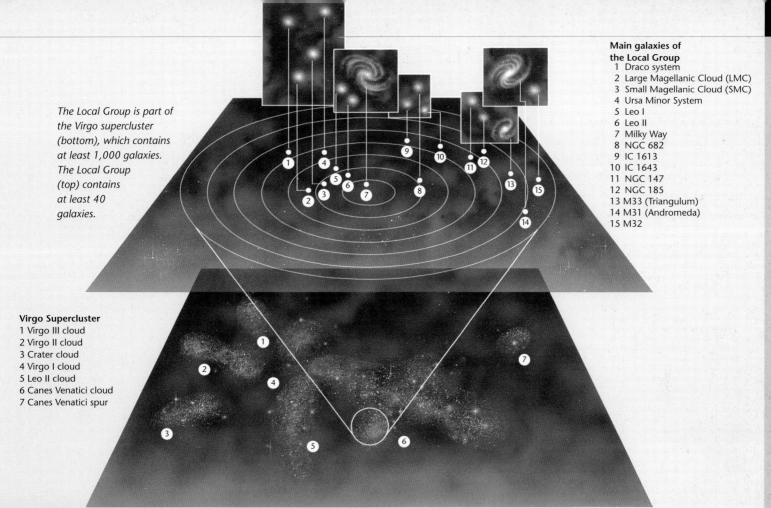

The Local Group is part of the Virgo supercluster (bottom), which contains at least 1,000 galaxies. The Local Group (top) contains at least 40 galaxies.

Main galaxies of the Local Group
1 Draco system
2 Large Magellanic Cloud (LMC)
3 Small Magellanic Cloud (SMC)
4 Ursa Minor System
5 Leo I
6 Leo II
7 Milky Way
8 NGC 682
9 IC 1613
10 IC 1643
11 NGC 147
12 NGC 185
13 M33 (Triangulum)
14 M31 (Andromeda)
15 M32

Virgo Supercluster
1 Virgo III cloud
2 Virgo II cloud
3 Crater cloud
4 Virgo I cloud
5 Leo II cloud
6 Canes Venatici cloud
7 Canes Venatici spur

A cluster of brilliant massive stars (bottom right) in the Tarantula Nebula.

Small Magellanic Cloud

The Small Magellanic Cloud (SMC) is the fainter and more irregular of the two Magellanic Clouds. It may once have been a disk shape, but as it orbited the Milky Way it was slowly pulled around and distorted. This is an example of "tidal" gravitational forces—one side of the galaxy feels the pull of the Milky Way more than the other, and as the cloud moves, the gravitational force slowly pulls stars and gas out of the cloud and into intergalactic space. Both Magellanic Clouds sit in a stream of cold gas, known as the Magellanic Stream, which is pulled out of the galaxies as they orbit the Milky Way.

The Magellanic Clouds are not the closest galaxies to us. The Sagittarius dwarf elliptical galaxy is closer, and it is currently merging with the Milky Way. The Magellanic Clouds are, however, the closest galaxies of which we have a clear view. They are examples of dwarf galaxies because they are small and have no clear structure or shape.

The distance to these galaxies is known very accurately, because we can see individual stars. Some stars, called Cepheid variables, have periodic changes in their luminosity, and the speed at which they vary (the period) depends on how bright they are. The period–luminosity relationship was discovered by American astronomer Henrietta Leavitt in 1912. Her discovery enabled astronomers to determine the distance to these stars: by measuring the period, astronomers can work out how bright a star actually is. By comparing that with how bright it appears in the sky, they can calculate how far away it is.

DATA

Distance: 210,000 ly

Constellation: Tucana

Embryonic stars embedded in nebula NGC 346 in the Small Magellanic Cloud.

Andromeda Galaxy—M31

The Andromeda Galaxy is the farthest object that is visible to the naked eye. On dark nights it appears as a faint fuzzy blob in the north of the constellation Andromeda. It is half as large again as our Milky Way and is the dominant member of the Local Group—the 40 or so galaxies in the vicinity of the Andromeda Galaxy and the Milky Way.

From inside our own galaxy, it is difficult for us to see how the Andromeda Galaxy behaves. However, it is close enough for telescopes to be able to resolve individual stars, and trace out some of its history. It is a spiral galaxy, similar to our own, and has two companion galaxies—M32 and M110 (also called NGC 205). Long-exposure observations show that the galaxy extends far beyond what is visible at first glance. The galaxy has a very bright extended disk of stars at its nucleus, estimated to be more than 220,000 light-years in diameter.

Average-sized galaxies such as Andromeda (200,000 light-years in diameter) and the Milky Way (100,000 light-years in diameter) are thought to grow by "galactic cannibalism"—the pulling apart and absorption of smaller galaxies by larger ones.

The Milky Way and the Andromeda Galaxy are gravitationally bound together. They are attracting one another and moving toward each other. In about five billion years they could collide. When that happens there will only be a handful of stellar collisions, but the gas and dust particles will collide. This will cause many stars to form, and could activate a massive black hole at the center of the resulting galaxy.

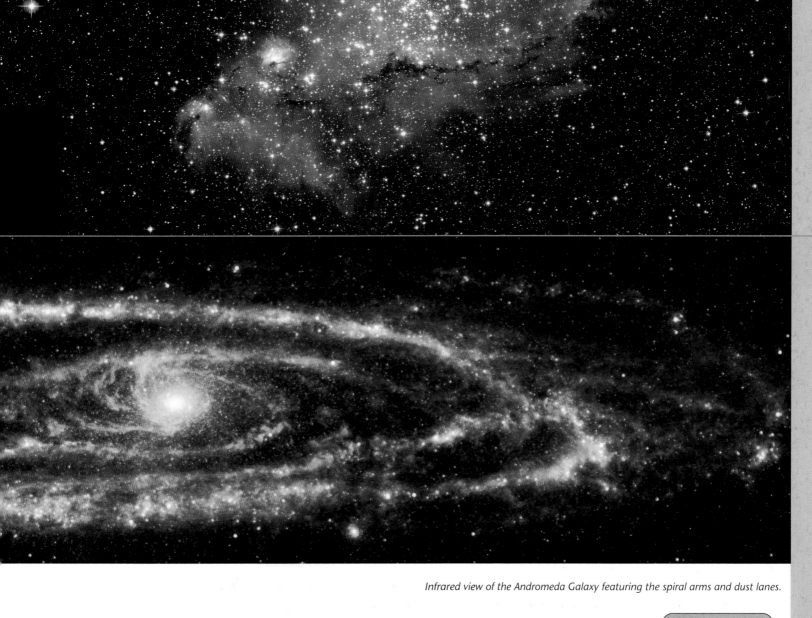

Infrared view of the Andromeda Galaxy featuring the spiral arms and dust lanes.

DATA

Distance: 2.5 million ly

Constellation: Andromeda

Triangulum Galaxy—M33

DATA

Distance: 2.3 million ly

Constellation:
Triangulum

The Triangulum Galaxy is the third largest member of the Local Group. With a diameter of about 46,000 light-years it is small in comparison with the Andromeda and Milky Way galaxies, but is similar in size to other spiral galaxies in the universe.

It was the first galaxy whose motion relative to the Milky Way was measured using maser emission from its nucleus. (Masers—an acronym for *m*icrowave *a*mplification by *s*timulated *e*mission of *r*adiation—are like lasers, but they exist in the microwave region of the electromagnetic spectrum. They have very specific frequencies that allow motions to be calculated precisely.) M33 is known to be moving at 118 miles per second (190 km/s) toward the Andromeda Galaxy. It may therefore be a large and distant satellite of the Andromeda Galaxy.

The Triangulum Galaxy has been cataloged a number of times since its discovery by Italian astronomer Giovanni Battista Hodierna in the 17th century. Early telescopes did not have lenses strong enough to discern the structure of what appeared to be "fuzzy blobs." During his life-long search for comets, the French astronomer Messier kept a note of the fuzzy blobs that did not move, so that he would not confuse them with comets. His journals became the Messier catalog.

In Ireland in the 1840s William Parsons, third Earl of Rosse, built "Leviathan," the largest telescope of its time, with a mirror 72 inches (183 cm) across. He could see more detail in the skies than anyone before. While surveying the nebulae, he noted spiral structures, and the Triangulum Galaxy was one of the first he observed. It took many years before these spiral nebulae were generally accepted to be galaxies—similar objects to the Milky Way, but seen from a great distance.

Local Group Detail

Left: *Sagittarius Dwarf Irregular Galaxy (SagDIG) is a very small Local Group galaxy. This Hubble Space Telescope image has revealed that many of the stars are old and that SagDIG is ancient, having formed early in the history of the universe.*

Below: *Double cluster NGC 1850 in the Large Magellanic Cloud is a young globular star cluster.*

M33, the Triangulum Galaxy.

Above: *Star-forming region (known as N11B) of the Large Magellanic Cloud.*

GALAXIES

Galaxies have been categorized according to shape into six main classes. The most photogenic are the spiral and the barred spiral galaxies, which are vast whirlpools in space. Elliptical galaxies on the whole have no clear structure and appear as diffuse clouds of stars. Dwarf galaxies are among the smallest star systems known and may have structures resembling either spirals or ellipticals. Lenticular galaxies form an intermediate class between spirals and ellipticals. The final class—irregular or peculiar galaxies—contains those galaxies that cannot be otherwise classified.

American astronomer Edwin Hubble devised a diagram relating to the major classes of galaxies. From its shape it became known as the "tuning fork" diagram. The handle of the fork is populated by elliptical galaxies, starting at the most spherically symmetrical and becoming flatter and flatter (classes E0–E6). The split in the fork is where lenticular (lens-shaped) galaxies occur (classes S0/SB0). The two prongs are the spiral galaxies and the barred spiral galaxies (Sa–Sd & SBa–SBd), in which the sequence a–d means that the arms become progressively less tightly wound. There are also Ir(regular) and dE (dwarf ellipticals) that do not fit on this sequence.

The sequence was once thought to be evolutionary, meaning that galaxies formed as E0 and evolved to become Sc or SBc, resulting in the terminology "early type" for ellipticals and "late type" for spirals. Galaxy formation is, however, a much more complicated process than this, but it represented the first real attempt to explain the range of galaxy shapes. It is now thought that some elliptical galaxies are the result of mergers, so ellipticals are "late type" in a different way. In recent years cosmologists have been able to simulate the growth of galaxies in the universe and produce computer-generated elliptical and spiral galaxies.

1

DATA

Types of Galaxies

Spiral, barred spiral, elliptical, dwarf, lenticular, irregular or peculiar

---

DWARF GALAXY
NGC 1705

Small galaxies can be active places, as shown in this Hubble Space Telescope image of the dwarf irregular galaxy NGC 1705. The satellite galaxies of the Milky Way, the Magellanic Clouds, are not as active as this galaxy. Stars are actively forming in the central regions. The light from these regions is dominated by emission from young hot stars that appear blue and are strongly concentrated toward the center of the galaxy. The older, cooler stars appear red and are spread farther out in the galaxy. Hot stars do not live long, and they end their lives in supernova explosions before they can travel far from the place of their birth. Cooler stars are much longer lived, by a factor of 10 or more, and can therefore travel much farther from the place of their birth.

It is likely that NGC 1705 has been slowly evolving, and that stars have been forming throughout its lifetime. However, the most recent peak in star formation probably started around 30 million years ago. This means that not all of the brightest and most massive stars have reached the end of their lives, and so the galaxy still appears bright.

This galaxy is too small to evolve into a shape with organized structures, such as spiral arms, so its appearance is amorphous. Using deep exposures (infrared images of a tiny area of space obtained by prolonged exposure) such as those from the Hubble Deep Field, astronomers can look back to the early universe and see how the first galaxies appeared. They are similar in shape to small galaxies such as NGC 1705: amorphous but forming stars fairly rapidly relative to their small size.

DATA

Distance: 17 million ly

Constellation: Pictor

Redshift 0.002

NGC 1705—a dwarf irregular galaxy.

2   3   4   5   6   7   8   9

*Hubble's tuning fork diagram classified galaxies as elliptical
(1, 2, and 3 above), spiral (4, 5, and 6), or barred spiral
(7, 8, and 9). Main picture (right) shows the center of the
barred spiral galaxy NGC 1512.*

DWARF GALAXY
I Zwicky 18

The universe is (currently) thought to be around 13.75 billion years old. Our Sun, along with the solar system, is comparatively young at 4.5 billion years. However, this image from the Hubble Space Telescope implies that the youngest stars of the galaxy known as I Zwicky 18 formed only 4 million years ago, which means that I Zwicky 18 is the youngest-known galaxy in the universe.

There are two major star-forming regions in this galaxy—the two bright knots in the center of the main image—in which the young hot stars are concentrated. Their extremely bright light illuminates the gas that makes up the outer regions of the cloud. The light from the stars is pushing the gas outward, and eventually the galaxy will stop forming stars because there is no new material from which they can be made.

The small collection of stars to the top right is a companion galaxy, striking in its difference to I Zwicky 18. There are few stars forming in this companion, and there appears to be little gas from which new stars can be formed. In contrast to the intense blue of I Zwicky 18, the fuzzy red blobs in this image are very distant background galaxies.

I Zwicky 18 is a galaxy that seems to occur at the wrong point in the evolution of the universe. It has more in common with the earliest galaxies than those in its immediate vicinity. However, its proximity has allowed astronomers to study in greater detail the processes involved in star formation within dwarf galaxies.

DATA	
Distance: 45 million ly	
Constellation: Ursa Major	
Redshift: 0.003	
Size: 3,000 ly	

Young galaxy I Zwicky 18 (left) with companion galaxy (top right). The red points are older stars, star clusters, and distant galaxies.

LENTICULAR GALAXY
NGC 2787

Galaxy NGC 2787 does not appear at first sight to have anything special happening at all, but it is still beautiful. Thin arms of dust wind around its center. It is of a type known as lenticular (lens-shaped), which falls between elliptical and spiral. Although galaxies of this type are flattened in the same way as spiral galaxies, they do not possess any clear spiral arms. NGC 2787 has a faint bar at the center, which is not visible in this image.

Many of the well-studied astronomical objects are unusual. They are the most studied because, on the whole, they are also the most interesting. However, it is occasionally useful to study what are regarded as "normal" objects in order to determine exactly what is normal and how abnormal the more exciting ones are.

The bright spots of light in the outskirts of NGC 2787 are globular clusters orbiting the galaxy. Many large galaxies have halos of star clusters like these. They are thought to be some of the oldest stellar structures that still exist in the universe.

Some elliptical galaxies are thought to form as the result of the merger of other, smaller galaxies. The collision causes most of the gas to be used to form stars, and there is no single direction of rotation. Such galaxies go on to evolve passively, since there is very little gas from which to form new stars and there are no events that could trigger the collapse of any clouds that exist.

DATA	
Distance: 24 million ly	
Constellation: Ursa Major	
Redshift: 0.0023	

NGC 2787 is a lenticular galaxy.

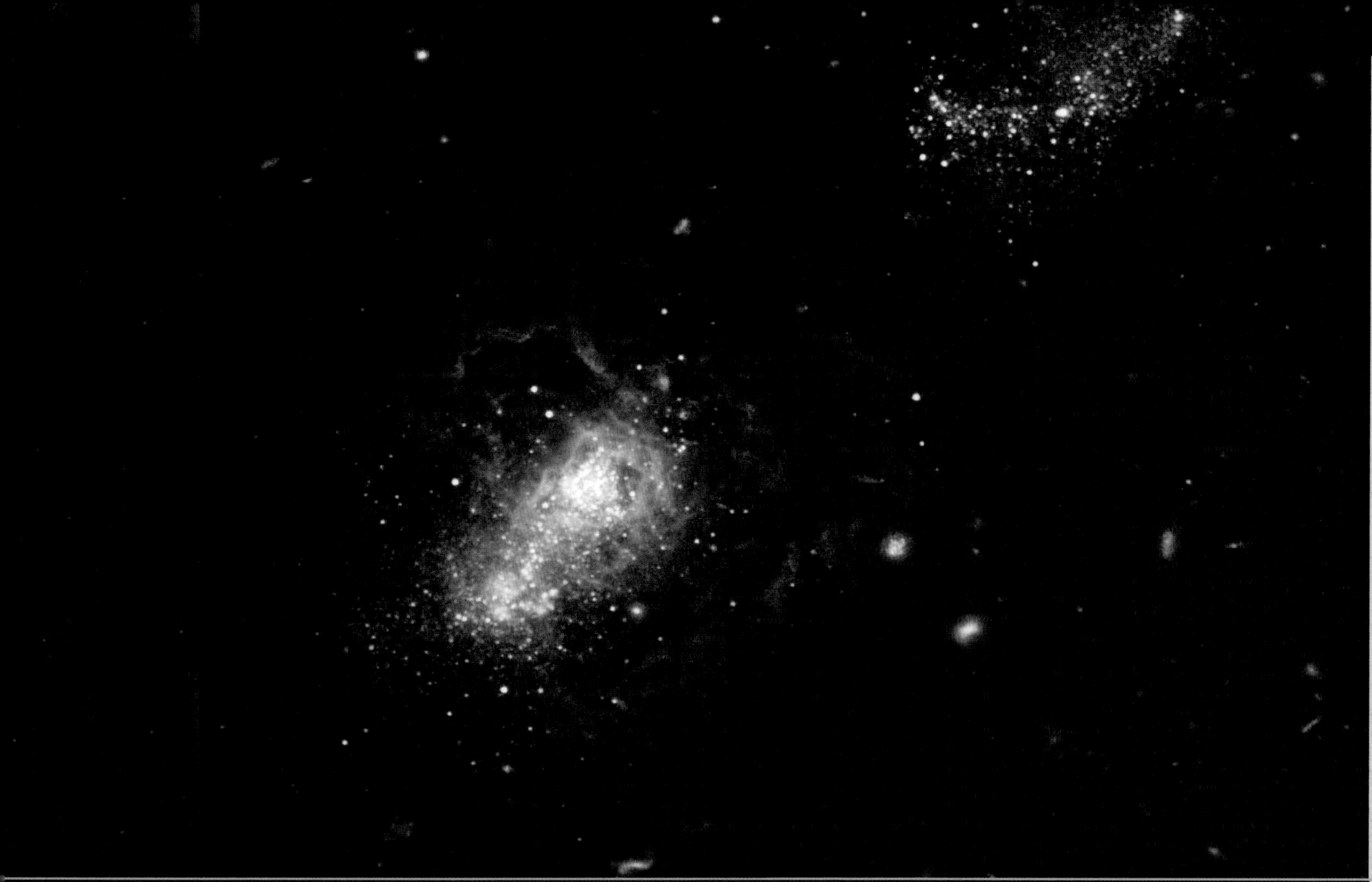

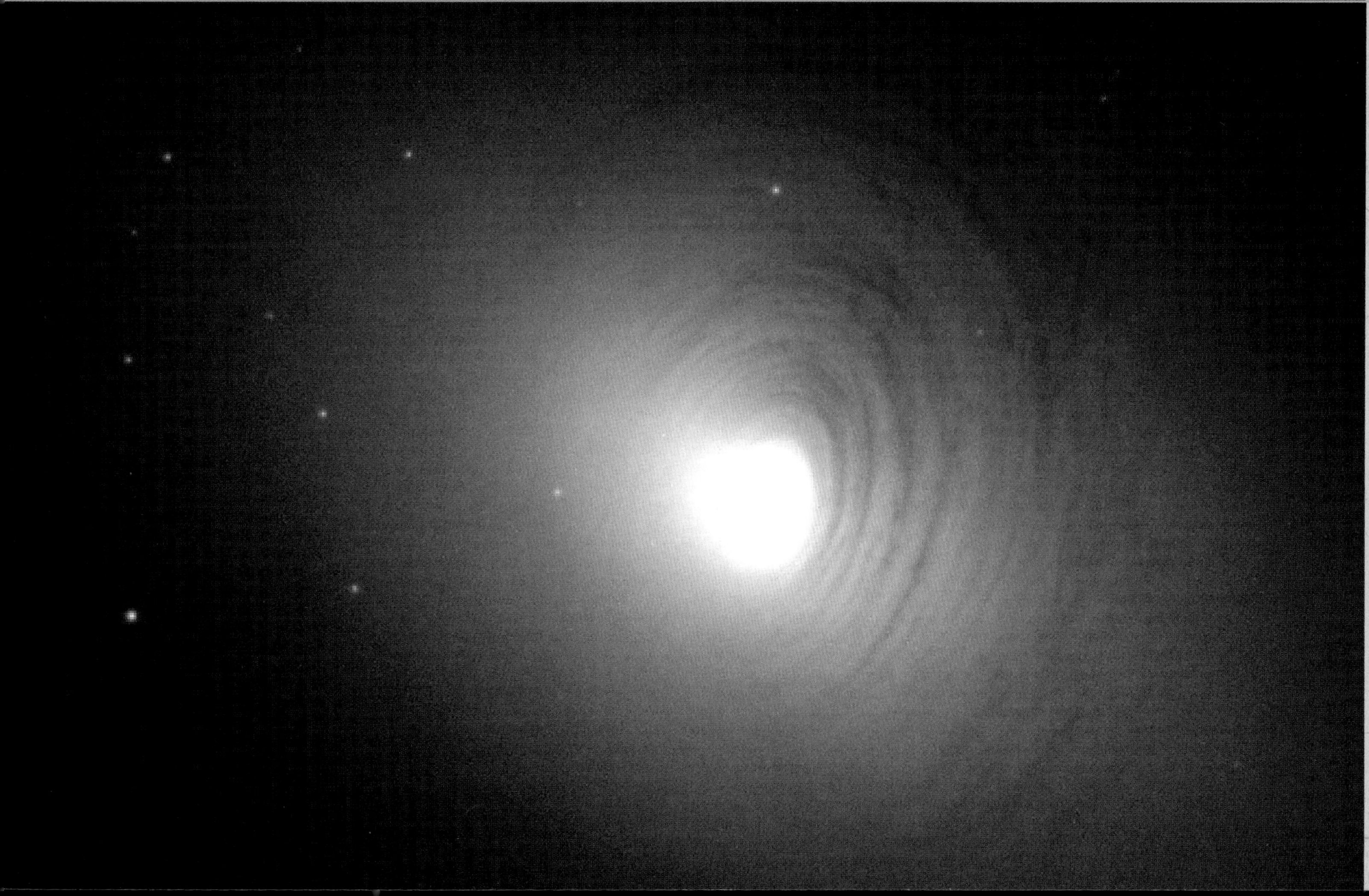

ELLIPTICAL GALAXY
NGC 1316

Galactic cannibalism is common in the universe, and various snapshots exist of different stages of the process. NGC 1316 is thought to be the final stages of a merger in which an elliptical galaxy has eaten a smaller spiral galaxy. NGC 1316 has the shape of an elliptical galaxy—the milky blur extends to the edges of this image—but it also shows dust lanes that are usually found in spiral galaxies. Elliptical galaxies generally show no structure and are thought to evolve passively.

This galaxy has been used in attempts to calibrate the size of the universe. Distances between the planets in the solar system can be determined using radar. Stars in the galaxy have their distance measured by parallax. For more distant galaxies astronomers use a "standard candle"—the distance of an event that always produces the same luminosity being determined by its brightness. If two identical candles were positioned so that one was twice as far away as the other, then the more distant one would appear four times fainter. Inverting this process, they could then calculate the distances by the brightness of the candles, if they are known to be the same.

NGC 1316 was found to have a high nova rate. Supernovae can be the explosions of white dwarf stars (in which case they are also known as supernovae Type Ia). They are all thought to have the same luminosity, because they are believed to be caused by the same process everywhere in the universe, since their masses are very similar. Therefore, assuming that the stars and the white dwarfs in NGC 1316 are the same as those in the local universe, then the distance to NGC 1316 has been found.

Dark dust clouds against the glowing nucleus of NGC 1316.

SPIRAL GALAXY
M74

Although usually known by its Messier number, this galaxy was discovered by French astronomer Pierre Méchain in 1780. It is a classic grand design spiral, classed as "Sc" (meaning it is a loosely wound spiral galaxy) using Hubble's tuning fork. The bright knots of emission from its arms are young star clusters and their surrounding nebulae, indicating that star formation is widespread in the disk.

In June 2003 a supernova was discovered in the outskirts of M74 by Rev. Robert Evans, a prolific supernova hunter who has found more than 30 since 1981. It was named SN 2003gd, and astronomers all over the world started to monitor the event. The supernova was identified as a type II-P because it had a plateau phase in its brightness. Comparing it to similar supernovae, astronomers estimated it had been discovered about 87 days after the explosion, when M74 was behind the Sun.

Then in January 2004 a group working in Cambridge, England, announced that they had identified the star that produced the supernova. Images from the Hubble Space Telescope and the Gemini Telescope on Hawaii that were taken approximately 200 and 300 days before the explosion showed stars where the supernova was found. By identifying the star that caused the supernova and analyzing its color, the group confirmed that it was a red supergiant with a mass of about eight times that of our Sun.

Galaxy M74, the site of a recent supernova.

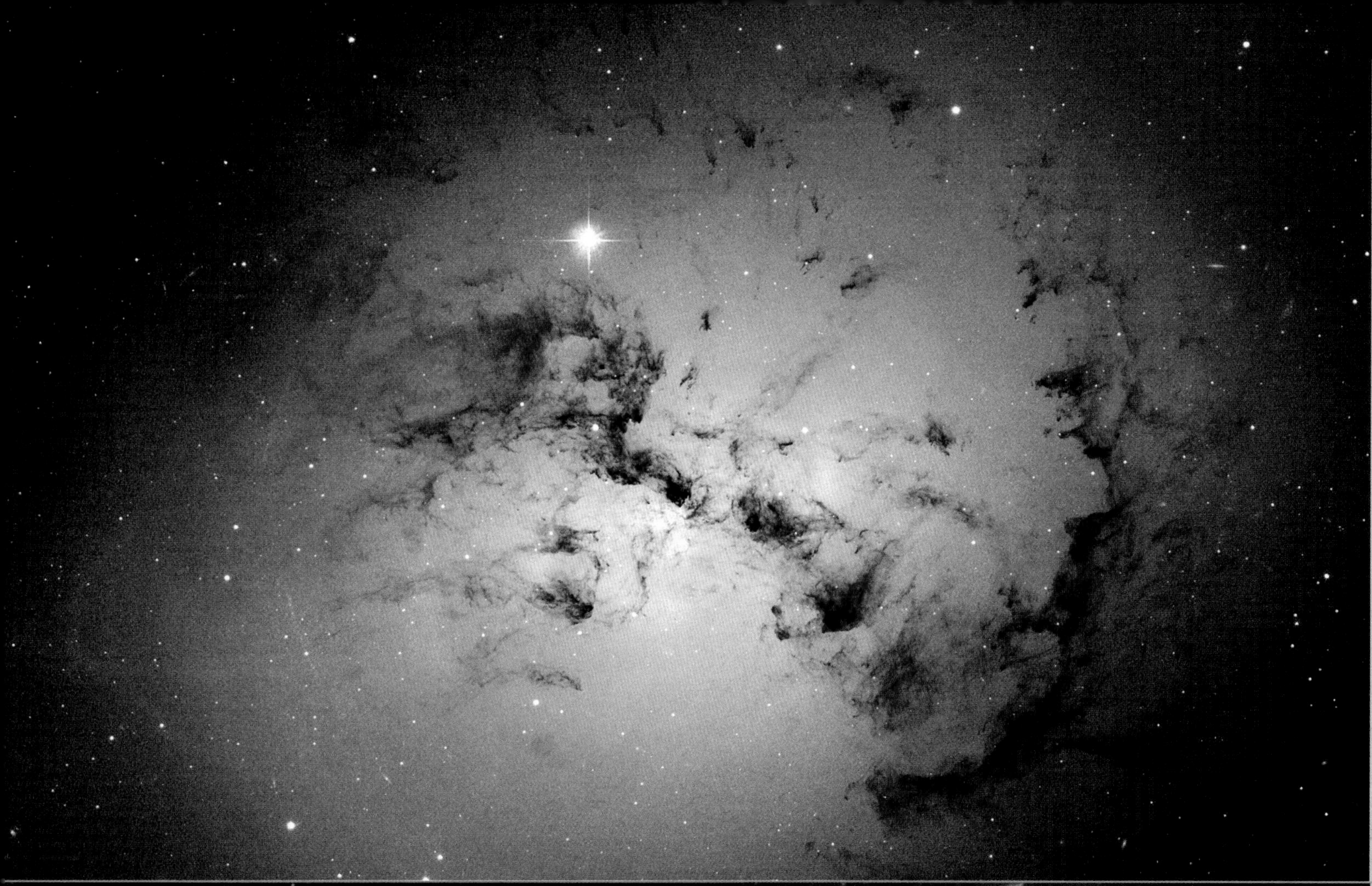

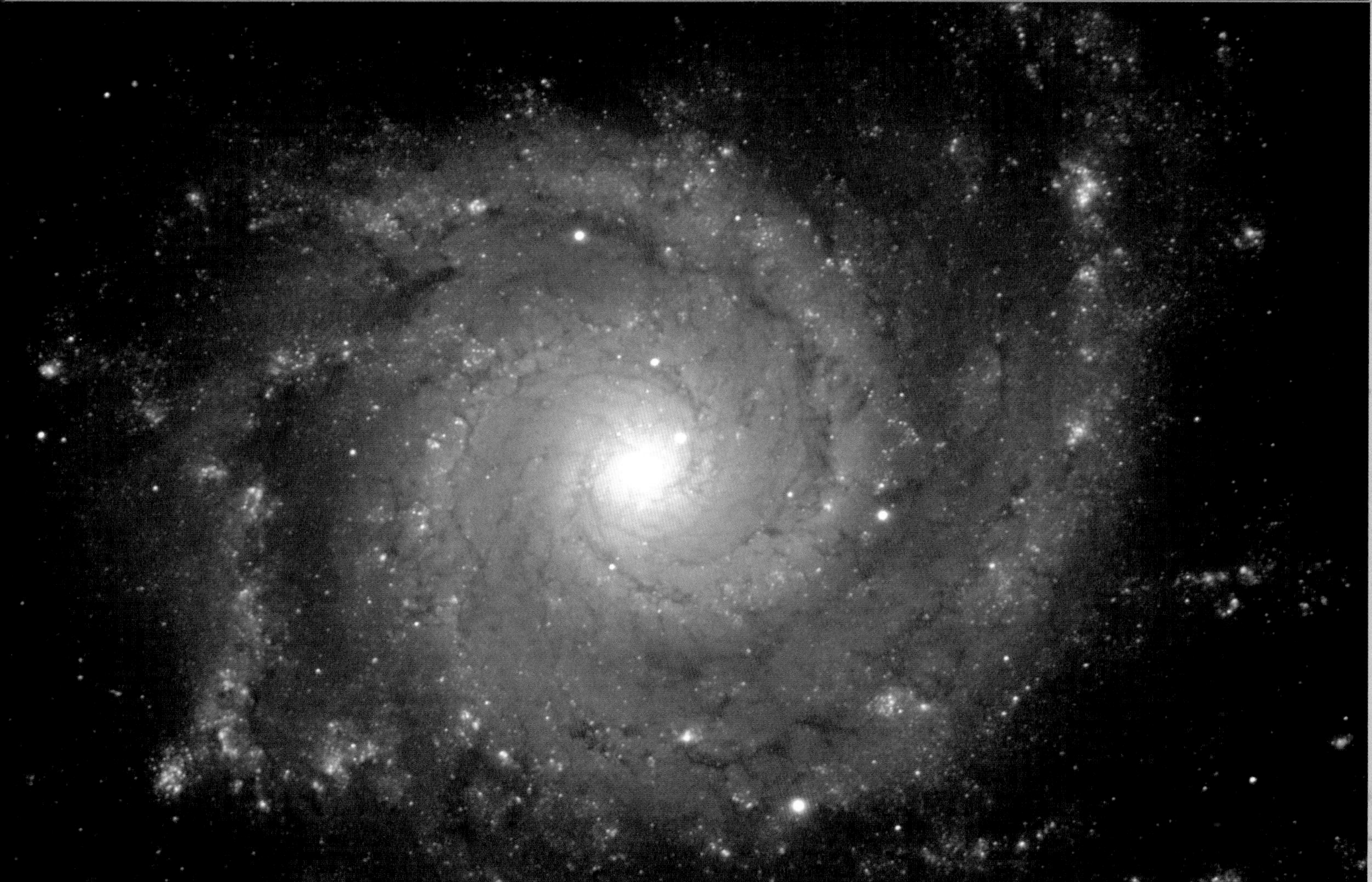

SPIRAL GALAXY
NGC 891

This galaxy is a beautiful edge-on spiral galaxy which clearly shows a dust lane along its middle. It is bright enough so that Charles Messier should have been able to see it. For reasons unknown he did not add it to his catalog and it was left for one of the first woman astronomers, Caroline Herschel, to discover it with her brother William in August 1783.

The Herschels made numerous discoveries and created extensive sky catalogs in the late 18th and early 19th centuries. They were originally from Hanover in Germany. Caroline joined her brother—who had already settled in Bath, England—in 1772 as his housekeeper. William was by this time a successful musician and conductor. He was also an amateur astronomer, and became the maker of the most powerful telescopes of the time. A grant from the King allowed him to give up his job to become an astronomer and telescope builder full time.

As this point Caroline became her brother's apprentice, initially taking notes for William while he observed. After acquiring her own telescope she began to work on her own more and more. In 1783 she discovered three new nebulae, and between 1786 and 1797 she discovered eight comets. She is thought to have discovered around 14 previously unknown deep-sky objects, including one of the satellites of the Andromeda Galaxy. After the death of her brother she returned to Hanover, 50 years after she had left. She died at the advanced age of 98. She has been honored by the naming of a crater on the Moon (Lucretia, for her middle name) and of an asteroid.

DATA
Distance: 27 million ly

Constellation: Andromeda

Redshift: 0.002

Edge-on view of NGC 891 (By permission Canada–France–Hawaii Telescope/J.-C. Cuillandre/Coelum).

SPIRAL GALAXY
NGC 3314a and NGC 3314b

It is rare for two relatively nearby galaxies to be close together on the plane of the sky and not be physically associated with each other. The image opposite shows an extraordinary chance alignment of a face-on spiral directly in front of a larger, oblique spiral.

The light from the more distant galaxy shines through the face-on spiral and shows in great detail the distribution of the nonluminous matter (dust and gas) in the foreground galaxy. This dust can be seen in many images of galaxies, but usually when the spiral galaxy is presented edge-on.

The dust comes mainly from stars similar to the Sun. It is blown from their outer layers toward the end of their life. It blocks light from nearby stars, allowing the gas mixed in with it to cool. In these protected regions of space the gas and dust can start to collapse under their own gravity, and form stars.

The nucleus of the background galaxy (NGC 3314b) can be seen as the reddish patch just to the bottom right of the core of NGC 3314a. It appears redder than other galactic centers as—in the same way as sunsets appear red—the blue light is scattered out of the line of sight more easily than red light. The dust particles in the foreground galaxy act like soot and dust in our atmosphere, and cause the nucleus to redden. The amount of reddening enables astronomers to estimate how much dust is present in these regions.

DATA
Distance:
NGC 3314a: 117 million ly
NGC 3314b: 140 million ly

Constellation: Hydra

Redshift: 0.0093

Galaxy pair—NGC 3314 and NGC 3314b.

M81

This spiral galaxy forms an interacting pair with M82. Both galaxies are part of the group of galaxies closest to the Local Group. It has been targeted by the Spitzer Infrared Space Telescope in order to study the structure of such a galaxy. In Hubble Space Telescope pictures it is seen as a beautiful object. The core takes on a soft peach color from the light of old stars, and the younger stars give the spiral arms a delicate bluish-gray color. The arms are also dotted with pink nebulae.

Observing this galaxy in the infrared shows the structure of the spiral arms in detail. The emission color-coded red is from the coolest material and is mainly due to dust. The blue light comes from the older cooler stars that dominate the center of the galaxy. When stars form, some are many times more massive than the Sun. These stars burn their nuclear fuel much faster, and are therefore much hotter than the Sun. This means that they emit more light at more energetic wavelengths (the blue end of the spectrum), and so emit much more ultraviolet light. The ultraviolet light is energetic enough to be absorbed by the dust particles, and heats them up. The dust then reemits the light at longer wavelengths (the red end of the spectrum), and usually in the infrared. The ultraviolet light also energizes the surrounding gas, causing the nebula to glow.

There are two main materials that form dust in a galaxy: silicates—similar to beach sand—and "polycyclic aromatic hydrocarbons" (PAHs)—molecules that are similar to coal and soot. Both dust types also absorb visible light, making infrared observations more effective in tracing the reservoirs within a galaxy available for star formation.

DATA

Distance: 11.8 million ly

Constellation: Ursa Major

False-color Spitzer image of the spiral arms of galaxy M81.

Pinwheel Galaxy—M101

M101, also called the Pinwheel Galaxy, was discovered in 1781 by French astronomer Pierre Méchain, a collaborator of Charles Messier, who added the galaxy to his catalog of nebulae and star clusters.

This image combines three colors. The red at the core represents visible light, the green shows near ultraviolet (UV) emissions, and the blue represents far ultraviolet emissions. Ultraviolet light is emitted mainly by objects that are very hot—much hotter than our Sun. (The Sun does emit some UV light, which causes exposed skin to burn, but most radiation is emitted in the visible as yellow-green.) Objects that are ten times hotter radiate mainly in the UV. They are massive stars which, because they burn their fuel at a greater rate, emit more light. Since they are hotter than the Sun, their characteristic radiation carries more energy, and is therefore more "blue" than yellow, which is why young massive stars appear blue in photographs.

This image shows that most of the massive stars are concentrated in lines that form the spiral arms of the galaxy. Massive stars will not move far from their birthplace before they die, and so they clearly indicate which regions are currently producing stars.

Most of the UV light that arrives at Earth is blocked by the atmosphere, much to the benefit of life on the surface. This means that ultraviolet astronomy has to be carried out from space. As a result, many discoveries have only been made within the last few decades.

DATA

Distance: 25 million ly

Constellation: Ursa Major

Redshift: 0.0008

Three-color image of M101.

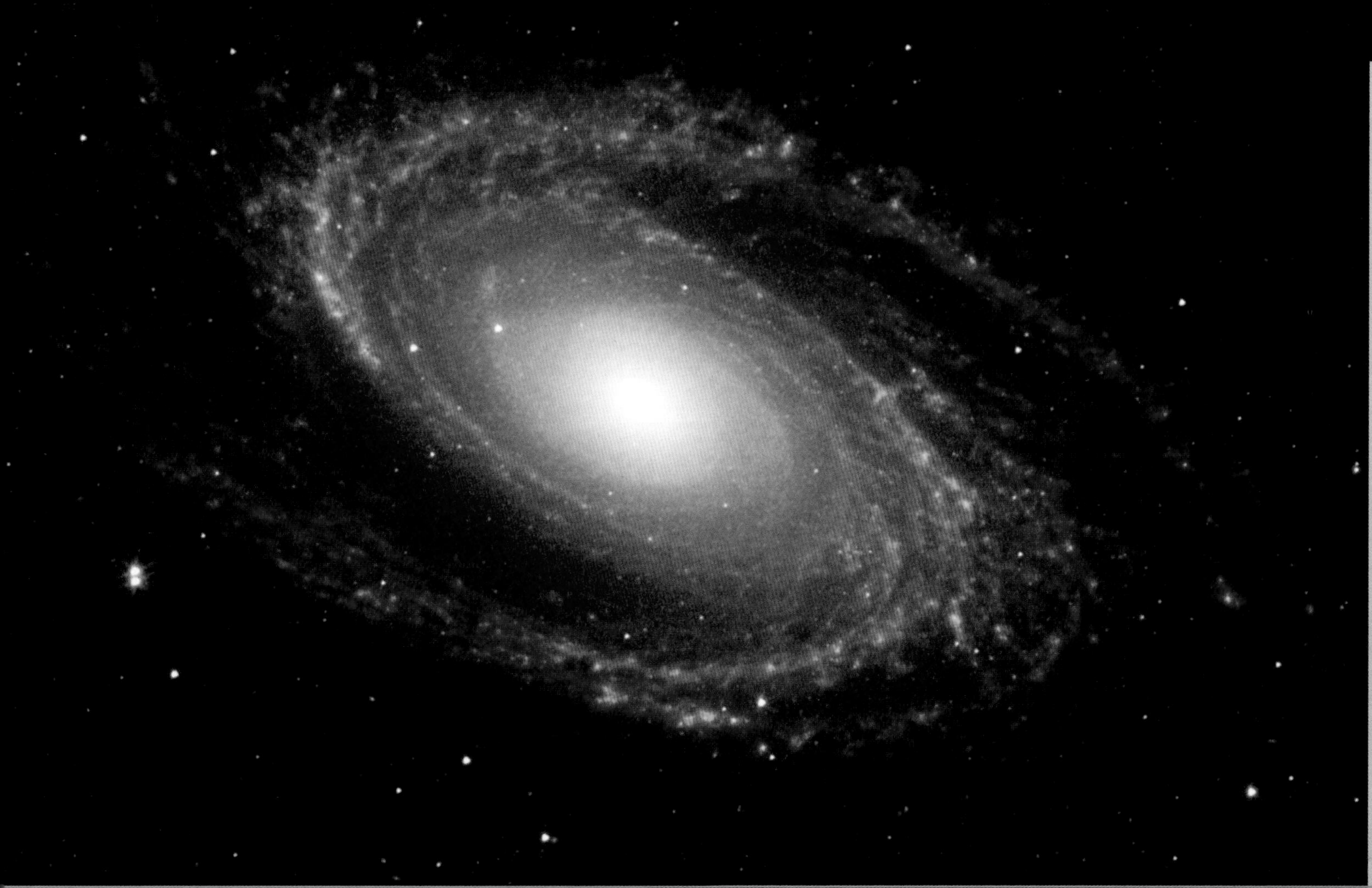

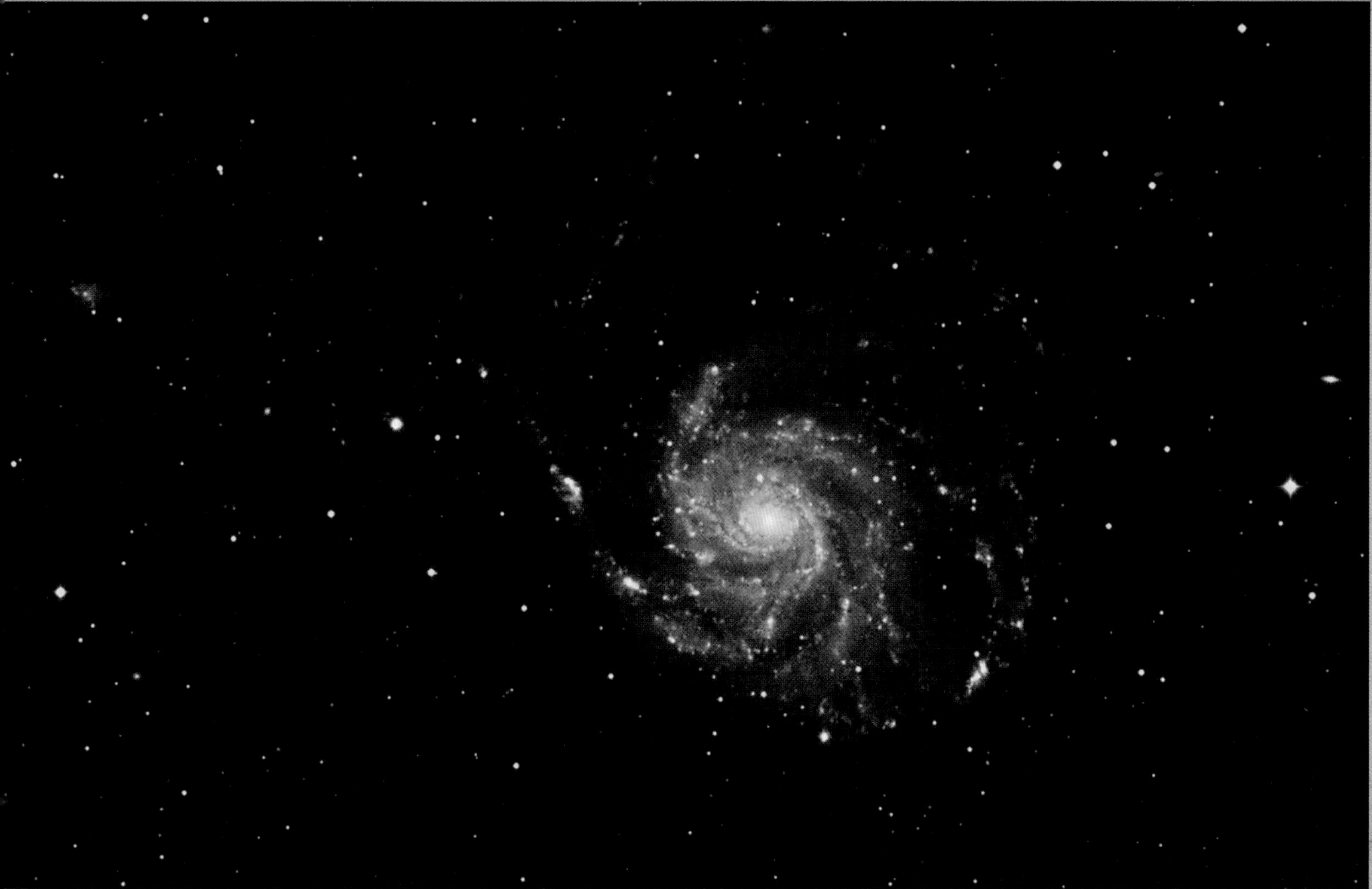

SPIRAL GALAXY
ESO 510-G13

All spiral galaxies have two parts. The outer part forms the spiral arms and contains, on the whole, young stars. All the stars usually lie in one plane, and this part of a spiral galaxy is called the disk. The inner region, called the bulge, is made up mostly of older stars and appears yellowish. To get an idea of the shape of a spiral galaxy, imagine two fried eggs back to back—the whites are the disk, and the yolks form the bulge.

Not all galaxies have a disk that is flat. ESO 510-G13's disk is dramatically bent in the image opposite. Disks result naturally from the collapse of diffuse material. As the material falls together, small differences in the density of the material result in small differences in the gravitational pull experienced by regions of the cloud. This causes a torque, which starts the rotation. Once it starts, it continues, and the rotation flattens out the disk. (This is the same reason why the planets in the solar system all orbit the Sun in the same plane.)

Disks are not solid objects, but are composed of billions of stars, all fairly loosely connected by their gravity. They all have different orbits about the galactic center, some oscillating between the "top" and the "bottom" of the disk. Warped disks are not uncommon and are seen in disks around new-born stars as well as in other spiral galaxies. The warps in galactic disks may be the result of interactions or even collisions with other galaxies.

DATA
Distance: 150 million ly
Constellation: Hydra
Redshift: 0.012

Warped galaxy ESO 510-G13.

SPIRAL GALAXY
NGC 4736 (M94)

This galaxy has a very bright core, which fades into tightly wound spiral arms. They in turn fade into a fainter amorphous outer disk that appears to have little structure when imaged in red light. When seen in more energetic wavebands such as the ultraviolet, however, the appearance of this galaxy changes dramatically. In the upper image the core is hardly visible, but there is a bright ring of emission that seems to occur at the edge of the spiral arms where they fade into the outer halo. The emission comes from young massive stars that are only ten million years old and are arranged in a ring about 7,000 light-years wide, encircling the nucleus.

One hypothesis is that there is an elongated bar in the center of NGC 4736. It is not visible in this image, but the bar is believed to be rotating and also to have generated the ring of stars. The hypothesis explains why the star formation occurs in a narrow ring—it could demarcate the ends of the bar. The symmetrical structure of spiral galaxies may be the reason why they form stars at a more rapid rate than ellipticals.

The startling contrast between the two images shows what "multi-wavelength" astronomy can discover: what would be hidden at one wavelength is unmasked at others. The combination of images and data from the whole range of the electromagnetic spectrum is likely to continue to reveal new and interesting phenomena.

DATA
Distance: 15 million ly
Constellation: Canes Venatici
Redshift: 0.0010

NGC 4736 in ultraviolet light (top) and in red light (bottom).

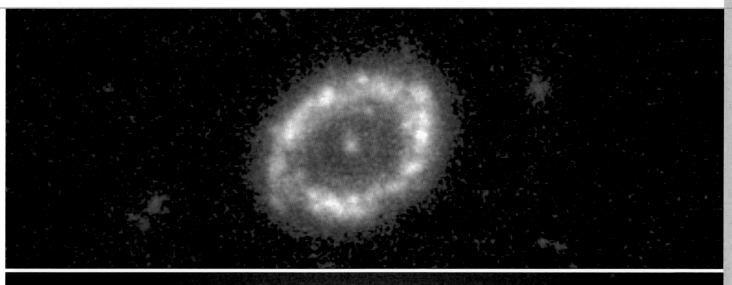

SPIRAL GALAXY
Sombrero Galaxy—M104

The Sombrero Galaxy owes its peculiar shape to the fact that we see it almost edge-on. It is a spiral galaxy, but it has a large central "bulge" of older stars that almost swamp the generally young disk stars. Many points of light in the visible-light image (top) are actually globular clusters. Because individual stars are too small to see, they form the diffuse halo around the center.

The lower image combines visible and infrared light. The galaxy looks different depending on what range of infrared light is observed. The shorter wavelengths come from stars that are older and cooler than our Sun. They are common in the central parts of the galaxy. At longer wavelengths, even these stars do not emit much light and so the emission is dominated by the interstellar dust and molecular gas. The dust and gas are concentrated in the spiral arms, where stars are most likely to form. The interstellar dust is very similar to soot. It is made up of carbon-based molecules known as polycyclic aromatic hydrocarbons (PAHs). They can clump together, forming dust grains.

When stars between one and three times the mass of our Sun come to the end of their lives they swell and cool, forming red giants. During this phase they lose up to 40 percent of their mass in the form of a strong stellar wind. At this stage they also contain a lot of carbon and other elements that are carried off by the wind and are lost in space, forming interstellar dust clouds. The amount of dust in a galaxy can therefore reveal how many of these types of stars have lived and died.

DATA

Distance: 28 million ly

Constellation: Virgo

Redshift: 0.0034

Sombrero Galaxy depicted in visible light (top) and in a combination of visible and infrared (bottom). The bottom image shows a ring of dust (red) and a disk of stars (bright spot) just visible at the center of the bulge of stars.

SPIRAL GALAXY
NGC 4013

This spiral galaxy is seen here edge-on. Although none of the spiral arms can be seen, this view enhances the dust in the galaxy as well as its thinness. Although it only shows about half the galaxy, the image reveals in detail the huge dust and gas clouds that form the galactic disk. The bright star to the top left of the image is a foreground star in our galaxy. The dust stands out in the image because it prevents the background starlight from reaching us. In the same way as smoke blocks out the sun's rays, the dust scatters the light out of our line of sight. Because less light reaches Earth, we perceive those regions as darker. On a dark night, away from city lights, you can see similar dark bands across the diffuse glow of stars that make up the Milky Way.

Different wavelengths of light are scattered by varying amounts, so the color of the light that reaches Earth and the amount by which it has been absorbed can give an insight into the size and composition of the dust clouds. The dust clouds in this image contain dust equivalent in mass to one million times that of our Sun. These interstellar dust clouds harbor the stellar nurseries in which new stars are born. The stars cannot be seen because, side-on, the dust is too thick.

By measuring the color of the starlight, astronomers can work out the speed at which stars are orbiting the center of the galaxy. Surprisingly, in the outer regions they are orbiting faster than would be expected if the mass of the galaxy came only from the visible matter (stars, gas, and dust). The extra mass required has been called "dark matter" because we cannot see it. We can only infer its presence from its gravity. About five times more mass in the universe comes from dark matter than from normal matter.

DATA

Distance: 55 million ly

Constellation: Ursa Major

Redshift: 0.0028

Edge-on view of NGC 4013.

NGC 4622

Not all spiral galaxies are easy systems to understand. One would assume that the spiral arms are trailing the rotation, so in this image the rotation of the galaxy would be counterclockwise. Closer inspection of NGC 4622 shows that it has an inner spiral arm which is wound in the opposite direction to the outer ones.

The junction of the two different spiral structures has produced a tidally generated ring. Most of the stars in this galaxy are probably old, since the light from the galaxy is yellowish. In the ring, however, there are young star clusters shining a brilliant blue. It is not currently known which set of arms is leading and which is trailing. Because the galaxy is face-on, astronomers cannot easily determine the rotation

from the Doppler shift of the light from the stars. So the outer arms may be opening out or winding in—the question is still unanswered. Simulations of this type of galaxy imply that the inner arm is leading. However, because NGC 4622 is an almost unique example, we cannot be absolutely certain that this is the case.

An interaction between this galaxy and a small companion galaxy might be the cause. A smaller galaxy (not visible) off the bottom left of the image is interacting with this galaxy. The outer arms of NGC 4622 are lopsided and indicate that there has been some interaction in the past. If the companion collided close to the nucleus of NGC 4622, it could have caused this strange motion.

DATA
Distance: 131 million ly
Constellation: Centaurus
Redshift: 0.0146

"Backward" spiral galaxy NGC 4622.

NGC 7331

This spiral galaxy is thought to resemble the Milky Way so much that it is sometimes known as our galaxy's twin. The galaxy, which is inclined to our line of sight, was discovered by William Herschel in 1784. Herschel also discovered infrared light, so it is appropriate that the Spitzer Space Telescope took an image of galaxy NGC 7331.

It is also one of the 14 "nebulae" that Irish nobleman William Parsons, the third Earl of Rosse, listed in his 1850 paper as "spiral or curvilinear." He wanted to have the largest telescope in the world and started work on one with a mirror 72 inches (183 cm) in diameter. His ambition caused problems—he had to build a foundry to create the mirror on site and also had to assemble a structure that could contain the telescope tube, which was 58 feet (18 m) long. Eventually the telescope was mounted between two parallel 70-foot-

(21-m-) high walls, built in a north–south direction, which would allow tracking of objects for around two hours.

This telescope could see fainter objects than any before and it allowed the earl to distinguish two types of nebulae: those that we now call nebulae and those that turned out to be galaxies outside our own. The latter type was found to have one of two shapes—elliptical or spiral—and NGC 7331 was one of the first to be described.

Working for the earl was Danish astronomer J. Dreyer, who began to compile the NGC (New General Catalogue) while at Parsontown in Ireland. The NGC is still in use today, although there are many others, and most objects in the sky have several names and appear in many catalogs. The telescope, located at Birr Castle, Ireland, was restored in the 1990s and is now open to the public.

DATA
Distance: 38 million ly
Constellation: Pegasus
Redshift: 0.0027

Infrared image of NGC 7331.

SPIRAL GALAXY
Black Eye/Sleeping Beauty/Evil Eye Galaxy—M64

DATA

Distance: 24 million ly

Constellation: Coma Berenices

Redshift: 0.00136

The collision of one galaxy with another resulted in a "black eye" for the remnant of the merger—M64. The collision left a thick dark band of dust across the galaxy's bright nucleus. The galaxy is visible through binoculars, and it was first cataloged by the French astronomer Charles Messier in the 18th century to avoid confusing it with a comet.

Although at first glance it appears to be an ordinary spiral galaxy, albeit somewhat dusty, it is not as simple and as peaceful as it seems. All the central stars and the dust are rotating in the same direction—clockwise in the image. The interstellar gas in the outer regions of the galaxy is orbiting the other way. Where these two parts of the galaxy meet, the shear forces cause the gas and the dust in the region to be compressed, and so they start to collapse under their own gravity. These new nebulae start to form stars that light them up from the inside. The newly formed stars emit mainly blue light, whereas the gas, mainly hydrogen, emits a reddish-pink glow. These nebulae appear as the points of light on the swirls of dust, rather like jewels on a carpet.

The opposite motion of the outer regions of this galaxy is likely to have arisen from the collision of a satellite galaxy with M64 around one billion years ago. The collider must have hit the outer regions of the galaxy in a counterclockwise direction in order to set the gas in motion. The collider galaxy has now been completely destroyed and assimilated by M64, its only trace being the strange motion of the gas.

The spectacular dark band of gas in front of its bright nucleus gives M64 the nicknames Black Eye and Evil Eye Galaxy.

SPIRAL GALAXY
Whirlpool Galaxy—M51

DATA

Distance: 23 million ly

Constellation: Canes Venatici

Redshift: 0.00154

With its sweeping spiral arms, the Whirlpool Galaxy is one of the most beautiful galaxies in the sky. It is almost exactly face-on, giving us a very clear view of its dynamic parts. The pink and red regions in the image from the Hubble Space Telescope (inset) are places where stars are being born—similar to the Orion Nebula in the Milky Way and the Tarantula Nebula in the Large Magellanic Cloud. The galaxy's nucleus has a yellowish tinge, because the stars there are mainly old, but in the spiral arms the stars are young, hot, and blue.

The infrared image shows the difference between M51 and its irregular companion in space, NGC 5195. The infrared image is color coded, the hotter parts being in blue and the cooler parts in red. The emission that is seen in the visible and the near infrared (blue) is mainly starlight. The infrared emission from the cooler parts (red) comes from gas and dust. The spiral galaxy's arms show a great deal of emission from dust as well as from stars. This means that there are large reservoirs of gas and dust to form new stars. Some of these star-forming nebulae are seen as the pinkish regions in the Hubble image, and where the northern arm passes in front of NGC 5195, the dust lanes can also be seen. NGC 5195 shows only emission from stars, and so there are not many regions in this galaxy that can form stars.

Although the companion looks as if it is attached to the end of one of the Whirlpool's arms, it is, in fact, gliding behind it. Its gravity may be what causes the Whirlpool to have many places where stars are being formed. As NGC 5195 drifts by, it causes pressure waves in the galaxy, enhancing those from the spiral arms, which aid the formation of stars.

Whirlpool Galaxy: infrared (main image), and visible light (inset).

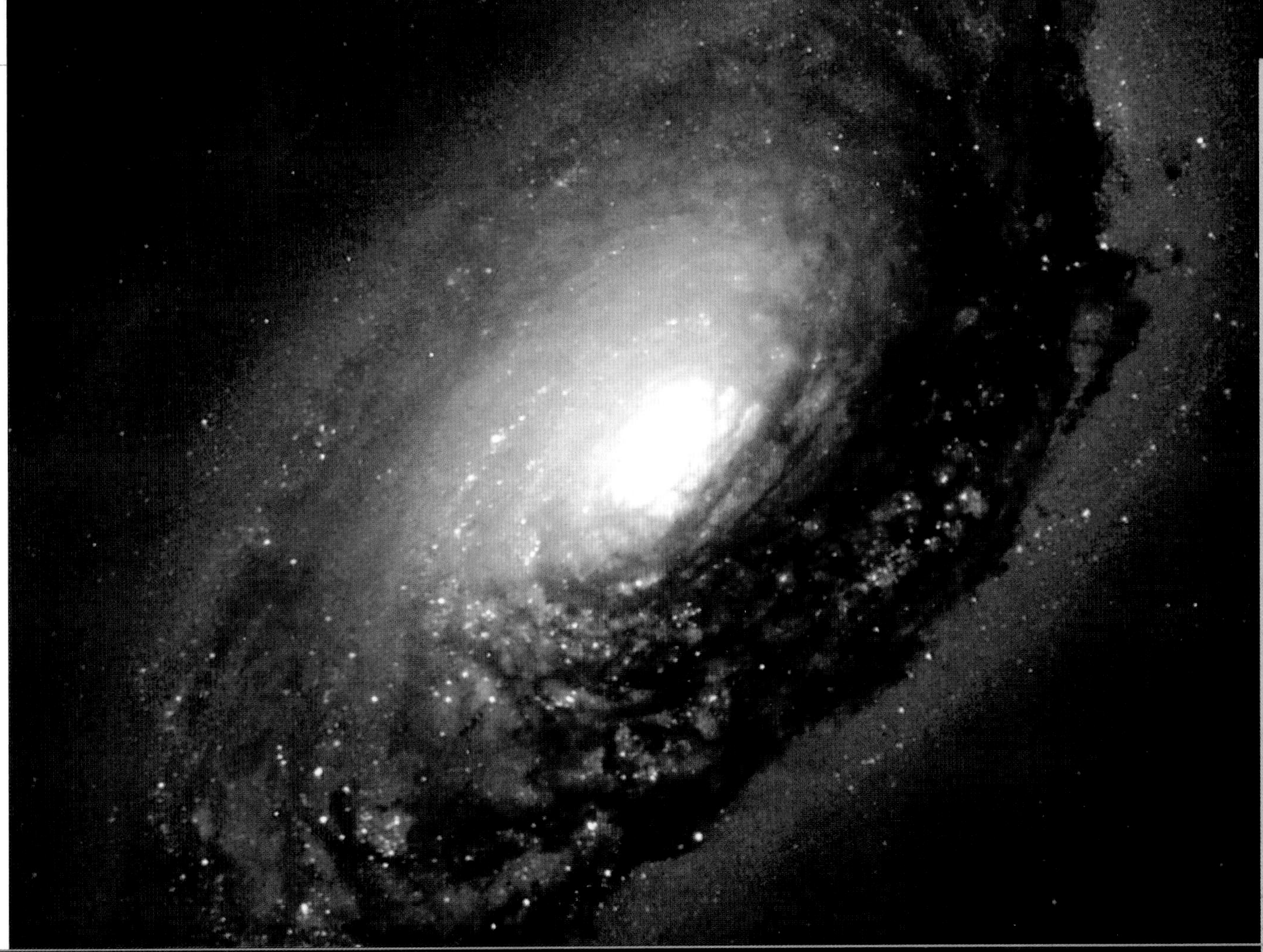

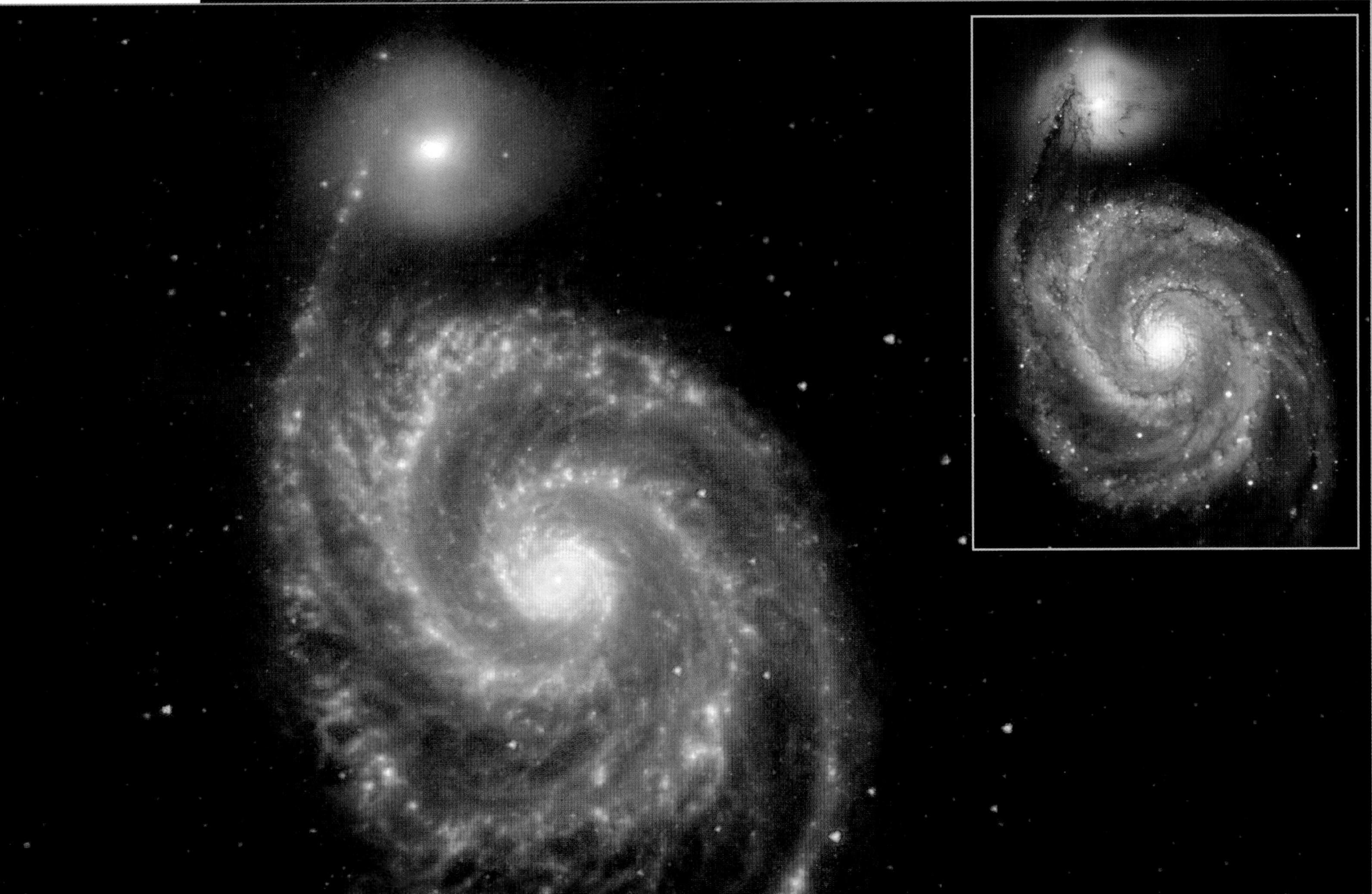

SPIRAL GALAXY
M66

M66, along with M65 and NGC 3628, is part of a conspicuous triplet of galaxies called the Leo Triplet. This galaxy is strikingly asymmetrical because it is so close to the other members of the Triplet. M65 has remained remarkably free of distortion, but NGC 3628 has a warp in its disk. The gravity of the other galaxies has combined with the rotation of M66 to pull its stars away from the nucleus. If the stars that are farthest from the nucleus fully separate from the main galaxy, they may have enough mass to collapse to-gether and form an irregular dwarf satellite of M66, although this will take many millions of years to occur.

In contrast to many of the images taken in visible light in this book, the image opposite was not taken by the Hubble Space Telescope but by a ground-based telescope.

The galaxy itself was cataloged in the 18th century by the French astronomer and comet hunter Charles Messier. He was interested in cataloging all the "fuzzy blobs" that he saw in the night sky so that he would not confuse them with comets, whose positions in the sky change from week to week. Over the course of 20 years and in three publications he cataloged the positions of 110 galactic nebulae and galaxies. The entry for M66 in his catalog states "Nebula discovered in Leo; its light is very faint & it is very close to the preceding [M65]: They both appear in the same field [of view] in the refractor. The comet of 1773 & 1774 has passed between these two nebulae on November 1 to 2, 1773." Messier most probably did not see them at that time because of the light of the comet.

DATA
Distance: 37 million ly

Constellation: Leo

Redshift: 0.0027

The spiral galaxy M66. (By permission Canada–France–Hawaii Telescope/J.-C. Cuillandre/Coelum).

SPIRAL GALAXY
NGC 2403

NGC 2403 is a conspicuous object that was (unusually) missed by Charles Messier when compiling his catalog. It is an outlying member of the M81 group. M81 is one of the closest grand design spirals outside the Milky Way, and its neighbor M82 is the brightest infrared source outside our galaxy. Strictly speaking, NGC 2403 is intermediate between an elliptical and a spiral galaxy because the bulge in the center is much larger than in most spirals, and the arms are a little faint.

Galaxy NGC 2403 was discovered in 1785 by William Herschel, who was working in England. He had moved there from Hanover, Germany, in 1759. In 1772 he went to live in the southwest of England, where he became interested in astronomy. He soon became an expert in making telescopes

and started to survey the night sky. In 1781 he discovered what he at first thought was a comet but was, in fact, the planet Uranus.

Herschel revisited the objects listed in Messier's catalog, studying them with his own superior telescopes, and began an extensive survey of the northern sky. He located about 2,500 new "nebulae" during this period. In 1787, six years after discovering Uranus, he discovered two of its moons—Titania and Oberon. He came close to determining the spiral nature of some of these nebulae, but this work was left to William Parsons, the third Earl of Rosse, in Ireland. William Herschel's son (John) also became an astronomer, continuing his father's work in the Southern Hemisphere, based in Cape Town, South Africa.

DATA
Distance: 13.9 million ly

Constellation: Camelopardalis

Redshift: 0.00043

Galaxy NGC 2403.

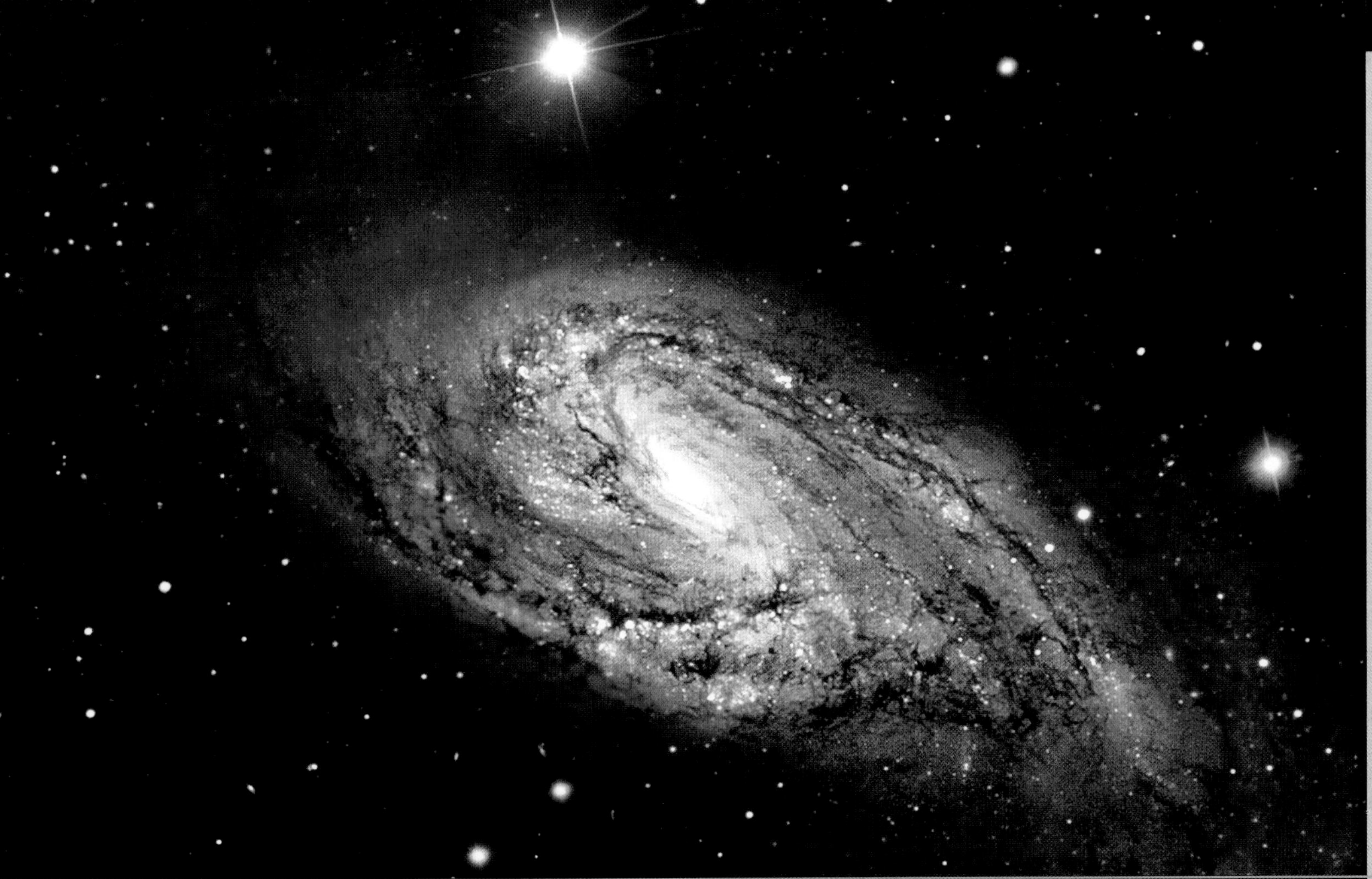

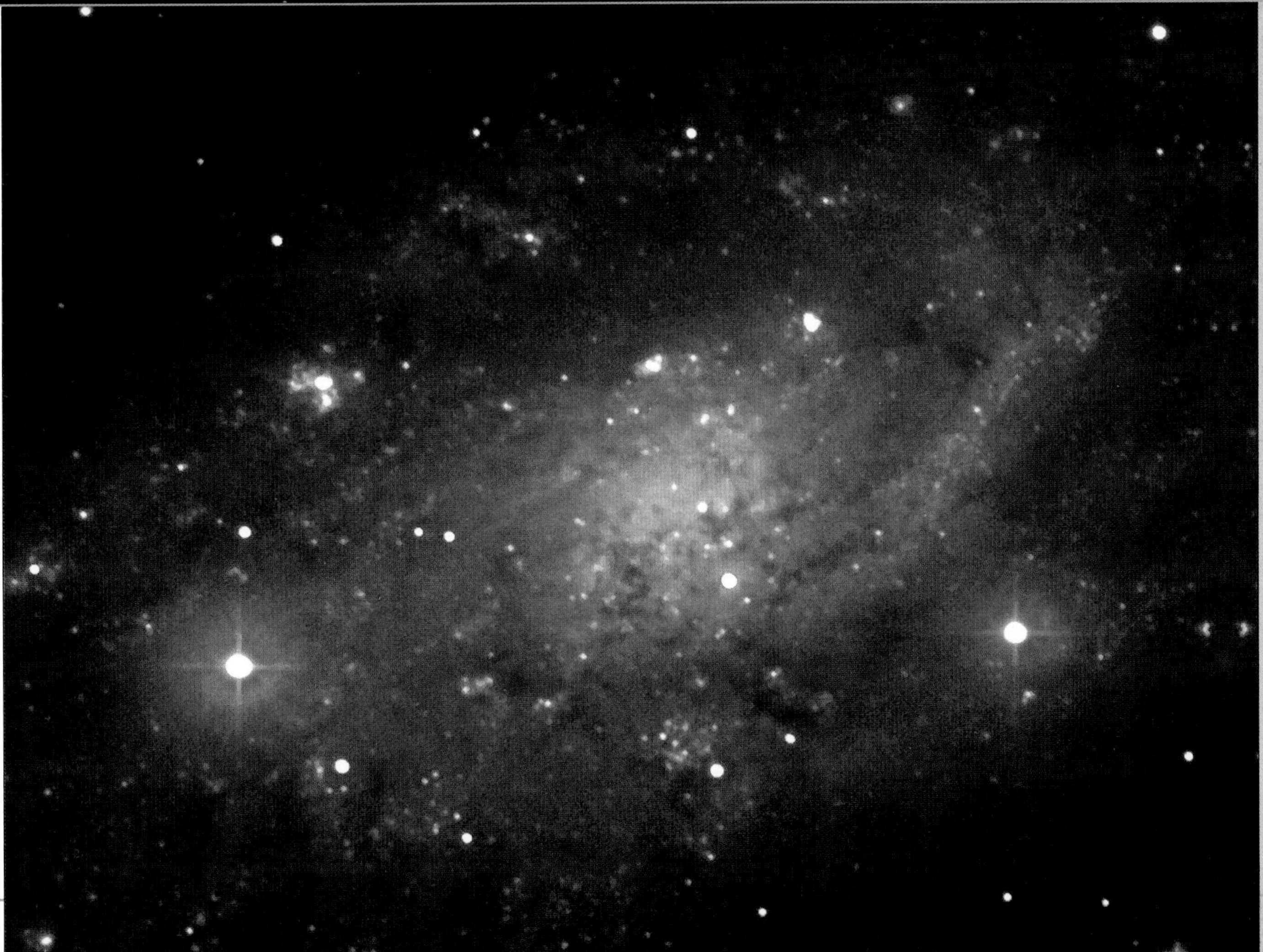

SPIRAL GALAXY
Other Spiral Galaxies

Left: *M65 is a spiral galaxy with a prominent central bulge and tightly wound spiral arms. It is 35 million light-years away, located in the Leo Triplet, or M66, group.*

BARRED SPIRAL GALAXY
NGC 1300

This archetypal barred spiral galaxy is shown in its full glory in the Hubble Space Telescope picture. Barred spiral galaxies, as can be clearly seen in the image, have arms that do not spiral right down into the center of the galaxy. Instead, they are connected by a straight bar that contains the nucleus.

The parts of the galaxy at the very center of the bar then show what appears to be a grand design spiral structure only 3,300 light-years across. This has been seen in other barred spiral galaxies. It is thought that the bars in this type of galaxy funnel material (gas and dust) inward toward the center. Close to the center of the galaxy the material forms the spiral structures as it falls. If the central nucleus is an active nucleus—one that contains a super-massive black hole that is accreting matter—then this material could eventually feed it.

DATA

Distance: 61 million ly

Constellation: Eridanus

Redshift: 0.0053

Starlight, glowing gas, and silhouetted dark clouds of interstellar dust in NGC 1300.

Right: *The spiral galaxy NGC 1232 has millions of bright stars along its spiral arms and contains dark dust between them.*

BARRED SPIRAL GALAXY
NGC 2903

Although it does not resemble one in this image, NGC 2903 is a barred spiral galaxy. Our own Milky Way is thought to have a central bar and is of a similar size to NGC 2903. There are many knots of bluish emission in the outer arms of this galaxy. These are young star clusters, similar to the Pleiades. The core of the galaxy is known for hosting many "hotspots," giving it its speckled appearance. The Hubble Space Telescope showed that most star formation does not take place in these hotspots, as had been assumed previously, but in large gas clouds found close to the hotspots.

The bar is the reddish band in this image. Its color arises from the large amounts of dust it contains. In the same way as the setting Sun appears to turn red, the dust changes the color of the light reaching Earth. It has been suggested that the bar causes the flow of gas into the center of the galaxy, triggering star formation. The rotation of the bar sends spiral pressure waves into the gas, causing both the gas and the dust to start to collapse. This action has formed a circular star-forming region around the nucleus of the galaxy some 2,000 light-years wide.

The galaxy is visible in the Northern Hemisphere and, surprisingly, is one of the more conspicuous objects missed by French astronomer Charles Messier when he compiled his catalog of nebulous objects. It was discovered by Willam Herschel in 1784.

DATA
Distance: 30.6 million ly
Constellation: Leo
Redshift: 0.0019

Close-up of barred spiral galaxy NGC 2903.

BARRED SPIRAL GALAXY
NGC 1365

Images from the ground show NGC 1365 as a majestic giant barred spiral galaxy 200,000 light-years in diameter. The dramatic spiral arms wind away from the large central galactic bar. The Hubble Space Telescope (HST) has zoomed into the very core of the galaxy (main image), showing that the dust lanes reach right into its center. There are also blue knots of light, which indicate the presence of young star clusters at the center of the galaxy. This phenomenon is in contrast to unbarred spiral galaxies—their star formation tends to occur mainly in the outer arms, with the central bulge composed mostly of older stars. The Near Infrared Camera and Multi-Object Spectrometer (NICMOS) infrared image shows the emission from these new star clusters as well as others that cannot be seen in the visible HST image because their light is blocked by dust.

It is thought that the bars may be a temporary structure and that they form naturally from symmetrical spirals. After a time, the bar weakens and disappears, only to be replaced by another later in the galaxy's evolution. The bars observed in current galaxies may be the third or fourth bars that the galaxy has experienced in its lifetime.

The nucleus of this galaxy is bright in both images, suggesting that there is a massive black hole at the center that is feeding on nearby material. It is likely to be of similar mass to the one in the center of the Milky Way, but we do not see ours because it does not appear to be accreting matter at present. The black hole in NGC 1365 is accreting matter, and the barred structure of the galaxy is thought to play a central role in funneling matter into the inner regions of the galaxy, where it forms stars or falls into the black hole.

DATA
Distance: 60 million ly
Constellation: Fornax
Redshift: 0.0055

Visible-light image (main picture) and infrared image (inset) of NGC 1365.

BARRED SPIRAL GALAXY
NGC 4639

This ordinary-looking galaxy was one of the galaxies used to determine distances in the universe. In the outer regions of NGC 4639 there are clusters of newly formed stars (blue dots in the image opposite). Among them are examples of a special type of star, called a Cepheid variable.

These stars were identified by Henrietta Leavitt, while working as a volunteer at the Harvard College Observatory from 1895. Seven years later the director, Charles Pickering, appointed her to the permanent staff. She discovered 2,400 variable stars during her career—more than half of the total number known in her day. Leavitt's most important discovery was the Cepheid Period–Luminosity relationship, which enabled astronomers to measure the distances to nearby galaxies. As their name suggests, Cepheid variables vary in brightness periodically and the rate at which they change

brightness (the period) depends on how luminous they are. This means that by measuring the star's period, astronomers would know its luminosity, and by comparing that to its apparent brightness, its distance could be calculated.

Having determined the distance to NGC 4639 using the Cepheid Period–Luminosity relationship, astronomers then turned to the next "standard candle" in the "distance ladder"—a kind of supernova known as Type Ia. These supernovae are caused by the explosion of white dwarfs. Their mass is limited by fundamental laws of physics, so their luminosity should be the same everywhere in the universe. The distance to NGC 4639 was used as a calibrator for SN 1990N, a supernova that went off in 1990 in this galaxy. In turn, Type Ia supernovae have been used to calculate other distances.

DATA

Distance: 78 million ly

Constellation: Virgo

Redshift: 0.0034

Combined visible light and near infrared image of NGC 4639.

BARRED SPIRAL GALAXY
Other Galaxies

Right: *Barred spiral galaxy NGC 1097 has at its center a supermassive black hole. It is about 45 million light-years away.*

Right: *NGC 5195 is a close companion of M51. The dust lane of one of the spiral arms of M51 can be seen in front of NGC 5195 at bottom left. The close encounter between these two galaxies has distorted NGC 5195 to an irregular shape.*

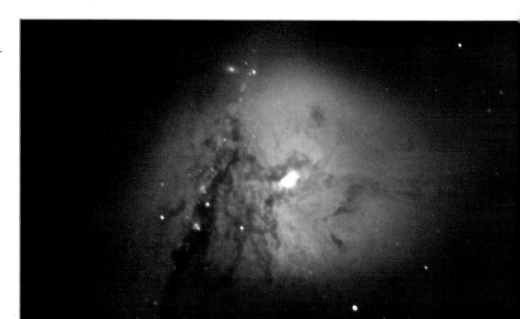

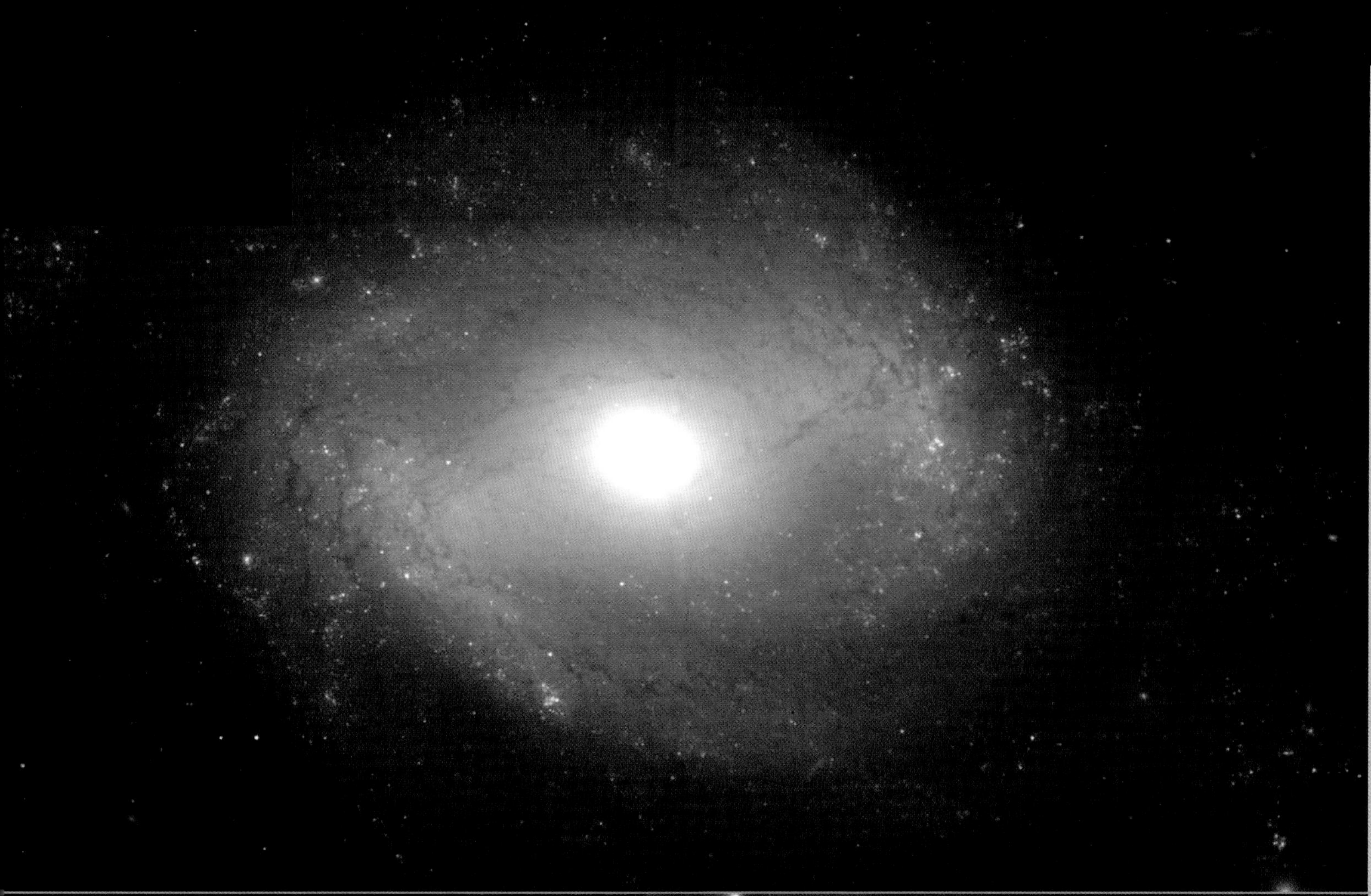

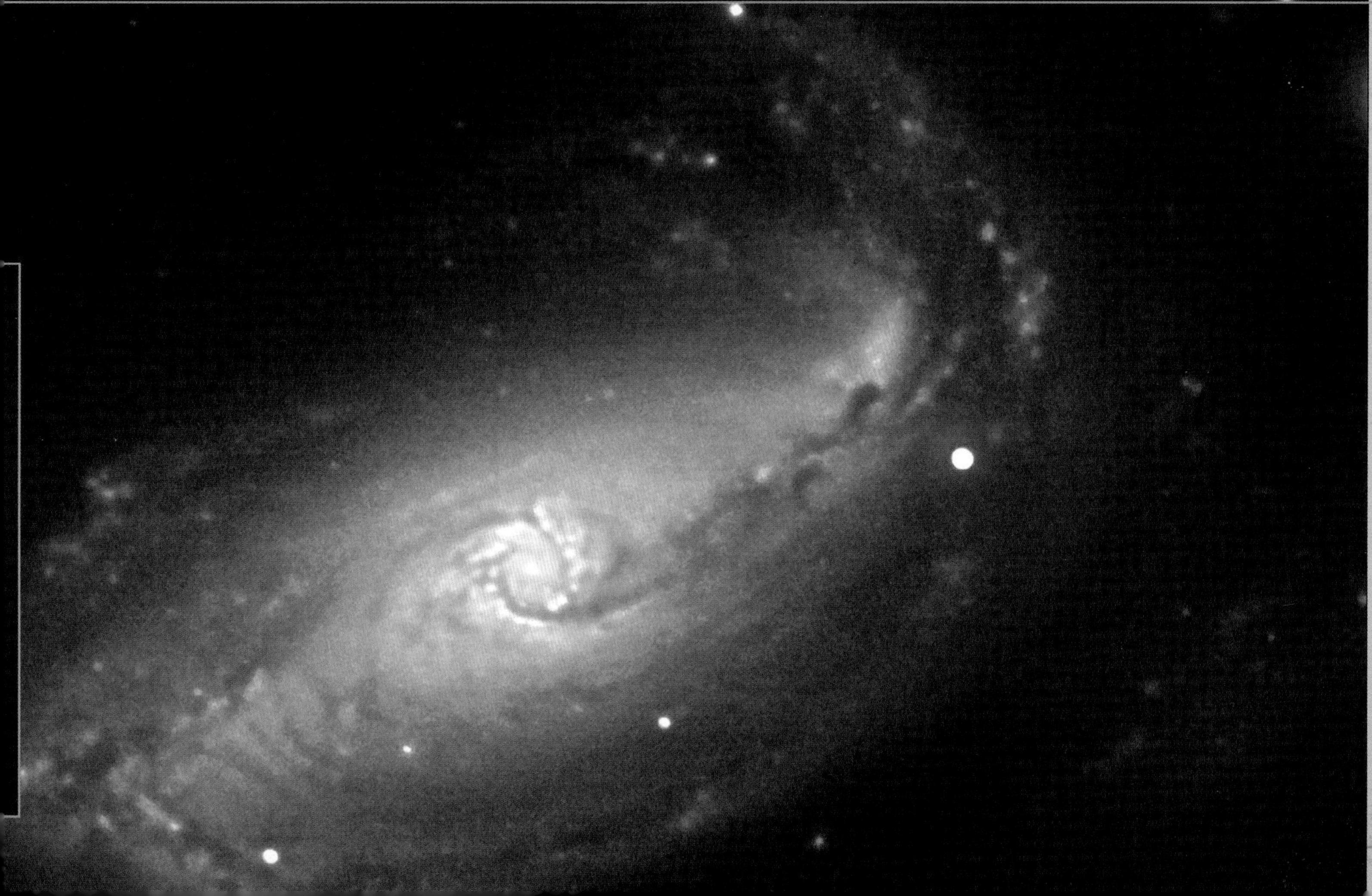

ACTIVE GALAXIES, INTERACTING GALAXIES, AND CLUSTERS

Most galaxies are relatively sedate, but some lead extremely active lives. Many names and classification schemes exist to describe such galaxies, and some may fall into more than one category. Many galaxies are found in groups. Within a particular group the galaxies are all relatively close to each other, bound by the gravitational pull they exert on each other. As a result, some galaxies collide, causing dramatic interactions—hence the term "interacting galaxies." If two colliding galaxies are of very unequal mass, one is highly disturbed while the other stays intact. When galaxies of similar masses collide, they disturb each other's stellar content. But it is the gas and dust that collide—the stars rarely do. This causes star formation in the resultant galaxy. In the collision stars are sometimes flung far outside the galaxy.

The onset of rapid star formation can trigger what is known as a starburst in the galaxy. This occurs when the star-formation rate of the galaxy is many times that of an average galaxy of its type. As an example, the Milky Way only forms a few solar masses worth of stars in a year, while starburst galaxies form hundreds. The deaths of the most massive stars formed in this way can cause large winds as the supernova explosion blows gas out of the galaxy.

It is now thought that most galaxies harbor a massive black hole at the center, ranging from millions to billions of times the mass of our Sun. These black holes are far too small to be seen (even if they were not black) but their effect on the surrounding galaxy and space can be detected. By imaging the center of a galaxy, we can see the motions of gas clouds and stars. The speed at which they orbit the center implies a very large mass at the galaxy's core, which can only be a black hole.

When the black hole starts "feeding" (in ways that are not yet fully understood), some material is not swallowed. Instead, it is ejected, sometimes in the form of jets. The nuclei of such galaxies are truly active and are therefore known as active galaxies or active galactic nuclei (AGN). There are many names for these objects. The least powerful, but also the most common, are known as Seyfert galaxies.

The more violent active galaxies include radio galaxies, quasars, and blazars. If we look at such an object using radio waves, the jets from the central black hole (known as the "engine") are clear. The material in the jets has been squirted over vast regions of space known as radio lobes (which can be much larger than the galaxy itself). In some nearby galaxies the jet is also visible at other wavelengths of light.

When a large number of galaxies are all in the same space, they are known as a cluster. Clusters are the most massive single objects in the universe, and they are still in the process of forming. They contain much more material than that associated with the stellar light from the galaxies. About ten times more material exists as a very hot, tenuous, X-ray-emitting gas in which the galaxies are bathed. This is still only a small fraction of the total mass in the cluster, which is dominated by dark matter. When clusters are observed in this light, they change character totally.

Some clusters have radio galaxies at the center. The jets from their central engines create large cavities in the X-ray gas and have a dramatic effect on the central regions.

Composite X-ray (purple) and optical (red) image of active galaxy NGC 1068. The hot X-ray-emitting gas is being blowing away from the galaxy center by a central massive black hole.

Left: *A long channel of material (dark streak) between NGC 1410 (on left) and NGC 1409.*

Right: *A black-hole-powered active Seyfert galaxy in the southern constellation Circinus.*

STARBURST GALAXY

M82

DATA

Distance: 12 million ly

Constellation: Ursa Major

Redshift: 0.00068

To our eyes—in visible light (seen left in the image opposite)—this edge-on spiral galaxy, also known as the "Cigar Galaxy," is fairly normal. In the infrared (center), however, it is the brightest galaxy in the sky. M82 is a starburst galaxy in which the rate of star formation is about ten times that of the Milky Way. The billowing clouds of gas resulting from such rapid explosion of stars can be seen in the images.

When gas clouds collapse and start to produce stars, they produce them with a range of masses—many small ones and a few big ones. Larger stars burn their fuel much faster than smaller stars, which means that they reach the end of their life much earlier. They die in spectacular supernova explosions that blast them apart from the inside, creating a wave of material rushing outward from the star. If many stars die in this way within a short space of time, the shock waves reinforce one another and can push material right out of the galaxy itself. The image taken in normal light used special filters to enhance the light coming from the filaments. The super-wind (gas rushing out from the center of the galaxy) stretches 10,000 light-years into space, but it will probably fall back onto the galaxy in the future.

The event that started the burst of star formation was probably a close encounter with its near neighbor, M81. M82 is the smaller of the two galaxies and therefore suffered more damage after the interaction. M81 has remained relatively unaffected and is a symmetrical spiral galaxy.

*Detail of M82 in visible light (left) and infrared (center).
The picture at bottom right shows the whole of the
visible-light image, with the white outline indicating
the area of the details shown in the other images.*

SEYFERT GALAXY

NGC 7742

DATA

Distance: 72 million ly

Constellation: Pegasus

Redshift: 0.00555

This ordinary-looking galaxy harbors something special in its core. In 1943 American astronomer Carl Seyfert noted that some apparently normal spiral galaxies had especially bright cores with peculiar spectra. He described the nuclei as almost starlike with very wide emission lines. Usually individual elements emit light at only certain frequencies (emission lines). For example, sodium emits an orange light as seen in street lamps. When the atoms that are emitting are moving very fast, the fact that light moves at a finite speed comes into play. If the atoms are moving away from us, the light they emit is stretched and becomes redder; if they are moving toward us then it is blueshifted. This is analogous to the Doppler shift exemplified by the sound of a racing car as it passes by.

This means that what would have appeared as an emission line is stretched and broadened out if the gas that is emitting is moving fast. So the wide emission lines at the center of Seyfert galaxies demonstrate that the gas there is moving fast. We now know that this is because the gas orbits a supermassive black hole. Seyfert galaxies are the least active of active galaxies, but they are useful because they are more common than their more violent cousins and therefore have enabled astronomers to perform larger studies to find out how they work. Later research into Seyfert galaxies showed that there are two distinct subclasses depending on the exact shapes of some of the emission lines. The difference arises from the density of the gas that is emitting the light.

Galaxy NGC 7742.

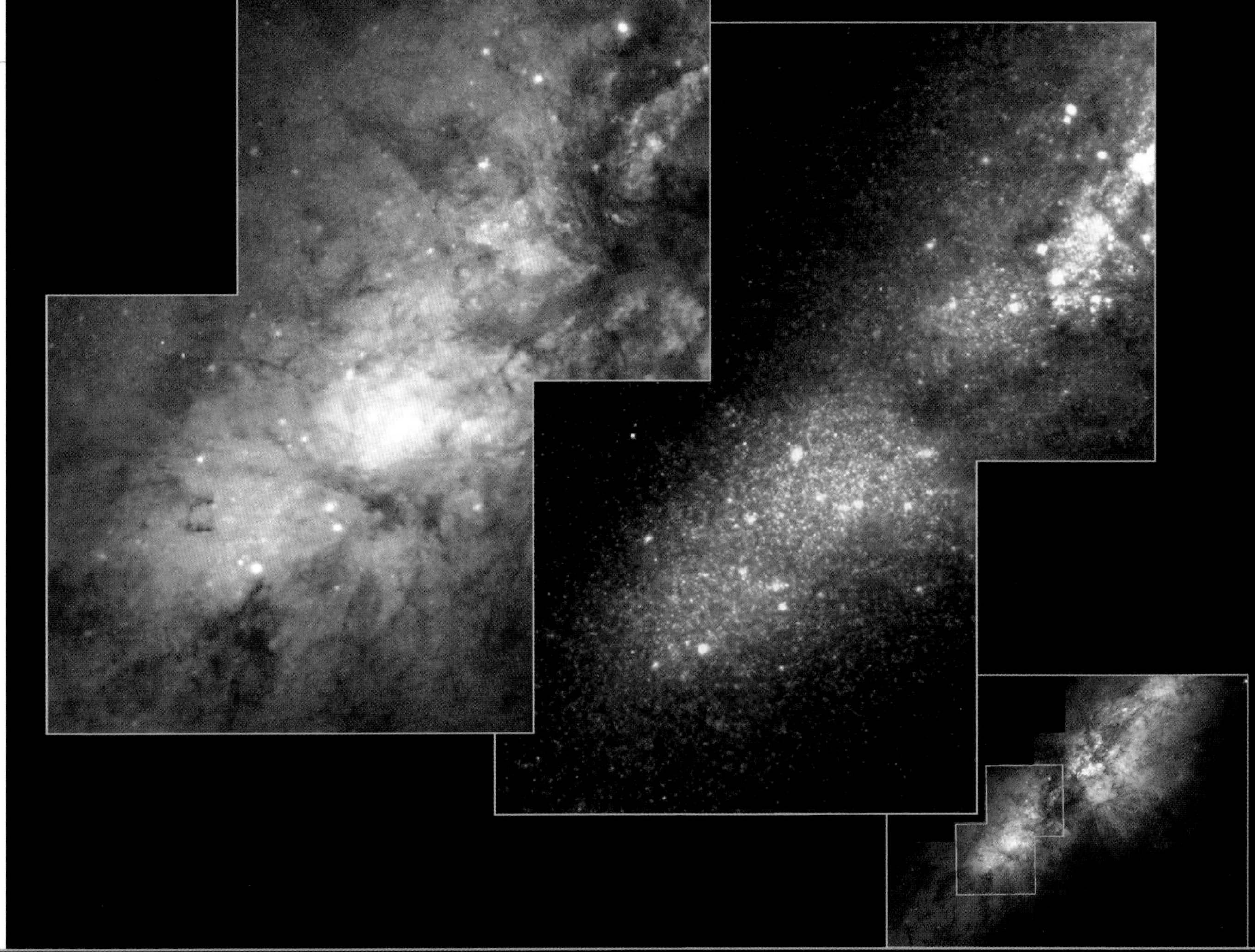

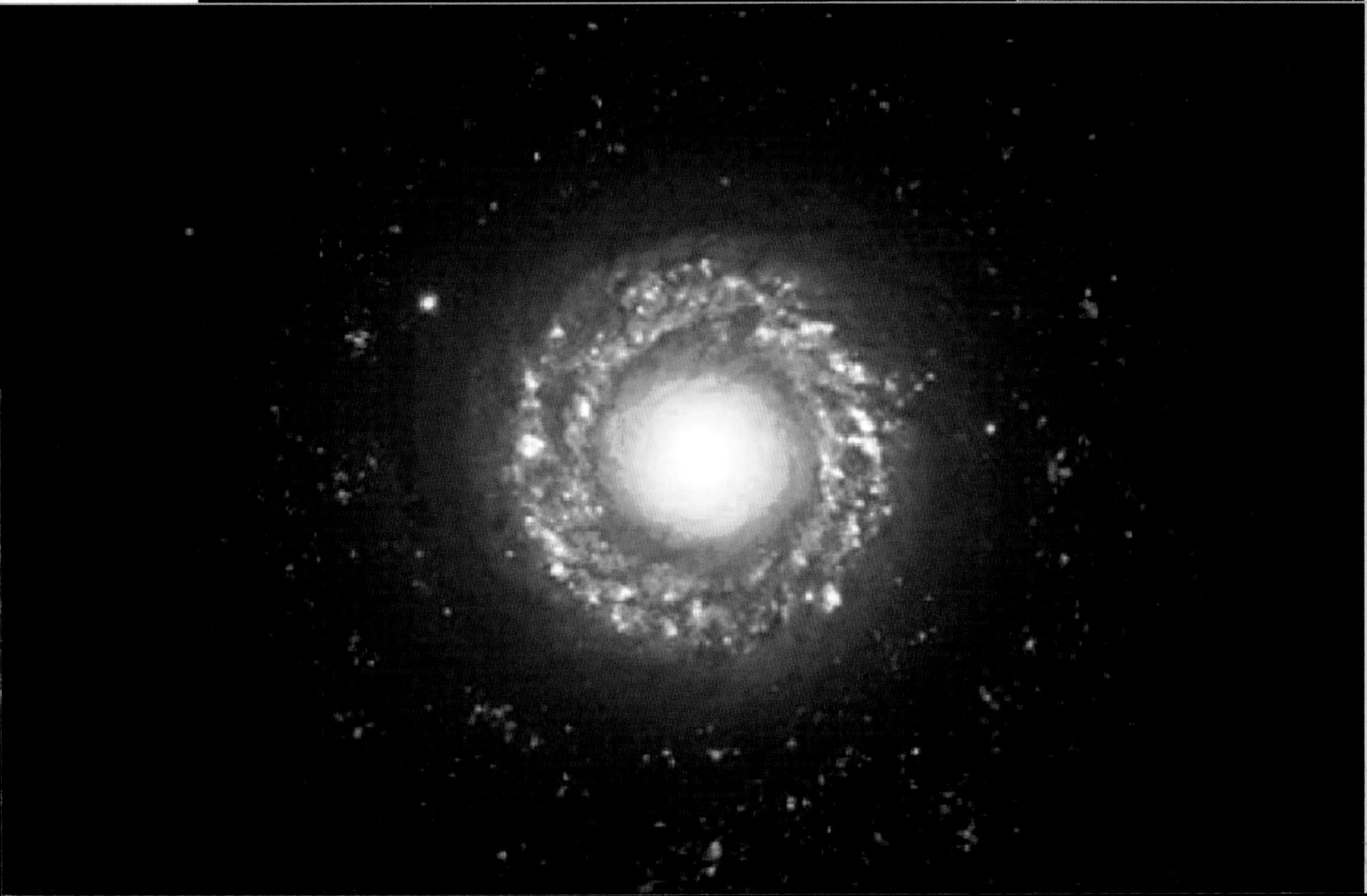

STARBURST GALAXY

NGC 4314

Ground-based images taken of the spiral galaxy NGC 4314 did not reveal anything special—just a large bulge with a bar across the center and two well-defined arms. There were no particularly bright parts, and stars were thought to be forming gently in a few places. Then the Hubble Space Telescope (HST) took a closer look at the core (the area covered by the Hubble image is shown by the white box in the inset picture). Hubble's spectacular resolution shows an intensely bright ring of young stars around the nucleus. The ring is the only place in the galaxy where stars are currently forming. This ferocious activity started only about five million years ago. Stars in this type of galaxy usually form in the spiral arms and not so close to the nucleus.

The region imaged appears like a minature version of the galaxy, yet the ring is only 1,000 light-years in radius. Just outside the ring are two bluish spiral arms and two wispy dust lanes. They trail on into the galaxy's bar. The stars in this region are older than those in the ring, suggesting that the star-forming ring is moving inward toward the galaxy's center. Interaction with another galaxy stripped most of the star-forming material from NGC 4314, but some fell back onto the bar near the center, which is now the only place in the galaxy that is forming stars.

DATA	
Distance: 31.6 million ly	
Constellation: Coma Berenices	
Redshift: 0.0032	

NGC 4314 as seen by the Hubble Space Telescope (main image), compared with a ground-based image (inset).

STARBURST GALAXY

NGC 3079

At first glance this spiral galaxy appears run-of-the-mill (main image). Detailed images show dramatic events occurring at the center (inset). Vast filaments of hot gas (10,000 K) extend outward and upward from the nucleus, forming a horseshoe-shaped structure. These filaments are thought to arise from a superwind (gas rushing out from the center of the galaxy). It carved a cavity in the gas surrounding the galaxy, but also stripped gas from the sides and stretched it up into the filaments. The hot gas, which is seen in X-rays, also heated the cool galactic gas so that it glows in the visible.

High-speed streams of particles, known as superwinds, are thought to be important in the evolution of galaxies. They are responsible for the expulsion of gas from the galaxy, and they may regulate the formation of new stars. It is not known exactly how they are created, but they may be caused by either supernova explosions or by activity surrounding the supermassive black hole at the center of the galaxy. Massive stars at the ends of their lives blow themselves apart. If several explosions occurred in a short space of time, the outward motions could cause the expulsion of the gas. Massive black holes sometimes produce jets of material that, in the early stages, may form a superwind.

DATA	
Distance: 52 million ly	
Constellation: Ursa Major	
Redshift: 0.0037	

The bubblelike structure in the center of NGC 3079 (main image) is composed of hot gas. The close-up (inset) shows the horseshoe-shaped filaments.

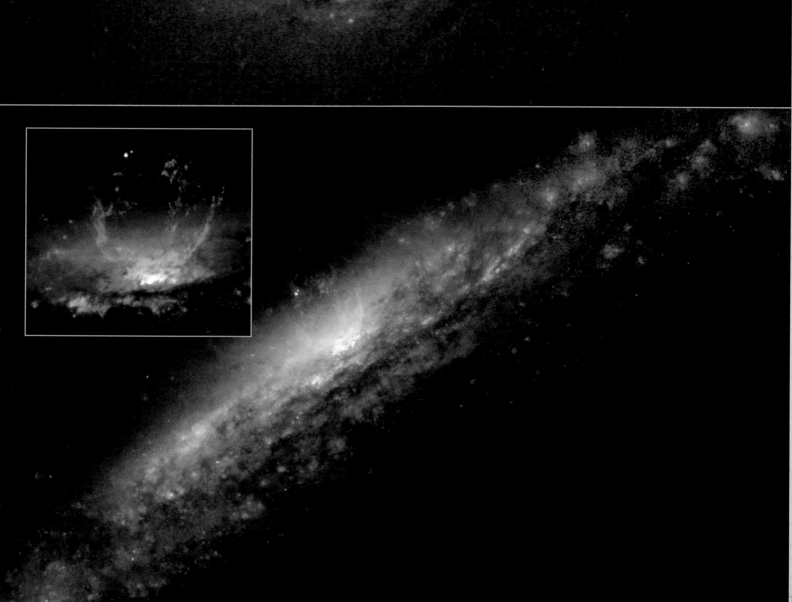

STARBURST GALAXY

NGC 3310

NGC 3310 is a special type of galaxy called a starburst galaxy. Starburst galaxies, as the name suggests, are forming stars much faster than ordinary galaxies, up to hundreds of times faster. They appear much brighter than other similar-sized galaxies because there are many more young stars present. The process of star formation produces stars over a wide range of masses. There are far more smaller stars formed than massive stars, but the massive ones are many times brighter and dominate the light. The rapid star formation causes the spiral structure to appear ragged and brilliant white.

The onset of this extreme rate of star formation was probably caused by the collision of NGC 3310 with another galaxy about 100 million years ago. The shock waves that the collision caused in the gas started the collapse of parts of the gas clouds. Eventually the clouds became dense enough to start nuclear fusion at their centers, and stars were born. This caused the arms of the galaxy to become brilliantly lit.

Some regions of the galaxy are winding down their star formation. Other clusters of stars are young and are likely to continue forming stars for many millions of years to come. This came as a surprise—it had previously been assumed that galaxies do not remain in a starburst phase for long before they run out of fuel to create new stars.

DATA	
Distance: 59 million ly	
Constellation: Ursa Major	
Redshift: 0.0033	

Starburst galaxy NGC 3310.

BLACK HOLE

M84

Trace from Hubble Space Telescope's space telescope imaging spectrograph (main image) across the core of galaxy M84 (shown inset).

Is this colorful zigzag the signature of a black hole? M84 is an elliptical galaxy (inset) that was for a long time suspected of harboring a black hole, since it is a powerful radio galaxy. The Hubble Space Telescope has produced more evidence in support of black holes.

At the center of this galaxy there are clouds of gas among the stars. Atoms in the gas emit light at certain frequencies (colors). The frequency changes depending on whether the gas is moving toward or away from us. Gas that is moving toward us has its light "blueshifted" and gas moving away has it "redshifted." Astronomers know at what frequency the light would be if it were not moving. The S-shaped spectrum (right) shows the change in frequency across the center of the galaxy (indicated by the box on the inset image). It shows that gas on one side of the nucleus is moving toward us and on the other side is moving away from us—this is orbital motion. This means that the gas clouds are orbiting the center of the galaxy, which isn't in itself a new discovery. The speed of the gas clouds is approaching 880,000 miles per hour (400 km/s). For the gas clouds to be moving so fast, they must be orbiting an object about 300 million times the mass of our Sun. It could be a very massive star cluster, but the stars would all collide and merge very quickly, which makes it unlikely. What is more likely is that a black hole lurks at the center.

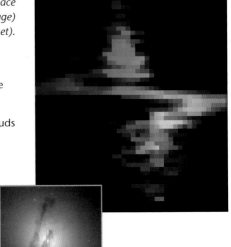

DATA	
Distance: 60 million ly	
Constellation: Virgo	
Redshift: 0.0035	

BLACK HOLE
NGC 7052

A dust disk (main image) encircles a black hole at the center of NGC 7052 (shown inset).

The extremely powerful resolution of the Hubble Space Telescope allows us to peer right into the center of NGC 7052. Wide-field images show an unremarkable spiral galaxy in the visible. In the core, however, we find a dusty disk surrounding a bright pointlike nucleus. This disk is not the accretion disk (material just about to fall into the black hole), since it is too far from the galaxy's center—the disk is 3,700 light-years in diameter. The reason why such a thick disk occurs in the center of the galaxy may be that a smaller galaxy collided with NGC 7052 at some time in the past, injecting large amounts of dust and gas into the inner regions. Just visible, peeping out from the center of the disk is a bright point of light. This is the light from material very close to the black hole around which the material in the disk is orbiting. The material in the dusty disk loses energy, and sinks toward the black hole.

This galaxy is a strong source of radio waves and has two jets, pointing in opposite directions, emanating from its nucleus—presumably from the black hole. The direction of the jets is not perpendicular to the disk, which indicates that the central black hole may not share the symmetry of the dust disk.

The disk is orbiting the center of the galaxy at a speed of 341,000 miles per hour (150 km/s). From this speed, astronomers have calculated that the mass of the black hole in the center is 300 million times that of our Sun. Although it is only 0.05 percent of the mass of the galaxy, it dominates the behavior of the inner regions.

DATA

Distance: 214 million ly

Constellation: Vulpecula

Redshift: 0.0156

BLACK HOLE
NGC 4438

NGC 4438 is known as a peculiar galaxy because of its unusual shape. It is another galaxy with an active core. In the large image the core is seen as much brighter than the surrounding stars. The Hubble Space Telescope's powerful resolution allows us to zoom into the very center of the galaxy and observe the effect of a supermassive black hole in the center of an otherwise normal galaxy. The black hole is blowing huge bubbles of very hot gas into space. The one coming up from the disk of the galaxy is clearly visible in the inset image. Its companion is behind the band of dust.

Black holes are massive objects, and so material travels toward them attracted by their gravitational pull. Objects in space tend to have angular momentum, so the material swirls around the black hole. As it reaches the black hole, not all of it is lost to the universe. It is not known exactly how, but some material is squirted up the rotation axis of the black hole and out into space at very high speeds. In this galaxy the jets act like high-pressure hoses, pushing material out in front of them and creating bubbles about 800 light-years in diameter. The collision of the jet with the normal galactic gas has heated it to temperatures where it glows and has pushed the material out of the plane of the galaxy. The bubbles will not last long—they will expand and eventually disperse because the black hole in this galaxy is not very powerful.

Bright core of NGC 4438 (top) with detail showing bubbles of hot gas appearing from its large black hole (bottom).

Distance: 50 million ly

Constellation: Virgo

Redshift: 0.0035

ACTIVE GALAXY
M87

Like a cosmic searchlight, a black-hole-powered jet of electrons can be seen clearly in the X-ray (bottom left) and visible-light (right) images of M87. M87 is the dominant elliptical galaxy in the nearest cluster of galaxies—the Virgo Cluster. It is also a radio galaxy with large radio lobes. It is one of the few galaxies where the jet produced by the black hole is visible on a number of wavelengths. This makes it an ideal target for investigating the interaction of black holes with their environment.

There are three almost separate structures in the radio image (bottom right). The inner (red) region contains the jet and the black hole. Beyond them are older structures (dark yellow swirls) that extend far into intergalactic space. The left one appears like a rising smoke ring. Above and below (paler yellow) are the oldest structures that surround the galaxy, possibly the result of separate outbursts from the black hole.

DATA

Distance: 54 million ly

Constellation: Virgo

Redshift: 0.0035

Images of M87: X-ray (bottom left); radio (bottom right); visible light, rotated 45 degrees and zoomed in (right).

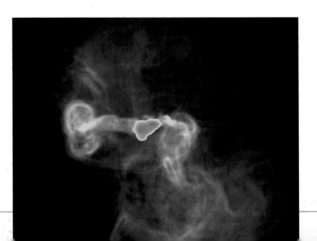

BLACK HOLE
NGC 4261

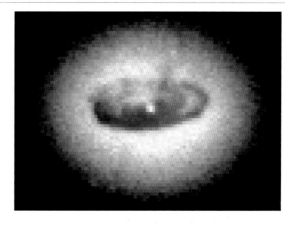

Galaxy NGC 4261.

The disk in the inner parts of NGC 4261 adds another galaxy to the list of those with known black holes in their core. In this case the black hole was not found in the exact center of the galaxy, but is offset. By measuring the changes in the light emitted by the gas in the disk as it orbits the core of the galaxy, the orbital speed was found. This means that the mass of the black hole has been calculated as 1.2 billion times the mass of our Sun. All this mass is concentrated in a volume of space not much larger than our solar system.

The galaxy was thought to have a black hole, as the galaxy is associated with strong radio emission. The dusty disk itself is about 300 light-years across, and is therefore likely to be connected to the inner accretion disk that feeds the central black hole.

One unsolved puzzle is what pulled the black hole 20 light-years away from the center of the galaxy. It may have moved as the result of the galactic collision that probably introduced the dust into the galaxy. Normally elliptical galaxies such as NGC 4261 do not contain much dust—most of it is thought to have been used up during star formation.

DATA

Distance: 45 million ly

Constellation: Virgo

Redshift: 0.0075

ACTIVE GALAXY

3C 31

DATA

Distance: 237 milion ly

Constellation: Pisces

Redshift: 0.0173

The radio source 3C 31 (NGC 383) is the 31st source in the third Cambridge catalog of radio sources observed at 159 MHz. The catalog was completed in the early days of radio astronomy, when most sources were just point sources because the resolution of the telescopes was not very high. The survey was revised in 1962 and is still in use today. Using more recent radio telescopes, such as the Very Large Array (VLA) in Socorro, New Mexico, more of the detailed structure of these sources has been seen.

Shown here is the whole galaxy (center) and a close-up of the very center (right), both produced by combining radio (red) and optical (blue) images. Most of the visible light comes from stars in the galaxy. Gigantic jets of material are seen rushing out of the center of the galaxy into intergalactic space, seen even more clearly in the radio image (far right). The jets are some of the largest single structures in the universe, stretching for millions of light-years.

These jets are caused by a supermassive black hole at the center of the galaxy. Almost all galaxies are presumed to have a large black hole at the center—with a mass millions or billions of times that of our Sun. When no material falls into the holes, we cannot see them. If material is swirling around a black hole, however, and is being accreted ("eaten") by it, we can see it because it is bright over large ranges of the electromagnetic spectrum—hence the bright point in the center of the galaxy. Not all the material falls into the black hole: some manages to escape and is accelerated to speeds close to that of light, and is ejected out of the galaxy. It forms the large jets and "lobes" most clearly seen in the radio image.

ACTIVE GALAXY

Fornax A

DATA

Distance: 80 million ly

Constellation: Fornax

Redshift: 0.0059

Fornax A is a relatively nearby radio source with very detailed structure in the radio lobes, appearing here as orange clouds. In contrast to 3C 31, Fornax A's "central engine" (the black hole that is creating the jet) is less powerful than the material with which the jet is interacting. This results in the lobes having a diffuse structure, because the particles that make up the plasma gently flow out into intergalactic space.

The jets that are the origin of this emission can just be seen in the center of the galaxy. The galaxy itself (yellow center and bluish white surround) is much smaller than the radio lobes it has produced. It is a giant elliptical galaxy, although it has distinct dust lanes in its inner regions. This implies that it has recently absorbed a dusty spiral galaxy, and that the merger is not yet complete. Faint shells of stars surrounding the galaxy also suggest that there has been some merger activity, which may have triggered the black hole in the center of the galaxy to start to produce jets of material.

The smaller galaxy to the top of Fornax A also shows features that give the impression it is interacting with the larger radio galaxy. Fornax A is the radio designation for NGC 1316. Galaxies that host supermassive black holes which emit jets and produce radio lobes tend to be elliptical galaxies. The current thinking is that the elliptical galaxies have black holes that are more massive than those in spiral galaxies. They produce more powerful jets that escape the galaxy and therefore produce the lobes.

Combined optical (white) and radio (orange) image of Fornax A.

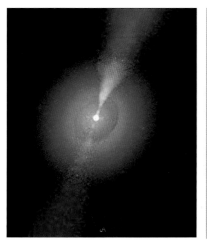

Combined optical and radio image of 3C 31 galaxy (above), with detail (above left). Radio image of jets (right).

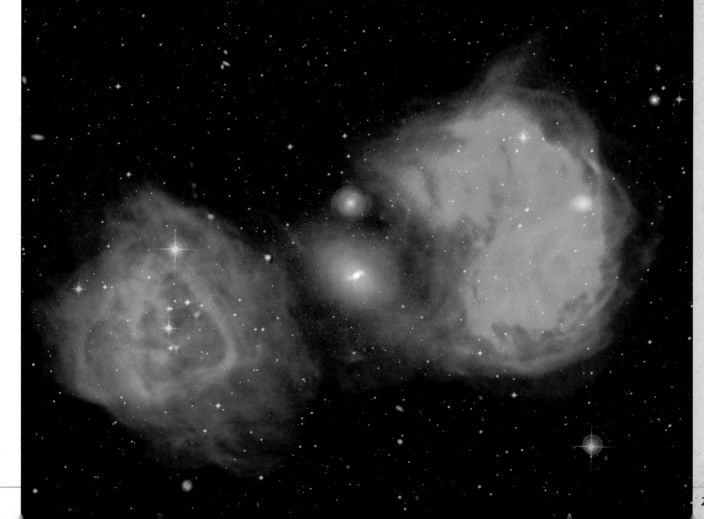

ACTIVE GALAXY
Centaurus A

Centaurus A is one of the closest active galaxies to us. In the combined optical image (far right) we observe a massive elliptical galaxy with a very prominent, thick dust lane. When seen at other wavelengths (smaller images, right) the galaxy takes on very different characteristics.

When imaged in X-rays by the Chandra X-ray Telescope (top), there is clearly something going on at the center of the galaxy. There must be a black hole in the middle, billions of times the mass of our Sun. Although no light escapes from a black hole, as material falls into it friction causes it to heat up. Eventually, it becomes so hot that it emits light, even X-rays. This is the cause of the bright spot of emission at the center. However, while all of the material that falls into a black hole is lost to our view, not all of the material near a

black hole ends up falling in; some gains a lot of energy through the rapid motion of the dense material and is squirted out along the rotation axis, which is what is seen to the upper left. Looking at the other end of the electromagnetic spectrum, in radio waves the effect of this jet is clear (center). Two regions of space at the end of the jet emit strongly in the radio band as high-speed electrons move in a magnetic field, causing synchrotron radiation.

If we look at the galaxy in the infrared (bottom) and the optical (far right), we see that there is dust present, which is unusual for an elliptical galaxy. It is possible that Centaurus A swallowed another galaxy. As well as creating dust lanes, the collision may have caused material to fall toward the central black hole.

DATA

Distance: 11 million ly

Constellation: Centaurus

Redshift: 0.00183

ACTIVE GALAXY
NGC 6240

NGC 6240 is a butterfly-shaped galaxy that has formed as a result of the collision of two smaller ones. These X-ray images show the hot gas in the galaxies and that there are two active supermassive black holes present. The fact that there were two nuclei in this galaxy had been detected in other wavebands. The Chandra X-ray Observatory was able to take spectra of the two nuclei. There are features in the spectra that indicate that they are, in fact, black holes—signatures from iron atoms in orbit around the black hole just before they are swallowed.

Currently the black holes are 3,000 light-years apart, but over the course of the next billion years they will drift toward one another. Because they are the most massive objects in this galaxy, they settle toward the center, where they will

merge and form one supermassive black hole. We do not fully understand how the black holes in the centers of galaxies reach their enormous size, but the discovery of a binary black hole supports the idea that mergers of black holes in the collisions of galaxies are the main method by which they grow.

The collision has transformed NGC 6240 into a starburst galaxy. The cause of the hot gas seen by the X-ray telescope surrounding the black holes may be the heat created by the frenetic activity. When observed in the Hubble Space Telescope, the central region of the galaxy is filled with streamers of gas, and the outer regions are a jumble of stars that were thrown out as the two galaxies collided.

DATA

Distance: 400 million ly

Constellation: Ophiuchus

Redshift: 0.0245

Chandra X-ray images of NGC 6240. The main picture shows a butterfly-shaped galaxy with a bright central core. The inset reveals the central region has not one but two active giant black holes.

Views of Centaurus A: X-ray (top), radio (center), infrared (bottom). Combined X-ray, radio, and optical (right).

ACTIVE GALAXY

Cygnus A

Cygnus A is the archetypal radio galaxy and the closest powerful one. When it is observed in the optical, all that can be seen is an elliptical galaxy at the center of a small group. In the radio band the galaxy—or specifically its effect on the surrounding space—is much more impressive. The bright spot at the center of the upper image is the result of emission from regions close to the black hole. The black hole is "feeding." In other words, it is accreting material, some of which is spewed out along its rotation axis, forming two jets. They are very thin and very straight, and they travel at close to the speed of light.

The jets collide with the gas that makes up the inter-galactic medium, where they form hotspots—the brightest regions in the radio emission. From there the particles that make up the jet "splash back" into the cavity carved out by the jets in earlier times.

When this galaxy is seen in X-ray light (bottom), the effect of the jets on the intergalactic medium is clear. The galaxy is surrounded by X-ray-emitting hot gas (red), but the jets have cleared large regions of space (yellow/light orange inner region). The scale of the jets and radio "lobes" is impressive: the galaxy, which is much larger than our own, fits in between the radio lobes.

DATA

Distance: 730 million ly

Constellation: Cygnus

Redshift: 0.056

INTERACTING GALAXIES

The Mice—NGC 4676

The long streamers of stars that have given this galaxy pair the nickname "the Mice" have been thrown out during a collision between the two. The encounter between the two galaxies has also caused massive numbers of stars to form. They are seen as the bright knots of blue emission in the left-hand galaxy. The long tail on the right also contains many regions of star formation. There are gaps between the star-forming regions, implying that stars have already formed from the gas and dust there and so they appear fainter.

The long straight tail on the right of the image is, in fact, curved. It only appears straight because we see it edge-on. Computer simulations show that we are seeing two spiral galaxies—originally almost identical—around 160 million years after their closest approach.

The two galaxies are likely to collide at least one more time before they fully merge and the resulting galaxy begins a more passive existence.

DATA

Distance: 300 million ly

Constellation: Coma Berenices

Redshift: 0.022

NGC 4676 captured by the Hubble Space Telescope's Advanced Camera for Surveys.

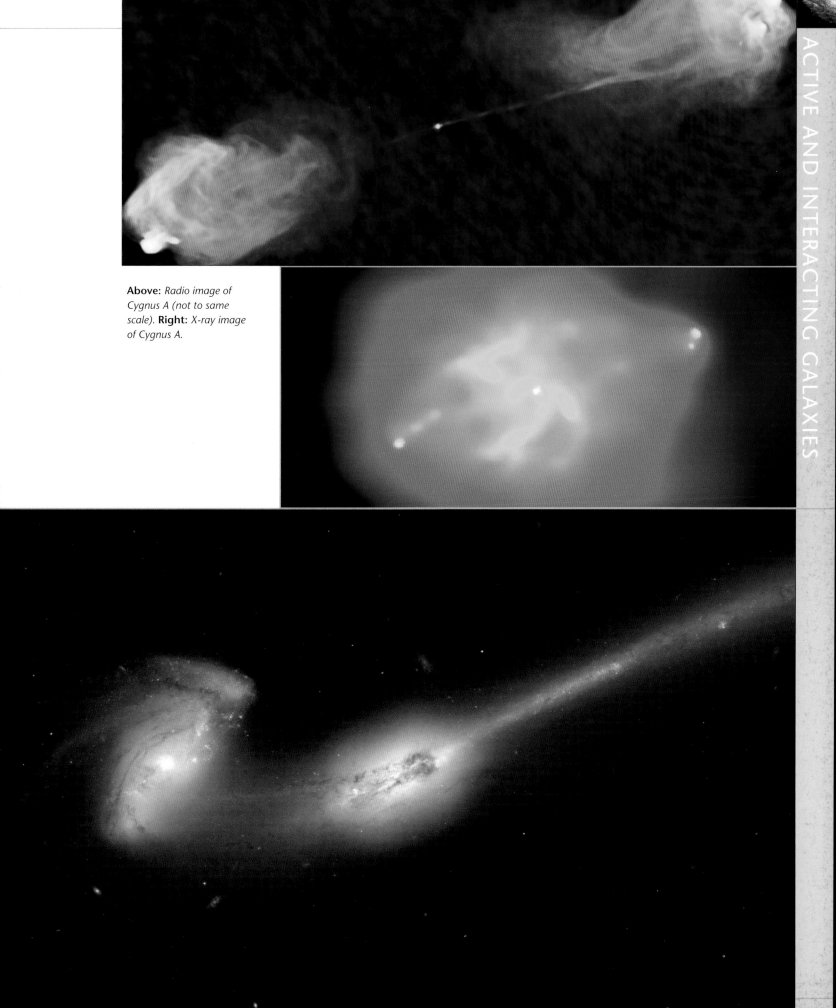

Above: *Radio image of Cygnus A (not to same scale).* **Right:** *X-ray image of Cygnus A.*

INTERACTING GALAXIES
NGC 1275

This galaxy is located at the center of the Perseus Cluster, the closest massive cluster to the Milky Way. It is a special type of galaxy known as a cD galaxy or a supergiant elliptical galaxy. Such galaxies are found only at the center of clusters of galaxies, and it is thought that they attain their truly massive size by merging with many other galaxies during the course of their lives. Because they are at the center of large concentrations of galaxies, mergers occur much more often than in the case of galaxies that are not in clusters. As a result of their massive size, they continue to attract matter and galaxies.

NGC 1275 is also an active galaxy—it has a black hole at its center and is a source of radio emission. Its effects can be seen in detail in the image of the Perseus Cluster in X-rays.

Silhouetted against the inner regions of the galaxy is another galaxy that is in the process of being torn apart by the massive gravity of NGC 1275. It was once a normal spiral galaxy, but it was caught in the gravitational pull of the Perseus Cluster that surrounds NGC 1275. During its passage through the center of the cluster, it was caught by NGC 1275 and on subsequent occasions when it has passed around the cD galaxy it has been slowly shredded. This is known as the "high velocity system," because it is moving fast relative to the central galaxy. We see mainly the dust in the system, since the light from NGC 1275 washes out its own starlight. The dust in this object prevents some of the X-rays from the cluster from reaching us.

DATA

Distance: 250 million ly

Constellation:
Perseus A

Redshift: 0.0183

Full-color image of NGC 1275.

INTERACTING GALAXIES
AM0644-741

The system known as AM0644-741 also goes by the more descriptive name of the "Lindsay–Shapley ring," because it was discovered in 1960 by Eric Lindsay, a student of American astronomer Harlow Shapley. This eccentric elliptical ring is situated close to the Large Magellanic Cloud. Their chance alignment caused problems in determining what this object was. Was it some planetary nebula in the very outer reaches of our companion galaxy, or was it a background object? Fourteen years later John Graham identified it as a "peculiar southern ring galaxy," and the object then became known as "Graham's Ring Galaxy."

In fact, at 150,000 light-years in diameter, the galaxy is larger than our own Milky Way. It is the result of a collision between two galaxies, very much like that of the Cartwheel Galaxy. In this case the "bullet" galaxy hit the target galaxy around halfway out from the center, whereas in the Cartwheel the collision was a "bullseye."

The stars in the two galaxies passed through each other, but the collision caused shock waves in the gas, triggering stars to form. The young massive stars emit more blue light than our Sun, giving the ring its striking blue color. As the shock waves spread outward, the star formation also moved outward. The shock waves also carried the gas and dust out from the collision site, causing the galaxy to have an asymmetrical, ringlike appearance.

DATA

Distance: 300 million ly

Constellation: Volans

Redshift: 0.022

Full-color image of AM0644-741.

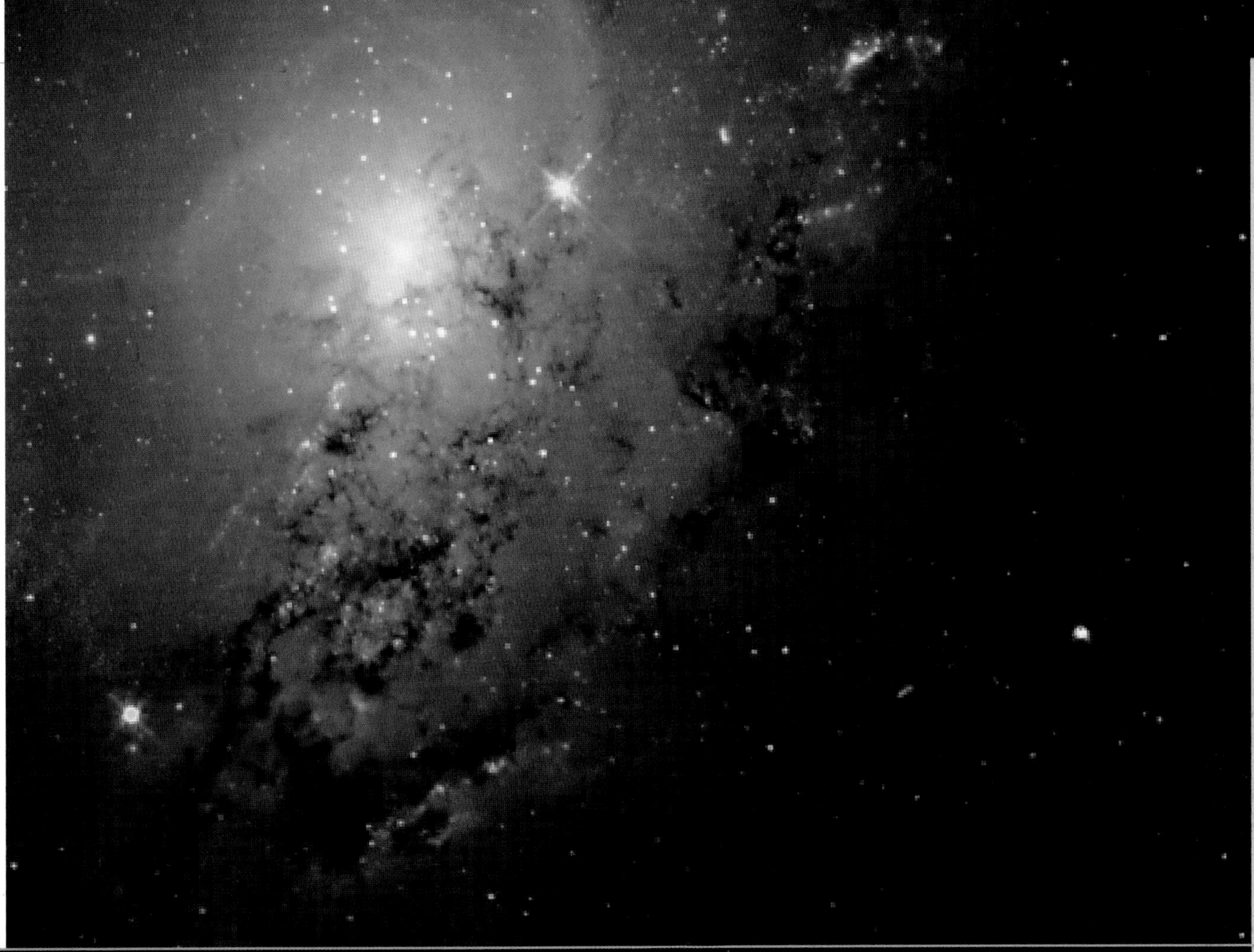

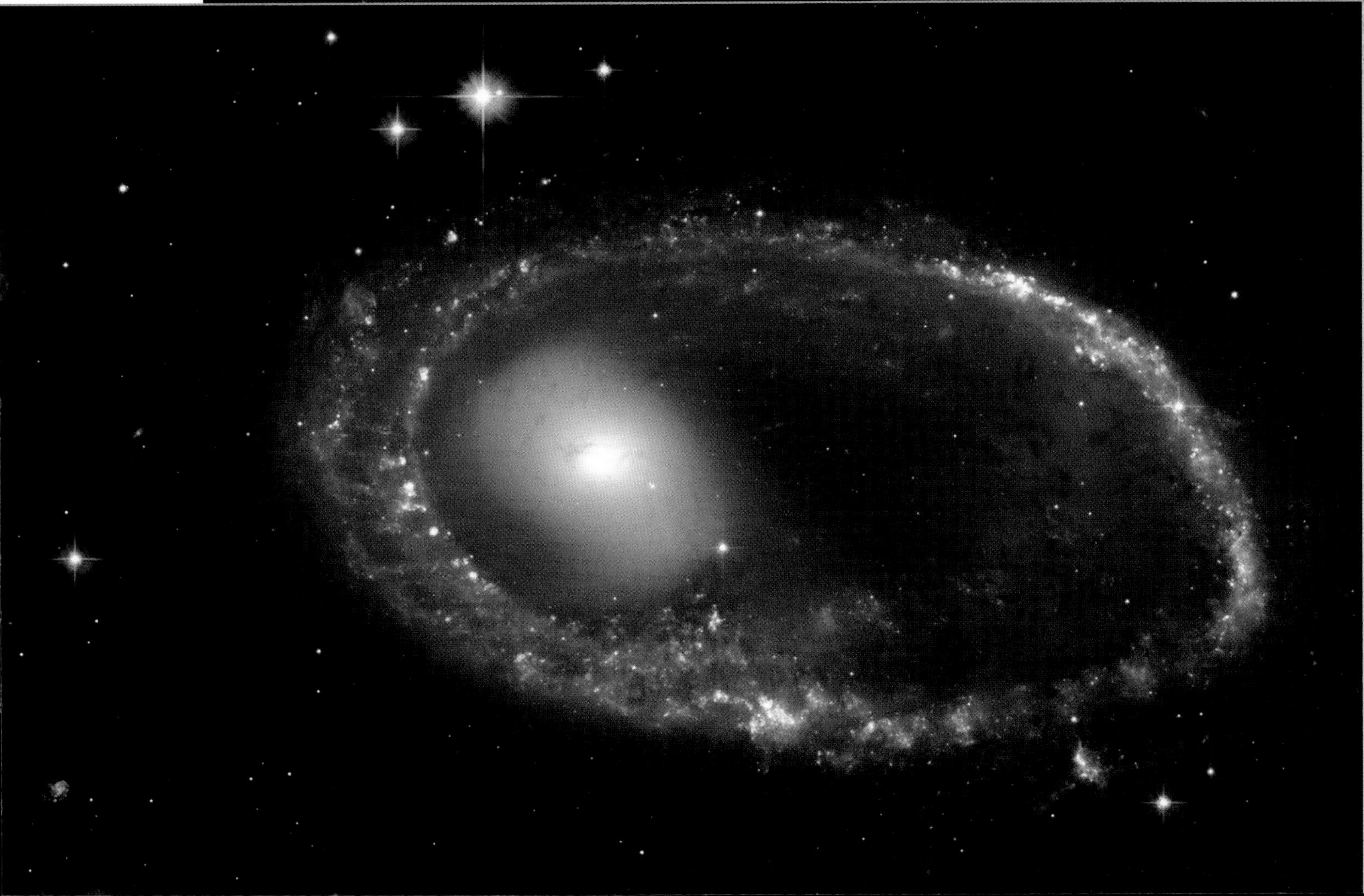

INTERACTING GALAXIES
Arp 220

Arp 220 is the brightest object in the local universe outside the Milky Way, but why does it shine so brightly? In the visible it is classed as a peculiar galaxy, but in the infrared it is known as an "ultraluminous infrared galaxy." High-resolution images taken by the Hubble Space Telescope show that Arp 220 consists of two colliding spiral galaxies, originally of similar size to our own Milky Way. The collision has caused massive amounts of stars to start forming. There are also likely to be two black holes in what is left of the cores of the two galaxies. The nuclei are only 1,200 light-years apart and are orbiting each other.

The furious activity at the center of Arp 220 has produced a "superwind," which can be seen in X-ray observations. It is probably the result of massive star formation in the remnants of the galaxies. Because many stars are forming simultaneously, many more massive stars are being created than is usual for galaxies of this size. These stars have very strong winds that stream off into space. When the stars die, the explosions that result also fling gas out into the galaxy and—if many explosions occur at the same time—out of the galaxy into intergalactic space.

The semicircular shape on the right is a cluster of around one billion stars, half of which is covered by dust that gives it the half-moon shape. The core of the other galaxy is the smaller bright region to the left of the semicircle. Its proximity to us has allowed astronomers to make detailed investigations into this active galaxy.

DATA	
Distance: 250 million ly	
Constellation: Serpens Caput	
Redshift: 0.0181	

Infrared image of Arp 220.

INTERACTING GALAXIES
NGC 4650A

NGC 4650A is an example of a rare type of galaxy known as a polar ring galaxy. In its center is what seems to be a small elliptical galaxy made up mainly of old stars, which is why it appears yellow. Seen here orbiting in an almost horizontal plane is a ring of younger bluish stars. Detailed measurements from the Hubble Space Telescope have shown that the ring of stars is, in fact, orbiting in another plane from the central elliptical component.

The most likely reason for the unusual shape of the galaxy is that it is the result of a massive collision between two galaxies some time around one billion years ago. The collision caused the gas and dust to be stripped from the smaller galaxy, which then formed the ring. The ring is distorted and it does not lie all in one plane, implying that the galaxy has not yet had time to recover from the collision.

Measuring the speed at which the stars in the ring are orbiting the galaxy leads to a surprising conclusion: if the only massive objects present are stars, gas, and dust, there is not sufficient mass in the galaxy to cause the stars to orbit as fast as they do. This has led to conjecture that a halo of dark matter surrounds this and other galaxies, creating the necessary extra mass. The ring of stars extends far out from the central parts of the galaxy, so the distribution of the dark matter can be calculated. Normal spiral galaxies are also thought to have such halos, since the outer regions rotate faster than would be expected if the glowing mass were the only mass present.

DATA	
Distance: 136 million ly	
Constellation: Centaurus	
Redshift: 0.010	

Multicolor image of polar ring galaxy NGC 4650A.

INTERACTING GALAXIES
The Antennae Galaxies

The Antennae Galaxies (NGC 4038/4039) are among the most dramatic objects in deep space. Ground-based images show two dense cores of stars from which two long arcs extend. The ultrahigh resolution of the Hubble Space Telescope (main image) allows us to see what has occurred to form this beautiful object.

Two spiral galaxies are colliding. Their centers are still separate (the red and yellow parts left and right of center). Their outer reaches, however, have been catapulted into intergalactic space. Spiral galaxies rotate, so when these two collided, the rotation caused the stars in the arms to be flung out into space, forming streams of stars. The tremendous collision sent shock waves through the interstellar gas clouds, which caused them to start collapsing. The two galaxies are experiencing an intense burst of star formation. The bright blue regions surrounding the cores of the galaxies are where most of the star formation is taking place.

When observed in X-rays (inset), these regions glow because the gas has been shock-heated to tens of millions of degrees. Some bright points—ultraluminous X-ray sources—are also visible. Astronomers are still investigating what the sources could be. The two current favorite theories are that they are intermediate-mass black holes of around 100 solar masses or that they are a short-lived phase of a special type of X-ray binary.

DATA

Distance: 43 million ly

Constellation: Corvus

Redshift: 0.0057

Optical (main picture) and X-ray (inset) images of the Antennae Galaxies.

INTERACTING GALAXIES
Hoag's Object

What caused this galaxy to evolve into its peculiar shape is currently a mystery. A perfect ring of young blue stars orbits (at a distance) a ball of older yellow stars. This type of galaxy—a ring galaxy—is itself rare. Visible through the gap is another, similarly shaped object in the background.

Since its discovery in 1950 by the American astronomer Art Hoag, numerous theories have been put forward to explain its origins. The initial explanation was that it was a result of gravitational lensing, meaning that the mass of the central object was bending the light of a background object into a perfect ring. However, since the two components of the galaxy are at the same distance from us this cannot be the case.

A collision may have created this shape (as in the Cartwheel Galaxy), but there is no trace of the "bullet" galaxy that would have collided with the progenitor of Hoag's Object. The collision may of course have caused the two galaxies to merge, and if that were the case the "bullet" would not be seen. The current favorite explanation is that it has formed from a barred spiral galaxy. The nucleus and the bar were rotating too fast for the arms to keep up, and so the two parts of the galaxy became detached. The nucleus lost its elongated shape and became spheroidal, and the arms merged to form a perfect ring. Some other galaxies share some of the features of Hoag's Object: a detached nucleus and a bright ring of young stars. However, the nuclei are usually elongated, and the rings show spiral structures.

DATA

Distance: 600 million ly

Constellation: Serpens

Redshift: 0.00425

Hoag's Object.

NGC 6745

In the center of this image is a large spiral galaxy that is feeling the effects of a close encounter with another galaxy. Almost off the picture in the bottom right is a smaller galaxy that collided with NGC 6745. Most galactic interactions are near misses, but in this case the galaxies actually hit. The knot of intense blue light on the lower right of NGC 6745 is a massive region of space where new stars are being formed. The upper portions of the galaxy are also bluer than the lower ones, which suggests the route taken by the intruder galaxy. This is because the collision has triggered star formation in the regions where the two galaxies interacted. Depending on the speed of the collision and the size of the galaxies, they may continue on their separate ways. Most near misses result in another collision later, when the mutual gravity pulls the galaxies back together. If this occurs, it may result in the eventual coalescence of the two galaxies.

When galaxies collide, the stars themselves very rarely collide. The stars are so minuscule compared to the size of the distances between them that the chances of a collision are small. The particles that make up the tenuous gas (around one atom per cubic centimeter) in between the stars do collide. This sets up shock waves in the denser gas clouds, which can cause regions in them to collapse and start to form stars.

DATA	
Distance:	208 million ly
Constellation:	Lyra
Redshift:	0.0152

Composite image of NGC 6745, using infrared, red, visual, and ultraviolet filters.

Tadpole Galaxy (Arp 188)

The Tadpole Galaxy is a well-known set of interacting galaxies, similar to the Antennae Galaxies. The close encounter between the two galaxies has caused them to tear pieces off themselves and create long wispy tidal arcs of stars, stretching out into space. The size and direction of the arcs depend on which way the progenitor galaxies were spinning before they collided, and also what their relative orientations were. The galaxy that caused the tail is thought to have come in from right to left and was slung around behind Arp 188.

During a collision some stars are accelerated as they swing around the other galaxy, in much the same way as some interplanetary probes accelerate around massive planets. They end up having speed so great that they almost escape from the galaxy. The collision also creates shocks in the interstellar gas in the galaxies. These shocks cause the gas clouds to collapse and to begin forming stars. Many interacting galaxies appear bluish because they have many more hot young stars. The tidal tail here is 280,000 light-years long. Over the course of the next few million years the tail will probably detach from the galaxy and clump together, forming satellites of the spiral galaxy.

Simulations of galaxies colliding can reproduce the shapes of these tidal arms, but they usually require there to be more mass present in the galaxies than just that from the stars. This is another indication that there is dark matter present that we cannot see, except by its interactions with gravity.

DATA	
Distance:	420 million ly
Constellation:	Draco
Redshift:	0.0314

Tadpole Galaxy (Arp 188).

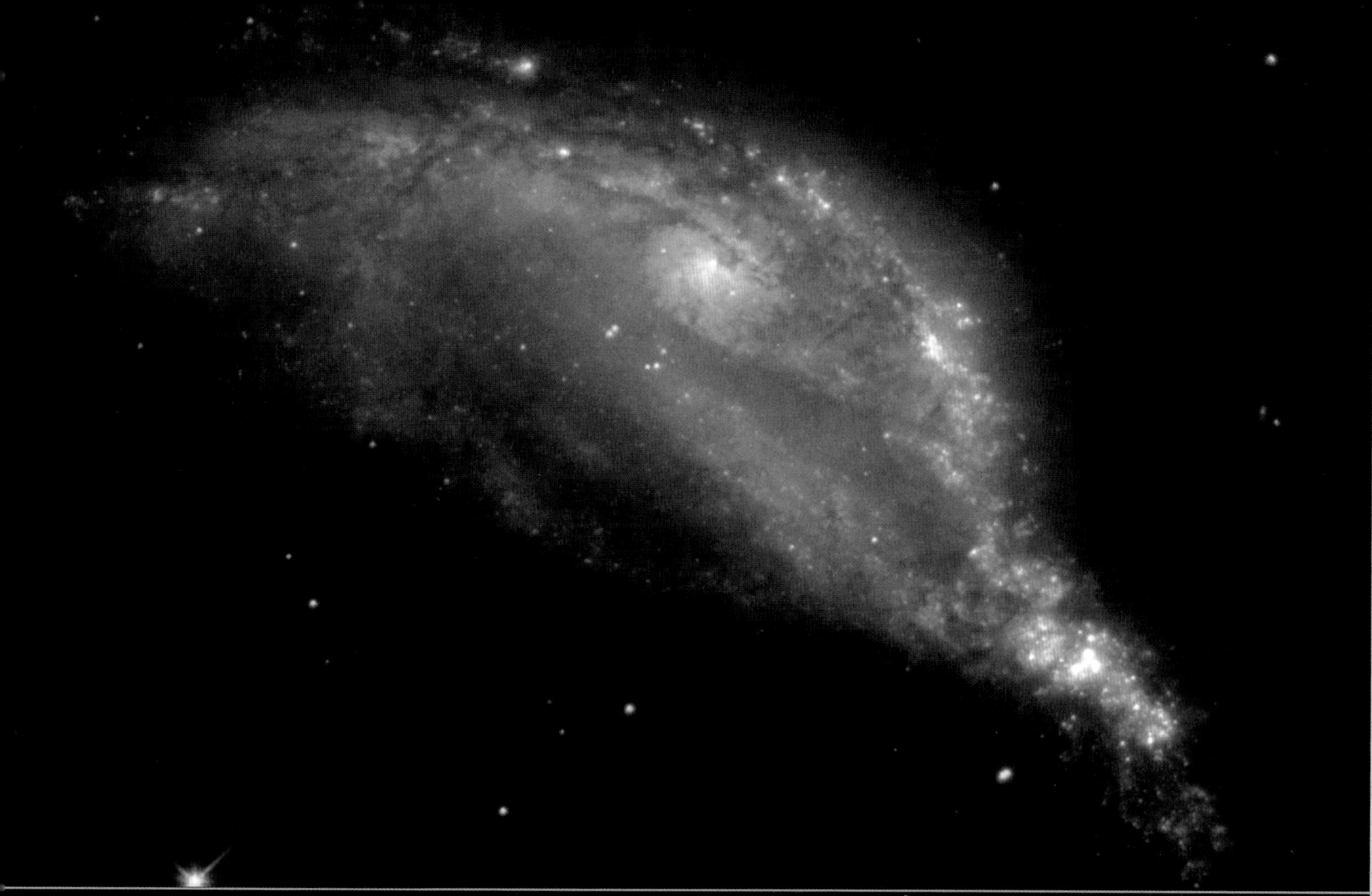

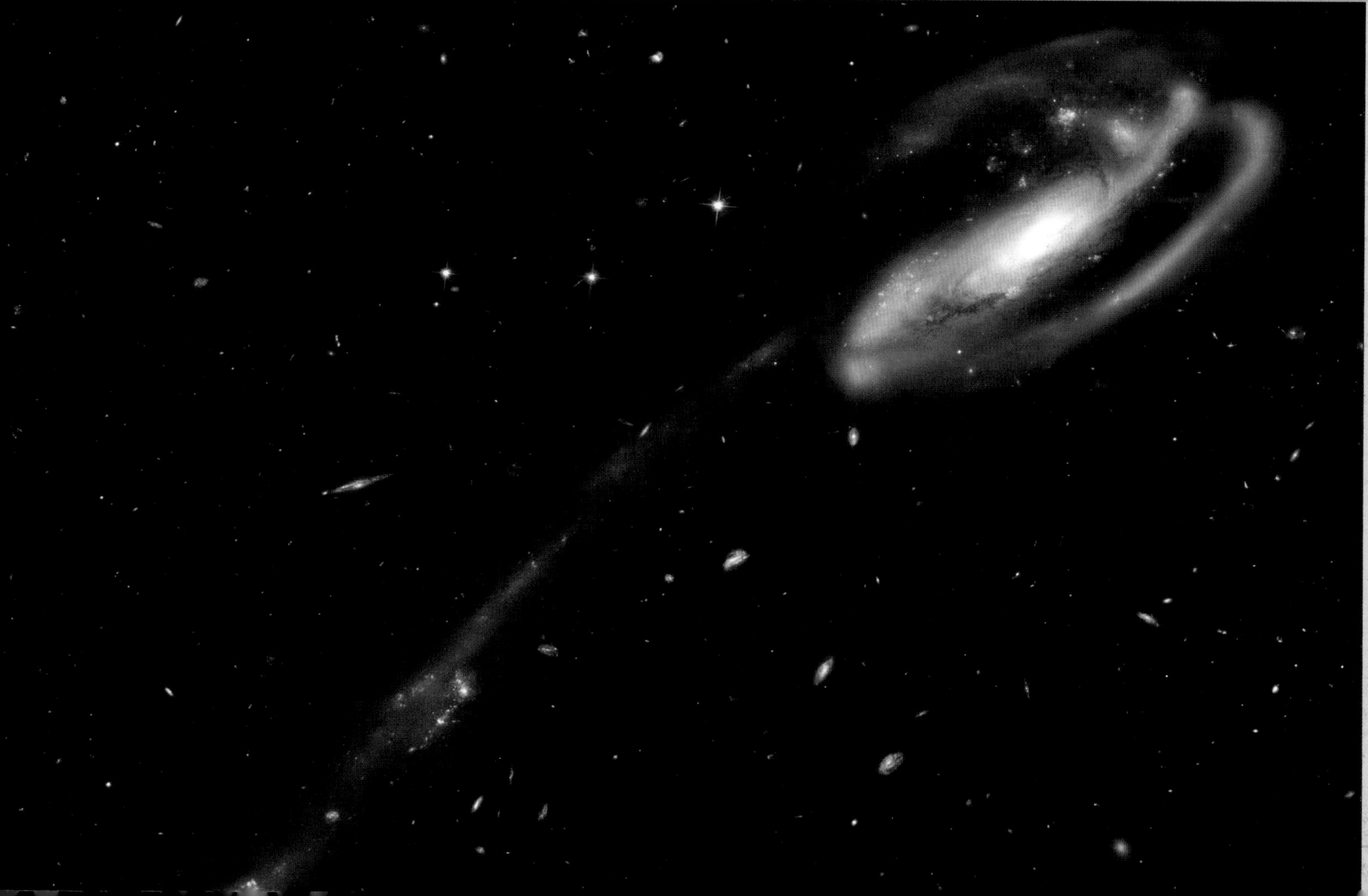

INTERACTING GALAXIES

Cartwheel Galaxy

The Cartwheel Galaxy is one of the most recognizable galaxies in the sky, and there are few others like it. A ring of hot young blue stars surround (at a distance) a core of yellowish older stars. What process could have caused a galaxy to result in a shape like this? We know that galaxies can collide, usually with spectacular results. In the case of the Cartwheel, one galaxy has passed straight through another, scoring a bullseye!

About 300 million years ago a small galaxy passed directly through the center of another galaxy. The collision sent shock waves out into the "target" galaxy, which slowly started to break the outer layers apart. As the shock waves traveled outward, they caused gas clouds to collapse and start to form stars. The ring of stars appears blue because most stars are young and very hot. It is formed in a similar way to the ripples on a pond when a stone is dropped into it, although it is many times the size—100,000 light-years in diameter—and is moving at 200,000 miles per hour (322,000 km/h).

With a bit of detective work, astronomers have discovered which galaxy was guilty of this intergalactic drive-by shooting. By following a trail of gas extending out from the Cartwheel, they have identified the intruder as being 250,000 light-years distant and looking remarkably unharmed considering the damage it caused.

DATA

Distance: 500 million ly

Constellation: Sculptor

Redshift: 0.030

INTERACTING GALAXIES

NGC 2207 and IC 2163

These two galaxies are performing a dangerous dance to the death in the depths of space. In billions of years from now only one galaxy will remain in this portion of the universe. Meanwhile, the gravitational fields of the two galaxies will pull stars and gas out of the galaxies, creating trails of matter across the space between them. NGC 2207—the larger galaxy on the left—will eventually incorporate IC 2163, which is currently moving away behind NGC 2207 after their recent close encounter.

Computer simulations can try to reproduce such collisions to see what the outcome of these interactions are. The simulations also show how these collisions occur at a leisurely pace. IC 2163 made its closest approach around 40 million years ago and is continuing its counterclockwise motion behind NGC 2207. IC 2163 does not have enough energy to escape the gravitational pull of NGC 2207 and will be pulled back into the larger galaxy in the future.

DATA

Distance: 53 million ly

Constellation: Canis Major

Redshift: 0.009

NGC 2207 (left) and IC 2163 (right).

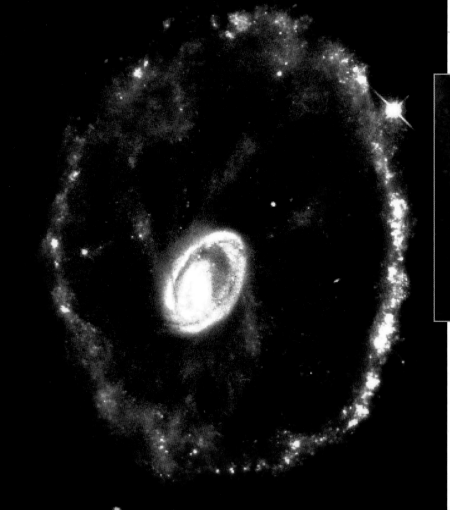

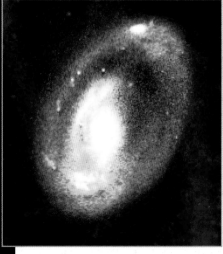

Above: *Core of Cartwheel Galaxy.*
Left: *True-color image of Cartwheel Galaxy, showing ring.*

Stephan's Quintet

The chaotic life of a group of galaxies is well illustrated in Stephan's Quintet, a group of five galaxies discovered by French astronomer Edouard Stephan in 1877. Groups of galaxies are intermediate between individual galaxies and clusters of galaxies. They are small groups that contain fewer galaxies but are more common than clusters. Some may be only chance alignments, in which the galaxies appear to be associated but in fact lie millions of light-years apart. In Stephan's Quintet this is also partly the case—the large galaxy to the bottom (F) is a galaxy that is not part of the group, but C, which is found off to the left, is part of the group.

Looking more closely at galaxies A and B, they appear to be interacting—there are two arms of faint stars (orange) pointing upward, reminiscent of the Antennae Galaxies. A collision is likely to have thrown off these tails and caused the burst of star formation that we can see.

The interaction can be seen more spectacularly when the group is seen in X-rays. An image taken with the Chandra X-ray Observatory clearly shows how galaxy B is the intruder in the group. Its rapid motion has shock heated the gas surrounding the galaxies to a temperature of six million degrees Kelvin, which causes it to glow in X-rays. The two galaxies that are interacting will probably merge to form an elliptical galaxy in the next few billion years. The remaining spiral galaxies may also merge with the resulting galaxy.

DATA

Distance: 300 million ly
Constellation: Pegasus
Redshift: 0.0215

In this optical image Stephan's Quintet (consisting of galaxies A, B, C, D, and E) can be seen at bottom right. Galaxy F is not part of the group. The detailed inset (blue, top) shows X-ray emissions of part of the group.

HCG 62

This is a Chandra X-ray image of the group of galaxies HCG 62. Groups of galaxies have halos of hot gas, many of which appear smooth and circularly symmetrical. This X-ray image of HCG 62 shows that some are not smooth. Two cavities (darker green patches at 11 o'clock and 5 o'clock) appear in the top and bottom of the hot gas that is emitting the X-rays. Something appears to be blasting out from the center of the galaxy and has pushed the gas out of the way.

This is, in fact, the most likely explanation. The group has a central elliptical galaxy. As in most large galaxies, it probably contains a black hole. When they are feeding, black holes can create jets of material that blast off into space at speeds approaching that of light. The jets have pushed away the tenuous gas that is emitting the X-rays, leaving the two large cavities at the upper left and lower right. A symmetrical arrangement has formed around the intense central X-ray region. The black hole is currently not feeding, and the jets have stopped, so the cavities have moved away from the center of the group.

The gas that is emitting the X-rays falls onto the group from the surrounding intergalactic space because of the group's gravity. The gas can then fall onto individual galaxies and provide new star-forming material. If the black hole at the center is too strong it could blow away all the material, and the galaxies would stop being able to produce stars. There may, therefore, exist a balance between the amount of material falling in and the amount that is blown out by a black hole.

DATA

Distance: 200 million ly
Constellation: Virgo
Redshift: 0.0137

False-color X-ray image of HCG 62.

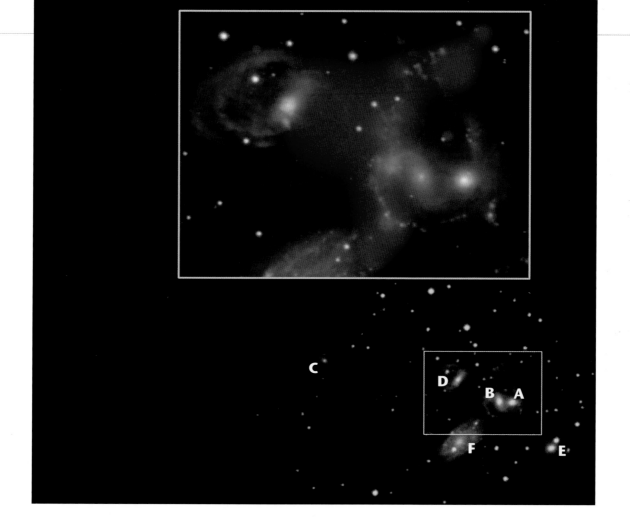

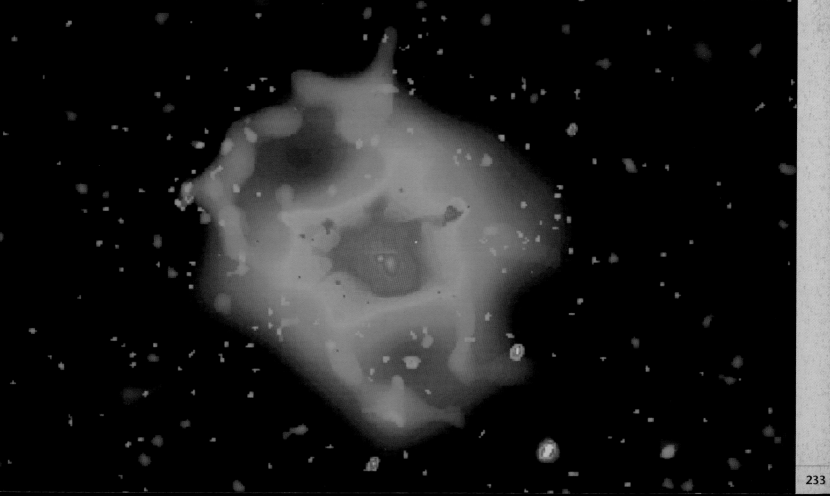

GALAXY GROUPS
Seyfert's Sextet

The group known as Seyfert's Sextet appears to be made up of six galaxies, but only four are interacting gravitationally. The face-on spiral galaxy to the right of center is nearly five times farther away than the other four, and is therefore not a member of the group. The sixth "galaxy" (the object farthest on the right in the image) is, in fact, simply a long tail of stars that have been pulled out of one of the galaxies by the gravitational effects of the other members of the group. The four galaxies are packed into a region of space only 100,000 light-years across—about the same size as the Milky Way. This makes the "Sextet" one of the densest-known galaxy groups. Deep images show that there is a halo of stars surrounding all the members of the group.

In the not-too-distant future the four galaxies will have merged to form one galaxy. It will probably be an elliptical galaxy because the rotations of the spirals will all be in different directions. The interactions among the members of this group will be spectacular, since it is likely that more than two galaxies will be colliding at once.

It is thought that most of the matter in the universe occurs in groups of galaxies that have between five and ten members, which makes them important in our understanding of the evolution of the universe. Some of the larger groups are reminiscent of clusters of galaxies because they can have halos of hot gas that glow in the X-rays.

DATA

Distance: 190 million ly

Constellation: Serpens Caput

Redshift: 0.0145

GALAXY GROUPS
HCG 87

There are at least three members of the group known as HCG 87: the edge-on elliptical spiral in the lower segment, the glowing elliptical to its right, and the spiral to the top left. The galaxy at the center of the image is a background galaxy and not connected to the group.

Most galaxies in the universe are neither isolated nor in large clusters, but are found in groups with about ten members. Most of the mass in the universe is also associated with these groups. Although the galaxies in the groups form individually, they exist in such close proximity—far closer than the distance between the Milky Way and the Andromeda Galaxy—that they influence each other as they evolve and age. If there are enough small galaxies and if

they are close enough, they may collide and eventually form a single large elliptical at what was the center of the group.

These galaxies are clearly interacting with each other. There is a stream of stars between the edge-on spiral and the elliptical, and both have active nuclei. The top spiral galaxy is forming stars at a more rapid rate then would be expected in a galaxy of its size and age.

In the 1980s, Canadian astronomer Paul Hickson compiled a catalog of compact groups (HCG = Hickson Compact Group). Each contained more than four galaxies that were in close proximity to one another and were of similar luminosities. Hickson's is one of the best catalogs dealing with relatively nearby groups.

DATA

Distance: 400 million ly

Constellation: Capricornus

Redshift: 0.0296

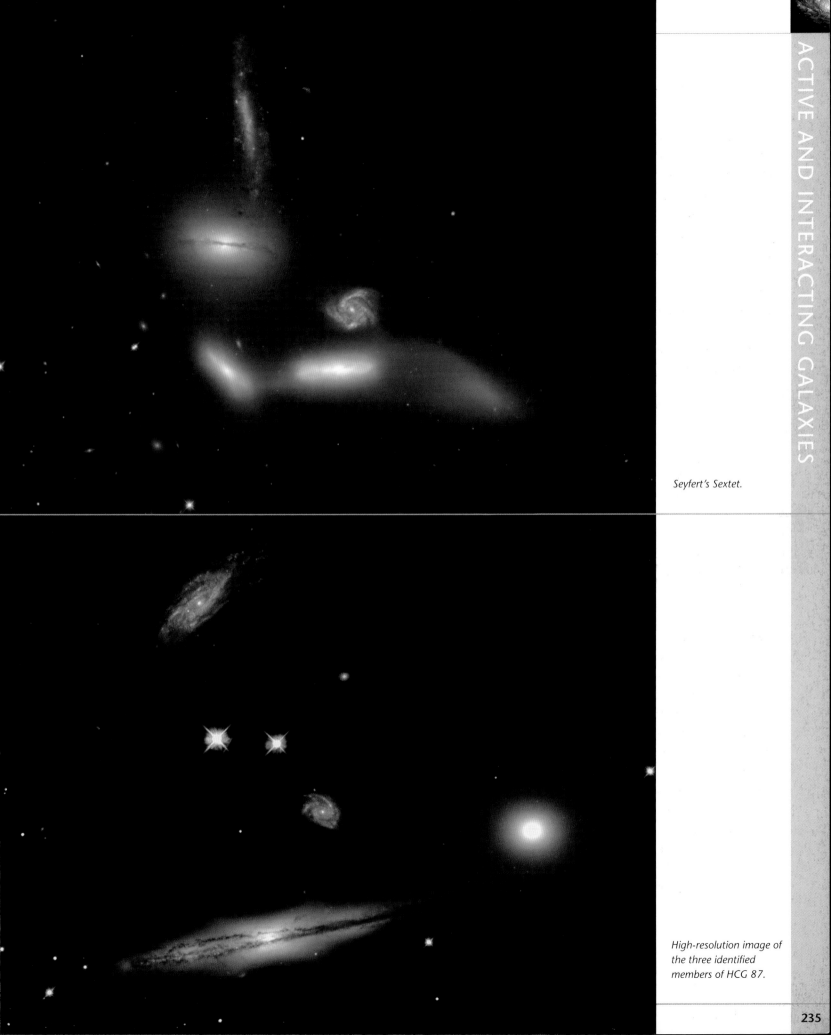

Seyfert's Sextet.

High-resolution image of the three identified members of HCG 87.

GALAXY CLUSTERS
Abell 2029

These images of Abell 2029, one from the Chandra X-ray Observatory (right) and an optical image from the Kitt Peak telescope (far right), show with startling clarity that there is much more going on in clusters of galaxies than the naked eye can see. Almost all of the points in the optical image are galaxies, some of which is similar to our Milky Way. Others, however, are much larger—the one in the center is many times more massive than our own galaxy.

In the X-ray image we cannot see the individual galaxies. What we are seeing is the emission from hot gas between them. The image shows that a large amount of other material exists between the galaxies. There is about ten times more material in the gas surrounding the galaxies than in the stars of the galaxies themselves.

DATA

Distance: 1 billion ly

Constellation: Serpens

Redshift: 0.0767

GALAXY CLUSTERS
Abell 2218

The Hubble Space Telescope image of the massive galaxy cluster Abell 2218 confirms some of the predictions of Einstein's general theory of relativity. Among the jumble of galaxies that make up this cluster are some streaks of light that do not appear to fit in with the known shape of galaxies. They are the result of the light from galaxies behind the cluster being distorted and stretched by the gravitational field of the cluster.

Einstein's general theory of relativity states that matter can bend space—we experience the result of this bending as gravity. An object with great mass will bend space far more and its effects will be felt much farther from the object than an object with less mass. Because light travels in straight lines in flat space, parallel beams of light will never meet. However, if space is bent—as a rubber sheet may be bent

by a heavy bowling ball—then beams of light that start off parallel may meet. In the image here, the cluster of galaxies plays the part of the bowling ball, and the light of a very distant background galaxy is bent and magnified so that we see it as streaks.

The background galaxy is so far away and so faint that, without the cluster acting as a fortuitous natural magnifying glass, we would not have detected it. If the distance to such a background galaxy can be found, then the shape of the streaks and arcs can help us estimate the mass of the cluster. This method of calculation is independent of the method employed by X-ray astronomers and therefore confirms estimates made of the mass from measurements of the X-ray emission.

DATA

Distance: 3 billion ly

Constellation: Draco

Redshift: 0.171

X-ray (above) and optical (right) images of Abell 2029 to same scale.

Hubble image of the galaxy cluster, Abell 2218.

GALAXY CLUSTERS
Centaurus Cluster

DATA

Distance: 160 million ly

Constellation:
Centaurus

Redshift: 0.0114

This is a Chandra X-ray Observatory image of the Centaurus Cluster, one of the closer clusters of galaxies. The false coloring that has been applied to the image is temperature coded—the sequence goes from red to blue with increasing temperature. It shows that the center of the galaxy cluster is cooler than the outer regions (this image shows only the inner regions). That is because, as the intracluster gas radiates in X-rays, it loses energy. As a result, it cools, much as a red-hot poker would cool, even in a vacuum. The cooler gas falls to the center of the cluster as it becomes denser.

The other interesting features of this cluster are the two cavities that sit left and right of the center (appearing as dark earlike patches). They are similar to those in the Perseus Cluster, among others, and have a similar origin. Features like the plume on the left of center, however, have been seen in very few clusters. Currently the formation of the plume is a mystery—there are various possible scenarios. It could be the result of a wake as another galaxy passed through the cluster or as gas cooled onto the moving central cluster galaxy.

The Centaurus Cluster is unique in it has a very high "abundance," that is to say, the concentration of elements heavier than helium in the periodic table is high—three times the concentration found in the Sun. Stars create heavier elements (termed "metals" by astronomers) when they fuse atomic nuclei in their cores. In this cluster large amounts of the "star dust" has entered the intracluster medium as the result of supernova explosions.

X-ray image of the Centaurus Cluster. The central galaxy fits in between the cavities.

GALAXY CLUSTERS
1E 0657-56

DATA

Distance: 4 billion ly

Constellation: Carina

Redshift: 0.296

This cluster comprises two parts (seen as pale violet areas). The main cluster in the center is vertically elongated, with a smaller "subclump" to its right. The X-ray image of this hot cluster shows a bow-shock (white arrowhead-shaped area) to the right of the subclump. The bow-shock is thought to arise from the passing of a group or subcluster of galaxies at supersonic speeds through the hot gas that forms the Intracluster Medium (ICM). The subcluster is traveling at about 6 million miles per hour (3,000–4,000 km/s) relative to the main cluster.

The shape of the shock wave is very similar to that of a bullet that travels supersonically in the air. As a result, the cluster has been nicknamed the "bullet cluster."

The subcluster passed close to the core, the densest part of the cluster, around 100 to 200 million years ago and is now in the final stages of being destroyed. As the subcluster falls through the ICM, its gas is stripped by the ram-pressure, rather like the way in which a strong wind removes leaves from a tree. The galaxies that form the subcluster actually sit to the right of the "bullet" seen in the X-rays. When the gravity of the cluster has caused the motion of the subcluster to stop, there will be no gas left in it, and the galaxies will slowly fall back into the main cluster.

Clusters, like galaxies, are thought to form by merging, and this cluster is rare because it has been caught in the act. The shock waves heat up the ICM, and since 1E 0657-56 is one of the hottest clusters, it is thought that a succession of mergers could have caused this.

X-ray image of 1E 0657-56.

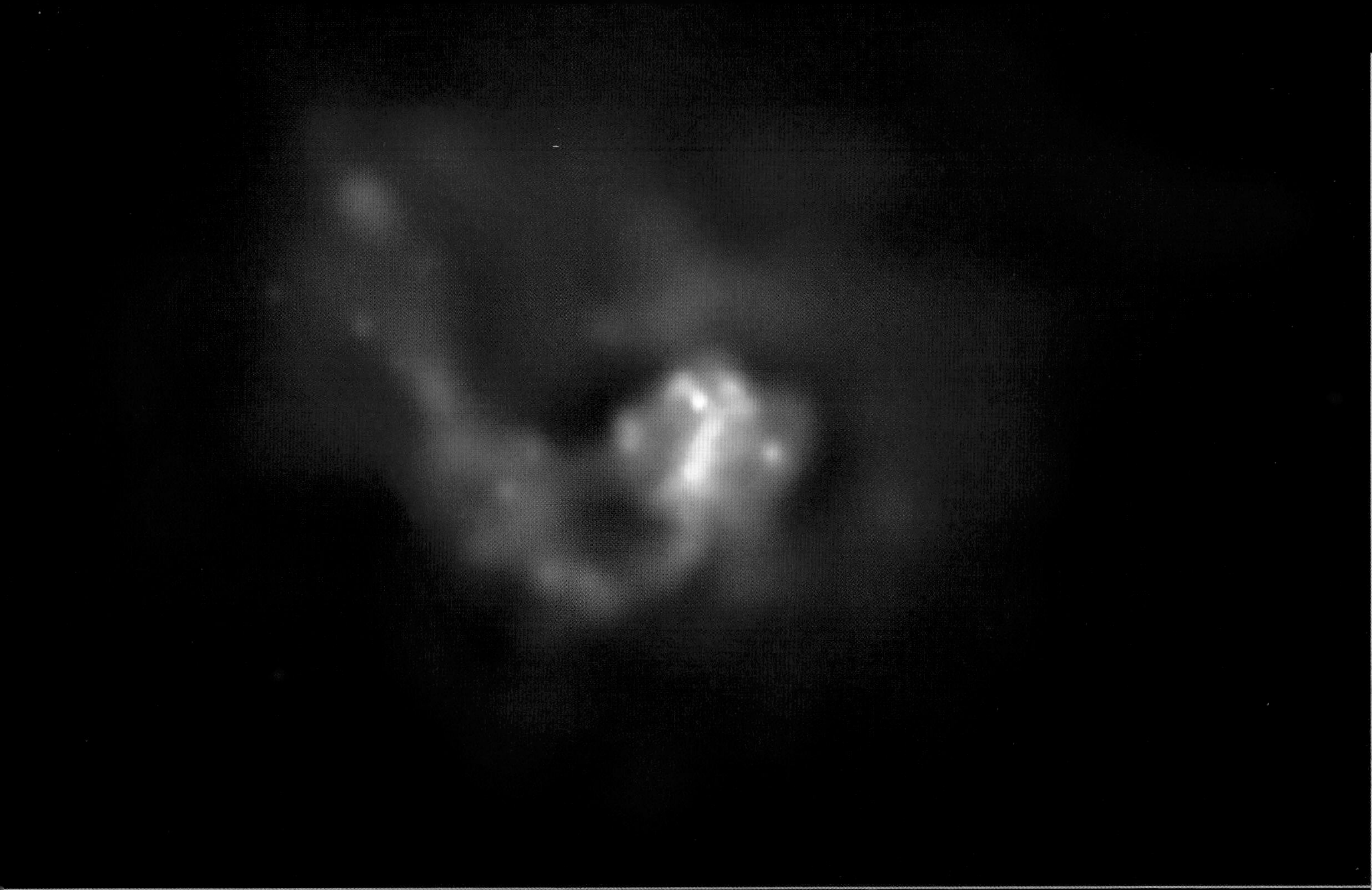

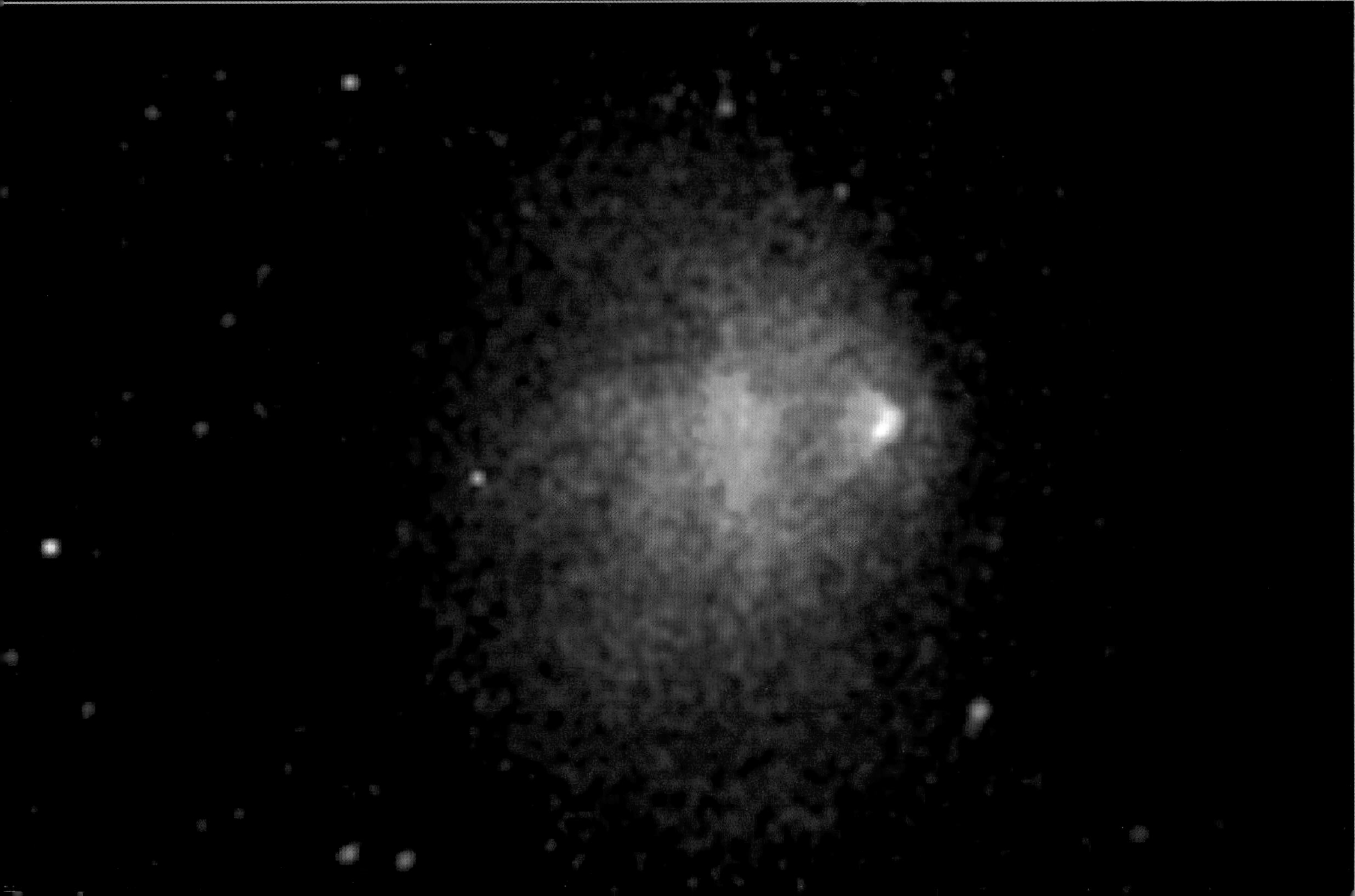

CL 0024+1654

This cluster of yellow elliptical and spiral galaxies clearly demonstrates that matter can bend light—or at least the space around it. Rather like a bowling ball on a rubber sheet, matter can bend the fabric of space. This means that, although light still travels in straight lines, the path can curve, and parallel beams of light are able to meet. Think of lines of longitude on the surface of Earth: lines of longitude that are parallel at the equator meet at the poles. The mass of the cluster distorts the geometry of space so that diverging beams of light are bent into converging beams. If Earth happens to be in line just when these light beams come into focus, then we see the distorted streaks, arcs, and copies of the background galaxy. (In this case there are five copies of a background galaxy seen as blue shaped objects.) As well as distorting the light, this process magnifies it.

Astronomers can use this serendipitous method to see farther into space and hence into the past than would be possible without the lensing. Some of the most distant objects in the universe have been discovered using this method. Seeing farther back into the universe means that we see objects when they were younger, and therefore at an earlier stage in their evolution.

The background lensed galaxy in this image is at a redshift of 1.67, which puts it almost twice as far away as the cluster CL 0024+1654, which is at a redshift of 0.40. More distant galaxies are expected to appear red because of the redshift from the expansion of the universe. This distant galaxy is very blue, because it is undergoing an intense burst of star formation.

DATA

Distance: 3.6/5.7 billion ly

Constellation: Pisces

Redshift: 0.40

Image of galaxy cluster CL 0024+1654 (yellow) with several arc-shaped copies of a far more distant galaxy (blue).

Perseus Cluster

In the X-ray view of the Perseus Cluster of galaxies (far right) the colors represent the different energies of the X-rays arriving at the telescope, red being the lowest and blue the highest. The amount of structure visible in this image of the hot gas bathing the galaxies in the cluster is impressive. At the center the bright point of light is the active nucleus of NGC 1275, a supermassive black hole that is producing jets of material. To the north and south there are "holes" (darker patches) in the X-ray emission—thought to be bubbles blown by the black hole—that sit either side of the central galaxy.

The material in the jets from the black hole does not mix with the hot plasma between the galaxies. Instead, it is constrained by it and so produces these bubbles. The radio emission from this source, known as 3C84, fits neatly into the holes in the X-ray emission. The darker streaks just right of center are the X-ray shadowgrams of the dust in the high velocity system—a galaxy in front of NGC 1275 that is being torn apart.

Farther out from the center are crescent-shaped dark patches, one to the top right and a fainter one to the bottom. They are thought to be older bubbles that have detached from the nucleus and are rising up through the gas. The fainter arcs are sound waves caused by the expansion of the bubbles. They are the lowest sound waves in the universe—a B-flat, but 57 octaves below middle C!

DATA

Distance: 250 million ly

Constellation: Perseus

Redshift: 0.0183

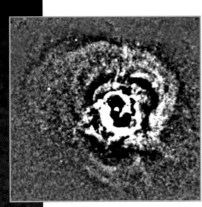

Images of the Perseus Cluster X-ray in color (left) and sound waves (above).

Abell 3627

The cluster Abell 3627 is in the plane of the disk of our galaxy, which makes it difficult to study because the dust and gas of the Milky Way blocks large amounts of its light. The image shows the large number of foreground stars in our galaxy that lie in its direction. The plane of the galaxy has been nicknamed the "zone of avoidance" over the years, since many surveys of the universe choose directions out of the galactic plane in order to avoid the obscuration caused by the gas and dust as well as confusion with foreground stars.

The cluster is thought to be among the most massive known. It is the prime candidate for what is known as the "Great Attractor." When the velocities of nearby galaxies were determined for the first time, many were found to be streaming in the direction of Centaurus. The Local Group as well as the Virgo Supercluster and the Hydra-Centaurus Supercluster are all moving at between 600 and 1,000 kilometers per second toward the Great Attractor.

Unfortunately, the large concentration of mass that must be pulling in all the galaxies of the Local Group and beyond lies in the plane of the Milky Way. It could not be seen through the stars, gas, and dust of our own galaxy, and so the name Great Attractor was used to describe this unknown mass. Abell 3627 lies close to the position where the Great Attractor should be found. Its mass has been calculated as around ten times that of an average cluster, so it could easily be responsible for the bulk motion of the local universe.

DATA	
Distance:	200 million ly
Redshift:	0.016
Constellation:	Norma

Galaxies and stars in the "zone of avoidance" towards the "Great Attractor." In this color composite the foreground stars of the Milky Way appear as whitish spots and the galaxies as yellowish green blobs.

MS 0735.6+7421

The results of one of the most powerful eruptions in the universe can be seen in this image of galaxy MS 0735.6+7421. The image was taken by the Chandra X-ray Observatory. It shows the X-ray emitting intracluster medium (red) of this distant cluster of galaxies, which contain two vast cavities—the darker patches within the globe (at 5 and 11 o'clock in the image). The phenomenon that caused this feature is the same as the one that caused the bubbles in the Perseus Cluster, but on a much larger scale. The cavities are 0.6 million light-years in diameter. The radio emission from the lobes, which are not shown in this image, neatly fills these holes.

Active galactic nuclei are caused by supermassive black holes that are accreting large amounts of material very quickly. Radio images show that some of this material is flung out into space in the form of huge jets. When the jets are hemmed in by the extremely hot gas in a galaxy cluster, they blow cavities in the gas, which are observed as X-ray holes. Many have been found in the relatively nearby universe, but this is one of the largest and one of the most distant.

Some clusters exhibiting this phenomenon experience the creation of cavities at a gentle rate. Others—like MS 0735.6+7421—have fewer but much more violent eruptions that disrupt the central regions of the cluster for tens of millions of years. To create cavities of this size, the black hole at the center of the cluster must have swallowed about 300 million solar masses of material over the course of the last 100 million years. That is as much material as exists in some entire galaxies.

DATA	
Distance:	2.6 billion ly
Constellation:	Camelopardalis
Redshift:	0.216

X-ray image of MS 0735.6+7421.

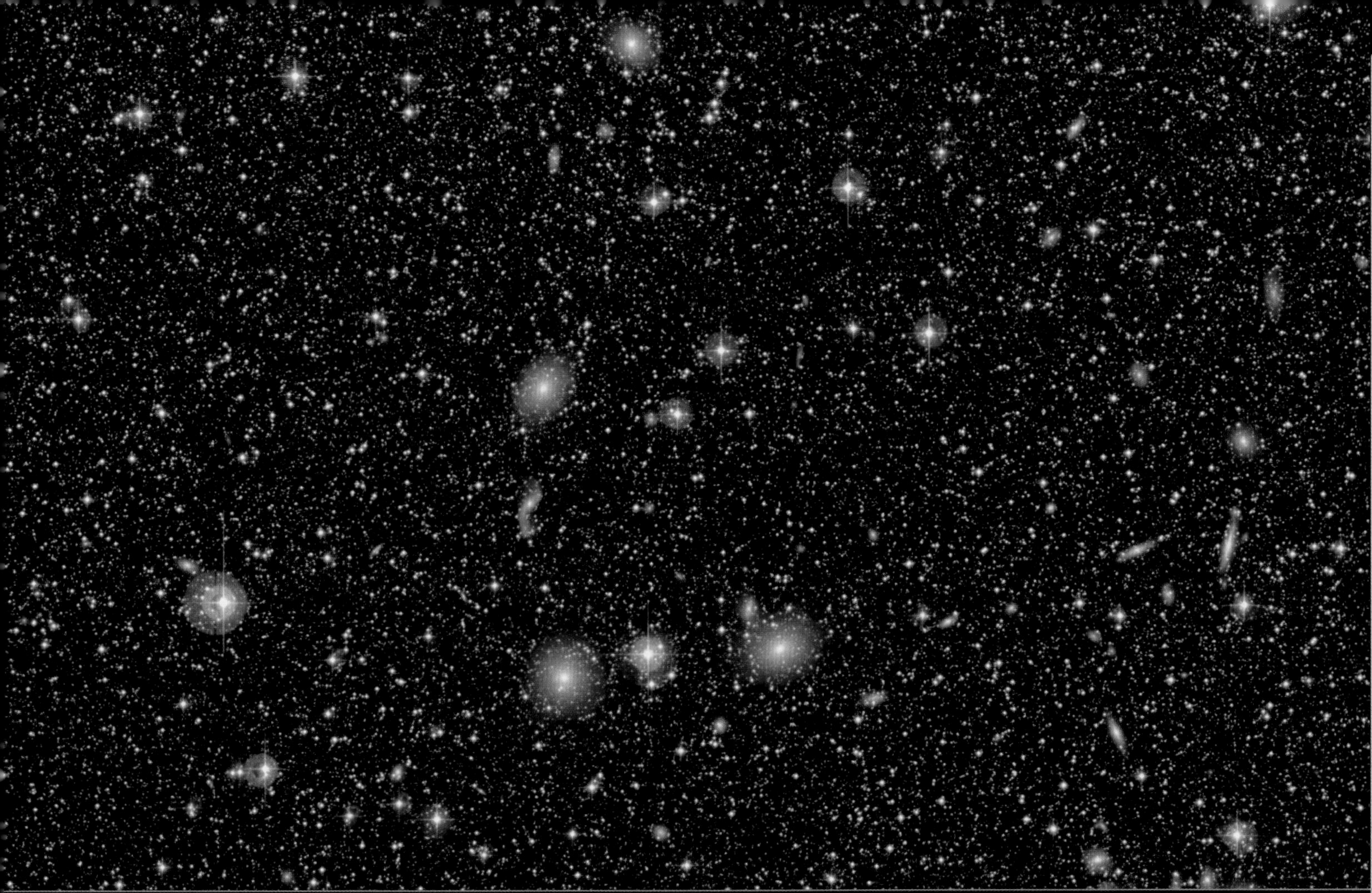

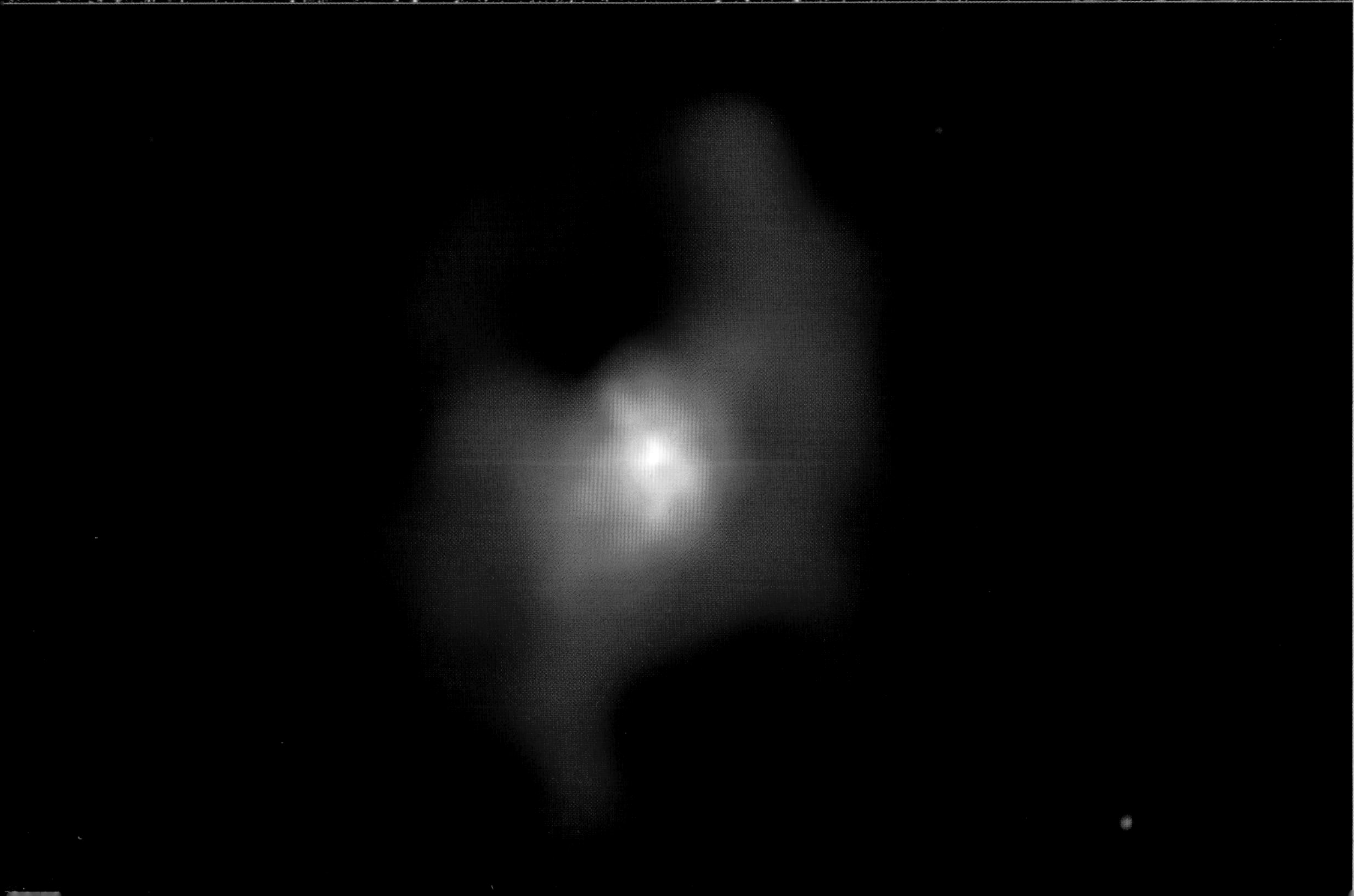

GALAXY CLUSTERS
Other Clusters

Left: *This Chandra X-ray Observatory image of galaxy cluster Abell 2125 reveals a complex of massive gas clouds in the process of merging. Ten of the pointlike sources are associated with galaxies in the cluster; the rest are probably distant background galaxies.*

Right: *A Chandra mosaic of images of the Fornax galaxy cluster, showing a vast gas cloud that extends for more than half a million light-years.*

COSMOLOGY AND ALL-SKY SURVEYS

As we explore areas of the sky that are farther away, present-day telescopes become insufficient to resolve individual objects with any great clarity, yet the diffuse glow from distant stars and galaxies combine together to show the largest-scale structure in the universe. Many satellite observatories have completed all-sky surveys. Depending on the satellite's sensitivity to different wavelengths of light, these studies investigate the structure of the Milky Way or the arrangement of galaxies far beyond our corner of the universe. By using a particular wavelength of light, scientists have been able to image the echoes of the formation of the universe (the big bang).

Another way of investigating the farthest reaches of the universe is to stare at one point on the sky for a very long time, gathering every last glimmer of light emitted from distant objects. In some cases what appears as a blank patch of sky is revealed to contain a multitude of galaxies, including the most distant galaxies known to date.

The image shows what the distant universe looks like. This Hubble Space Telescope image is looking back billions of years into the past and shows an area on the sky that is just a fraction of the size of the full Moon. A random variety of galaxies is shown: spirals, ellipticals, and interacting ones that have come too close to each other. The smallest and reddest ones are those that are still forming in the early universe. The pointlike objects (including the bright red one in the center) are foreground stars in the Milky Way.

The Hubble Space Telescope views hundreds of faint galaxies in the distant universe. The red or white points with diffraction spikes (crosses) are Milky Way stars in the foreground.

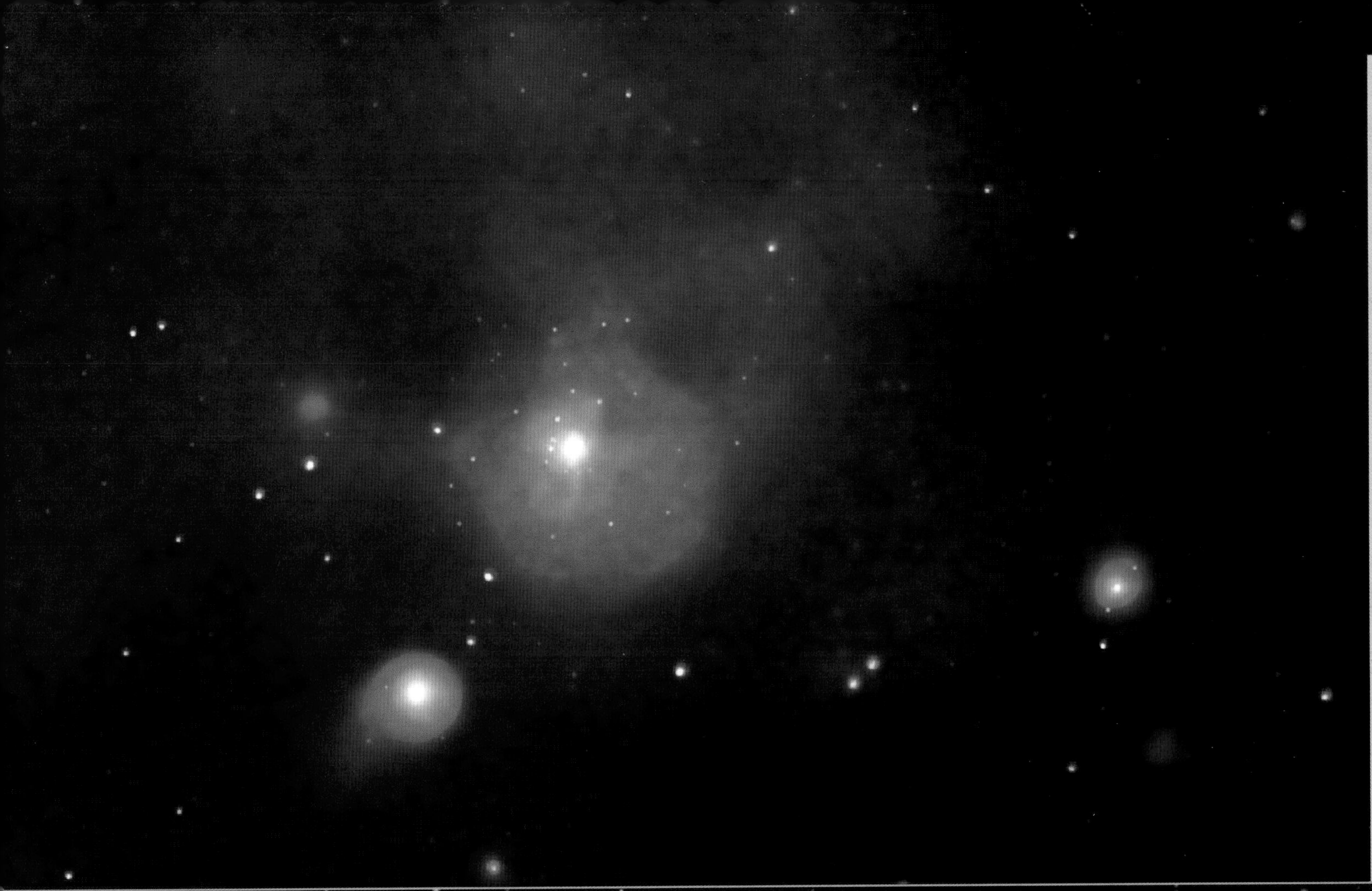

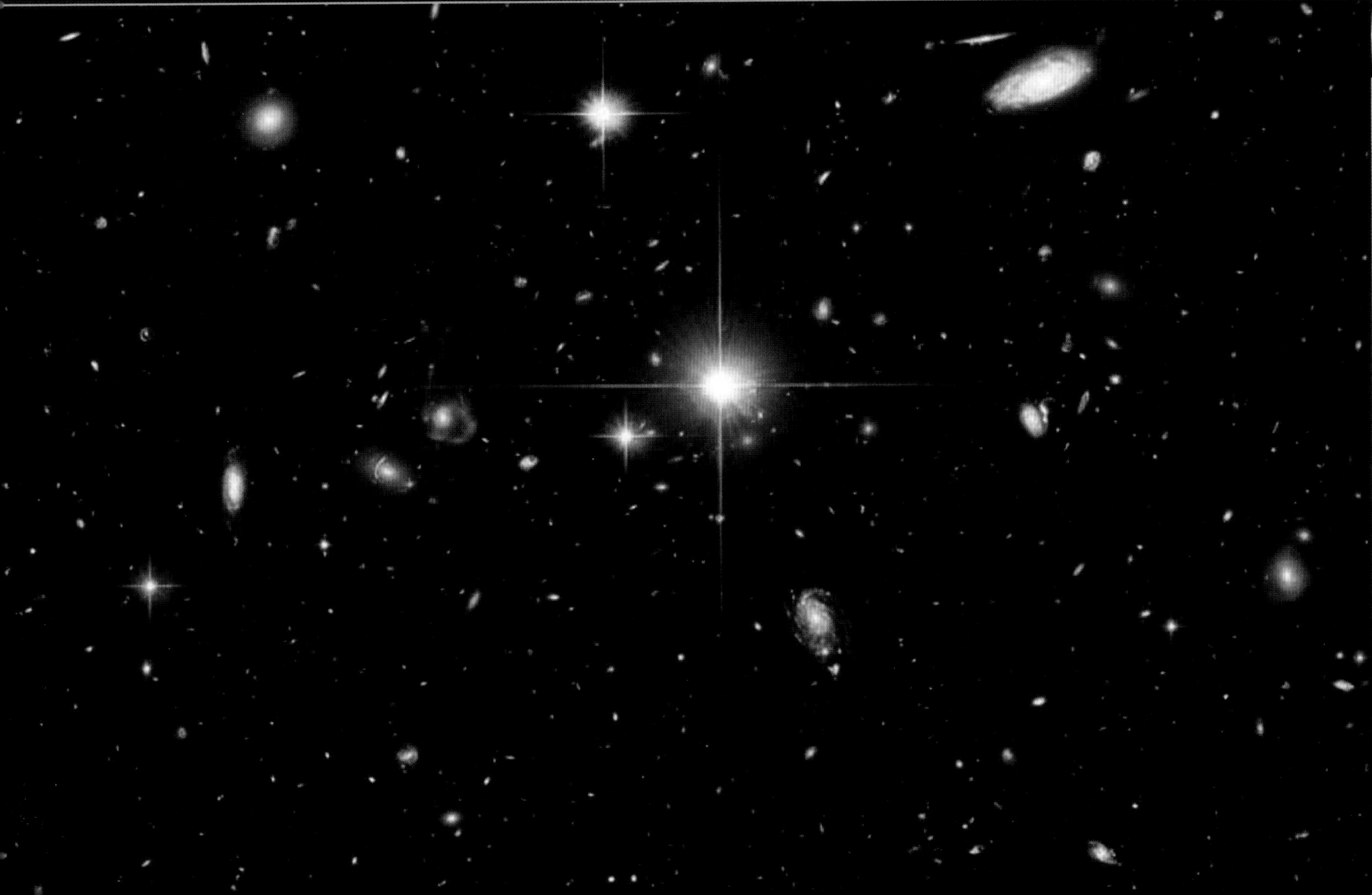

First Stars from Spitzer

The top panel to the right shows an image from the infrared Spitzer Space Telescope after looking at a patch of sky for ten hours. It shows objects familiar to astronomers—stars and galaxies. The lower panel shows what is left on the image when all the individual objects are masked out. There is a nonuniform diffuse glow across the image that could not be associated with any known object. Astronomers believe that it may be the light from the very first stars that formed after the creation of the universe or perhaps from hot gas falling into the first black holes.

Current understanding is that the universe and everything within it was created around 13.7 billion years ago in the big bang. The material was extremely hot to begin with, but cooled over time. The Cosmic Microwave Background (CMB) is what is left of the first light to stream across the universe around 300,000 years after the big bang. However, stars did not form immediately—it is thought that they lit up the universe about 200 to 400 million years later.

Images of the first stars from the Spitzer Space Telescope. The straight lines of dots around the brighter sources are artifacts.

EGRET All-Sky Map

Gamma rays are a form of light that has even more energy than X-rays. In space the most energetic objects emit gamma rays, as do some radioactive elements. Luckily, Earth's atmosphere stops these rays from reaching the ground.

The Energetic Gamma Ray Experiment Telescope (EGRET) took this all-sky image at some of the highest energies (>100MeV). It shows the diffuse emission from the plane of the galaxy—the center of which is in the middle of the image. It arises from cosmic-ray (high-energy protons and electrons) interactions with the interstellar medium.

There are a few notable points: 3C279 is the bright source that appears near the top of the image just to the right of center. The Vela, Geminga, and Crab pulsars are on the far right-hand side. 3C279 was initially identified as a radio source, hence its designation as the 279th source in the 3rd Cambridge catalog. It is, in fact, a type of blazar—one of the most powerful nonexplosive energy sources in the universe. The pulsars are the cores of exploded stars that are very hot and have intense magnetic fields. These two factors can combine in pulsars to produce gamma rays.

EGRET all-sky gamma-ray survey.

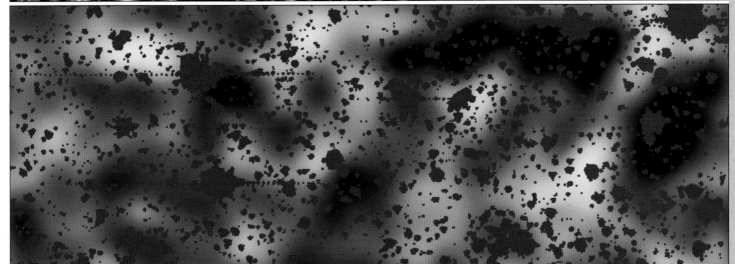

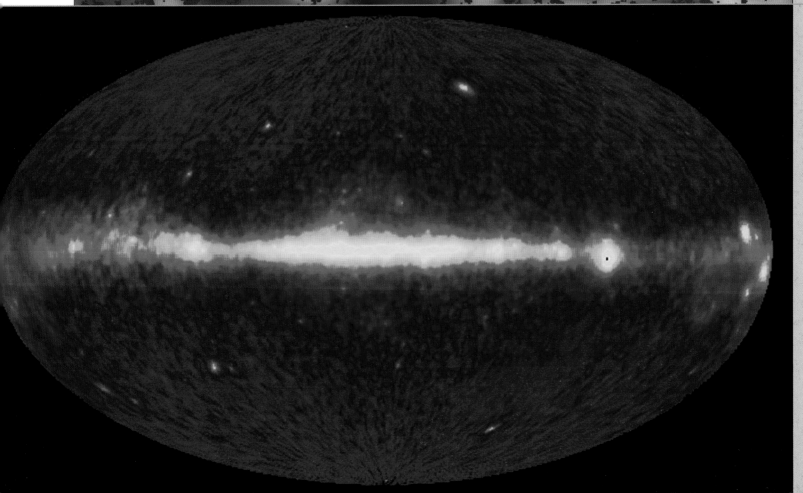

2MASS Map

The 2-Micron All-Sky Survey (2MASS) shows the entire sky at a wavelength of 2 microns. This wavelength is characteristic of objects that have a temperature of 1,000 to 2,000°C—much cooler than the surface of the Sun.

A wavelength of 2 microns was used because it is not scattered by interstellar dust in the way that optical emission is. Visible light is scattered and blocked by dust and gas that prevent an accurate map of the Milky Way from being plotted. The 2-micron radiation allows the true distribution of the luminous mass in the Milky Way and enables its largest structures to be revealed. It also allows the survey to see galaxies that are outside the Milky Way but that lie close to the plane of the galaxy.

The image here shows those galaxies that appear bright in the infrared and reveals that they are not uniformly distributed, but appear in clusters with filaments and voids. The vertical blue sash across the image arises from stars in our own Milky Way.

WMAP

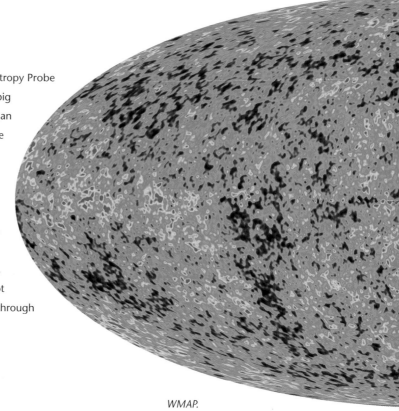

This map of the entire sky was produced by the Wilkinson Microwave Anisotropy Probe (WMAP) in the period from 2001 to 2003. The map shows echoes of the big bang, observed around 13.7 billion years after they occurred. The universe began with an explosion (the big bang) of space rather than in space. It created all the particles and energy that we see around us today. It was very hot—so hot that physicists cannot recreate even in particle accelerators what the conditions were like in the first few microseconds afterward.

As it began to expand after the big bang, the universe cooled, and the particles that make up the nuclei of atoms (protons and neutrons) were able to join up to form ions—the nuclei of hydrogen, helium, and a small amount of lithium. The electrons that surround the nuclei in ordinary atoms were still free, and the matter was still a plasma. These objects interact with light, so it was not possible to see across the universe (in the same way as it is not possible to see through a thick flame).

WMAP.

The 2-Micron All-Sky Survey (2MASS) map. It has been projected in a different way to the other maps, with the Milky Way running vertically up the center rather than along the "equator."

The explosion was not uniform. Some parts of the expanding universe were slightly denser than others. When the universe had cooled enough for the plasma to form a gas—about 300,000 years after the big bang—the light suddenly streamed out across the universe. The temperature fluctuations seen in the Cosmic Microwave Background (CMB) arise from density fluctuations in the early universe, which was not entirely uniform. However, denser regions are only denser by 1 in 100,000. The universe has expanded since the light started its journey, and the light has been stretched and cooled, and now corresponds to a temperature of 2.73 degrees above absolute zero.

Astronomers can use the details of this map to work out what the universe was like in its early life and also what its future may hold.

Hubble Ultra Deep Field

The Hubble Ultra Deep Field is the deepest image ever taken of the universe. It took 11.3 days of observation between September 2003 and January 2004 to produce. There are about 10,000 objects on this image, only a handful of which are stars. The patch of sky from which the image came is one-tenth the apparent size of the full Moon and from the ground looks to be entirely empty. Additional observations in 2009–2011 have added 87 hours to this exposure with Hubble's WFC3 infrared instrument. These observations provide significantly more information about the most distant galaxies, including a galaxy 13.2 billion light-years away that formed only 480 million years after the big bang.

In the image, there are some relatively nearby galaxies, which are very similar to the familiar ones found in the nearby universe, the spiral and elliptical galaxies. However, when you look at progressively smaller and fainter sources, the galaxies change. They become more irregular: some still have spiral arms, but they are incomplete or distorted; the elliptical galaxies are no longer smooth and egg-shaped.

As we look back in time, we notice how the galaxies that we see in the local universe today evolved into their current shape. Galaxies grow by merging together, a process which in the local universe tends to occur between large galaxies and their satellites, with a few spectacular exceptions, such as the Antennae Galaxies. In the early universe, when it was about one-sixth of its current age (800 million years after the big bang), galaxies had formed but they were small and irregular. They had not had enough time to assemble themselves into the majestic galaxies we see around us today. They are among the first galaxies that lit up the universe after the big bang.

In the past few years, observations using the WFC3 instrument have added more than 87 hours of exposure and provided significantly more information about the most distant galaxies. This includes a galaxy approximately 13.2 billion light-years away that formed only 480 million years after the big bang.

Chandra Deep Field North

To complement the Hubble Deep Field observations, the Chandra X-ray Observatory watched the same region of the sky for two million seconds (23 days). As you can see, there are not very many sources, especially when compared to the Hubble Ultra Deep Field. In fact, there are 503 objects in the image.

The objects at the edge of the image appear to be stretched, but this is a result of the telescope optics. The colors show what energy levels the X-rays had when they arrived at the satellite—red is the lowest, followed by yellow. Blue represents those with the highest energies. The extended red sources are distant clusters of galaxies, of which there are only a few in the image.

Most of the sources are distant supermassive black holes at the centers of galaxies. The galaxies themselves are not visible in this image. As material falls into the black holes, friction causes it to heat up, since some parts move faster than others. In the very central regions the material is so hot that it emits X-rays, which are what astronomers use to observe black holes. The expansion of the universe is a useful phenomenon when analyzing these data. It means that X-rays whose energy levels would usually be far beyond the capabilities of Chandra become stretched, and their lower energy levels are detectable.

The diffuse background was thought to be made up of sources that could not be detected with the instruments available at the time. Although some of the background is made up of distant X-ray sources, there is still some emission missing. It is now thought that this arises from sources that have large amounts of material around them. They cannot be detected from Earth because they are too faint for the current generation of satellites.

The Hubble Ultra Deep Field image.

Chandra Deep Field North image.

Einstein Cross—Huchra's Lens

Contrary to appearances, this is not an image of five different objects. It is, in fact, an image of just two: a quasar in deep space, which is almost exactly behind a large galaxy when viewed from Earth. The diffuse galaxy at the center is some 400 million light-years away. The four point-like images that surround it to the top, right, bottom, and left are of the same quasar that is about 8 billion light-years away. Light from the quasar is bent in its path by the gravitational field of the galaxy, which acts as a gravitational lens. The phenomenon of gravitational lensing creates streaks and arcs out of the lensed object, generally a background galaxy or quasar. However, if the alignment is just right (as in this example) and the lens is very symmetrical, then the image forms a so-called Einstein Cross. (This object is sometimes known as Huchra's Lens, named for American astrophysicist John Huchra, who first discovered it in 1985.)

The Hubble Space Telescope has zoomed all the way into the center of the galaxy. The quasar images are so bright that only the densest regions of the galaxy's nucleus are visible.

BATSE

This map of the sky was made by the Compton Gamma Ray Observatory's (CGRO's) Burst and Transient Spectrometer Experiment (BATSE) during the period 1991 to 2000. The CGRO was one of NASA's four great observatories—the others being the Hubble Space Telescope, the Chandra X-ray Observatory, and the Spitzer Space Telescope. The CGRO had four main instruments, but the BATSE was designed to watch the entire sky for intense bursts of gamma-ray emission (light more energetic than X-rays).

These bursts of gamma-ray radiation, known as Gamma Ray Bursts (GRBs), were discovered by military satellites after the signing of the limited nuclear test ban in 1963. The satellites were launched to monitor the radiation that would be released should an atomic device be set off. The satellites did detect bursts, but not of terrestrial origin.

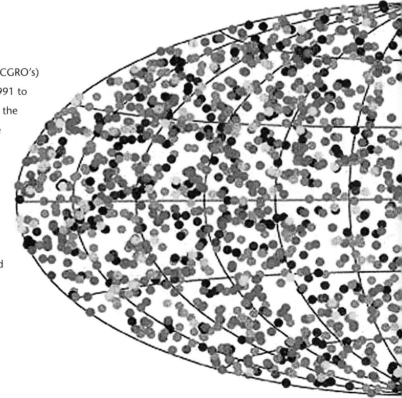

Image of Einstein's Cross taken by the Hubble Space Telescope.

These gamma-ray bursts are now known to originate outside the Milky Way and are the brightest objects in the universe. The distance to some of them has been measured, which was a challenge—the longest-lived bursts last less than one minute, some last only a few milliseconds, and they shine all the way across the universe. For the short time that they last, they are the most energetic form of light in the universe.

There are various competing theories that attempt to explain the nature of GRBs. The current favorite is that they are a special type of supernova (an explosion created by a dying massive star). The SWIFT mission, operating since 2004, has detected 500 GRBs to date.

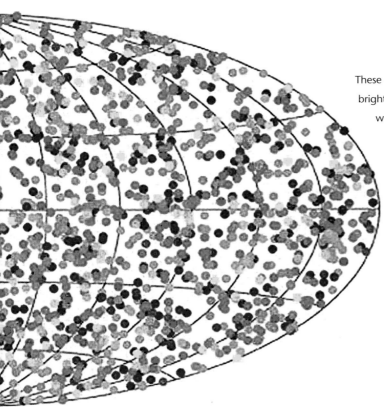

BATSE map recording 2,704 Gamma-Ray Bursts. Bursts are color coded according to duration and strength. For example, long-duration, bright bursts are red and short-duration, weak bursts are purple.

Glossary

Words in SMALL CAPITALS can be looked up elsewhere in the Glossary.

Abell, George (1927–83) American astronomer who analyzed CLUSTERS OF GALAXIES and in 1958 published the first catalog of clusters in the Northern Hemisphere. Today the extended Abell catalog is a comprehensive list of about 4,000 clusters.

absolute magnitude A measure of the true brightness of a STAR. It is the APPARENT MAGNITUDE a star would have at a distance of 10 PARSECS (32.6 LIGHT-YEARS).

absolute zero In theory, the lowest temperature it is possible to attain, equivalent to –273.16°C and zero degrees KELVIN.

accretion The process by which a body grows by the attraction of material as a result of its GRAVITY. Planets grow by accretion of dust and GAS in a PROTOPLANETARY DISK. BLACK HOLES grow by accretion; any material that strays too close is pulled into ORBIT, forming an accretion disk.

active galactic nucleus (AGN) A GALACTIC NUCLEUS in which vast amounts of energy are being liberated in the form of ELECTROMAGNETIC RADIATION (sometimes enough to outshine all the other STARS in the GALAXY combined). They are currently thought to be SUPERMASSIVE BLACK HOLES.

active galaxy A GALAXY in which an unusual, perhaps violent, phenomenon is releasing energy, usually from the GALACTIC NUCLEUS.

albedo The proportion of incident light (light that strikes an object) that is reflected from the surface of an object, expressed as a percentage. For example, the average albedo of THE MOON, although it appears silvery bright to observers on EARTH, is just 7 percent. Some ASTEROIDS are as black as soot and reflect less than 1 percent of incident light.

andesite A fine-grained volcanic rock, named for the Andes Mountains where it is common.

Andromeda Galaxy The nearest SPIRAL GALAXY to the MILKY WAY and the most distant object visible to the naked eye.

angular momentum The product of the moment of inertia and the angular velocity of an orbiting or rotating object.

aphelion The greatest distance from the SUN reached by an object, such as a planet, in an elliptical ORBIT around it.

apogee The point in its ORBIT where THE MOON or an artificial SATELLITE is farthest from EARTH.

apparent magnitude The brightness of a CELESTIAL OBJECT as it appears from EARTH.

asteroid Any of many rocky objects, most of which ORBIT the SUN between Mars and Jupiter.

atmosphere The layer of GASES that surround a planetary object.

aurora The glow that appears in the night sky at high northern and southern latitudes, commonly known as the Northern or Southern Lights. It is produced when charged particles (ELECTRONS and PROTONS) collide with atoms and molecules in EARTH's upper ATMOSPHERE.

barred spiral galaxy A form of disk GALAXY in which the SPIRAL ARMS are attached to the GALACTIC NUCLEUS by a straight bar of STARS.

barycenter The common center of MASS around which two or more bodies revolve.

big bang The theory that the UNIVERSE came into being in an instantaneous event about 14 billion years ago. The theory states that everything contained within the universe was created in that initial event and, as time passed, the universe expanded and its contents evolved into the STARS and GALAXIES of today.

binary star A pair of STARS that ORBIT one another about their BARYCENTER. They are held together by the force of their mutual GRAVITY.

black dwarf The theoretical cold remnant of a dead, low-mass STAR.

black hole A CELESTIAL OBJECT so dense that not even PHOTONS of ELECTROMAGNETIC RADIATION can escape from it.

blazar The most active type of ACTIVE GALAXY. The name is a combination of BL LACERTAE object and QUASAR. It is thought that the activity in blazars is caused by jets of GAS being expelled from the ACTIVE GALACTIC NUCLEUS at speeds close to the SPEED OF LIGHT.

BL Lacertae A type of ACTIVE GALAXY, the first BLAZAR to be discovered, thought to be the bright NUCLEUS of an active ELLIPTICAL GALAXY.

blueshift The shortening of the wavelength of spectral lines caused either by the motion of the source toward the observer or by the motion of the observer toward the source.

carbonaceous asteroid An ASTEROID composed mainly of carbon.

Cassini–Huygens A joint American–European space probe, sister to GALILEO, used to explore Saturn and its MOONS. Launched in 1997, Cassini is the orbiter and Huygens is a detachable PROBE that successfully landed on Saturn's largest MOON, Titan.

cD galaxy A huge ELLIPTICAL GALAXY found at the center of a CLUSTER OF GALAXIES. These galaxies are believed to have grown so large because they merged with other galaxies.

celestial object Any object, such as a STAR, GALAXY, or planet, that is projected onto the CELESTIAL SPHERE.

celestial sphere The imaginary sphere, centered on EARTH, upon which STARS and other CELESTIAL OBJECTS appear fixed; in reality they are distributed through the volume of space and are moving with respect to each other.

center of mass The point at which a system of masses would balance if placed on a pivot, and the point at which the MASS of the object can for many purposes be thought to be concentrated. This point need not be in the center of the object. (*See* BARYCENTER.)

Cepheid variable star A STAR that has left the MAIN SEQUENCE and begun to pulsate because it is unstable.

chromosphere The layer above the PHOTOSPHERE on a STAR.

cloud belt A persistent band of gaseous cloud in a planet's ATMOSPHERE, associated with the planets Jupiter, Saturn, Uranus, and Neptune.

cluster of galaxies Collection of GALAXIES held together by the force of their mutual GRAVITY; the largest objects in the UNIVERSE that are held together by their own GRAVITY.

coma The cloud of dust and GAS that surrounds the NUCLEUS of a COMET.

comet Relatively small icy objects that may have extremely long tails and that populate the KUIPER BELT and the OORT CLOUD. Some comets are in highly elliptical solar ORBITS that bring them close to the SUN, which drives off GAS and dust to form spectacular tails visible from EARTH.

core The central region of a CELESTIAL OBJECT, where the DENSITY and temperature are highest. In a planet the constituent chemical elements are expected to be in a metallic state. In stellar cores, NUCLEAR FUSION takes place, releasing energy and causing them to shine.

corona The outer part of a STAR's ATMOSPHERE. (*See* CHROMOSPHERE, PHOTOSPHERE, and SUN.)

cosmic background radiation Weak ELECTROMAGNETIC RADIATION that is believed to be a remnant of the radiation released during the decoupling of matter and energy around 300,000 years after the BIG BANG.

cosmic ray A subatomic particle, usually a PROTON, moving through space at close to the speed of light.

crater A depression formed during METEORITE impact or by volcanic explosion.

crust Outermost layer of a planet's LITHOSPHERE.

dark matter A form of matter that is thought to exist in the UNIVERSE in vast quantities. It is difficult to detect because it is nonluminous or of very low LUMINOSITY and only interacts via gravitational effects.

density An object's MASS-to-volume ratio.

diurnal Repeating daily; or relating to the rotation of the STARS across the sky.

Doppler effect A physical process that alters the wavelength of ELECTROMAGNETIC RADIATION or sound because the source of emission and the observer are in relative motion. When the source and observer move closer together, the wavelength becomes "squashed." In sound waves the result is that the pitch increases. In electromagnetic waves the PHOTONS become more energetic. When the source and the observer are moving apart, the wavelength is "stretched." Sound drops in pitch, and photons are shifted to the less energetic, red end of the ELECTROMAGNETIC SPECTRUM. (*See* REDSHIFT.)

Dwarf planet A SOLAR SYSTEM object massive enough to be spherical under its own GRAVITY but not enough to clear its orbital environment of other debris. Also, not a SATELLITE of a planet.

dwarf star A STAR lying on the MAIN SEQUENCE that is too small to be classified as a giant star or a SUPERGIANT star. (For example, the SUN is a yellow dwarf star.)

Earth The third planet in distance from the SUN, and a member of the four TERRESTRIAL PLANETS that make up the inner SOLAR SYSTEM.

eclipse The partial or whole apparent disappearance of one CELESTIAL OBJECT when it passes through the shadow of another.

ecliptic The path that the SUN appears to take across the sky every day. It is the plane of the SOLAR SYSTEM, projected onto the CELESTIAL SPHERE.

electromagnetic radiation The kind of radiation that consists of electric and magnetic fields and travels at the speed of light. Visible light is electromagnetic radiation, along with waves on the ELECTROMAGNETIC SPECTRUM from RADIO WAVES to GAMMA RAYS. Most objects we detect with telescopes emit some form of this radiation.

electromagnetic spectrum The range of ELECTROMAGNETIC RADIATION from the longest wavelength (RADIO WAVES) through MICROWAVES, INFRARED rays, VISIBLE-LIGHT rays, ULTRAVIOLET RAYS, and X-RAYS to the shortest and most energetic wavelength radiation (GAMMA RAYS).

electron An elementary particle that has 1/2000th the MASS of a PROTON and carries negative charge.

electron degeneracy pressure The great pressure exerted by ELECTRONS in a WHITE DWARF that keeps it from collapsing further. In white dwarfs the DENSITY of matter is so great that electrons can no longer exist in ORBITS around the NUCLEI. They become compressed and try to fill the lowest energy states. Because two electrons cannot occupy the same quantum states at the same time, additional electrons are forced to occupy states of higher energy. The atoms have reached their minimum size and, because they cannot get any smaller, the white dwarf cannot collapse further.

elliptical galaxy A GALAXY with an ellipsoidal shape, characterized by the absence of SPIRAL ARMS. Most are deficient in dust, and few possess regions of current STAR formation.

emission nebula A cloud of interstellar GAS that glows because it absorbs and then reemits ELECTROMAGNETIC RADIATION from hot STARS embedded within it.

equator An imaginary line that encircles EARTH at zero latitude and is equidistant from the poles, dividing Earth into the Northern and Southern Hemispheres. By analogy, other planets have an equator in the plane that is perpendicular to their axis of rotation.

equinox A time when day and night are of equal length. It occurs twice a year, when the SUN crosses the celestial EQUATOR. The equinoxes mark where the ECLIPTIC intersects the celestial equator.

false-color image An image that has been artificially colored in ways that do not match how the eye would actually see the object when it was photographed. Completely "wrong" colors may be used in order to highlight subtleties of detail that are otherwise too difficult to see. Sometimes false colors have to be used to represent wavelengths of radiation that are normally outside the visible range of the spectrum. (*See* TRUE COLOR IMAGE.)

galactic disk The flattened region of GAS and dust in which the arms of a SPIRAL GALAXY can be found.

galactic nucleus The central regions of a GALAXY. In SPIRAL GALAXIES it is also called the bulge, because it is thicker than the disk and is usually composed of older STARS. It is also where a SUPERMASSIVE BLACK HOLE probably resides.

galaxy A collection of STARS, dust, and GAS that are held together by the force of its constituents' GRAVITY.

galaxy cluster (*See* CLUSTER OF GALAXIES.)

Galileo Galilei (1564–1642) Italian astronomer and physicist who was the first person to view Jupiter through a telescope and record his observations. He made numerous important discoveries, including the MOONS of Jupiter. He supported the theory proposed by Nicolaus Copernicus of a Sun-centered UNIVERSE, for which he was accused of heresy by the Roman Catholic church.

***Galileo* space probe** One of the largest interplanetary space PROBES ever launched, *Galileo* is named for GALILEO GALILEI and was launched to orbit and study Jupiter in greater detail than possible before.

gamma rays The highest energy form of ELECTROMAGNETIC RADIATION. The wavelength of gamma rays is typically less than 10^{-12} m.

gas A phase of matter characterized by low density and viscosity; it undergoes relatively great expansion and contraction with changes in pressure and temperature, it has the ability to diffuse readily, and it tends to become distributed uniformly throughout any container. It usually occurs at higher temperatures than liquids or solids.

gas giant A planet made up of vast quantities of GAS. In the SOLAR SYSTEM there are four gas giants: Jupiter, Saturn, Uranus, and Neptune.

grand design spiral A SPIRAL GALAXY in which the spiral structure is clearly defined and well organized.

gravitational energy GRAVITY is weak universal force realtive to other fundamental universal forces at the shortest distance scales (nuclear and atomic), but dominates the UNIVERSE at cosmic distances (assuming we ignore dark energy). As a result of the mutual attraction between objects that increases proportionally to their MASSES, huge amounts of potential energy are stored inside them.

gravitational lens Any massive CELESTIAL OBJECT, such as a GALAXY, that distorts the SPACETIME CONTINUUM so that the light from more distant objects is magnified and distorted. This can result in more than one image of the background object being visible.

gravitational resonance A state in which an orbiting object is subject to periodic gravitational disturbances by another. The two objects are in a simple ratio to each other. (For example, for each ORBIT of Jupiter by its MOON Ganymede, Europa orbits exactly twice and Io exactly four times.)

gravitational slingshot A means of using the motion of a planet to alter the path and speed of a spacecraft. It enables spacecraft to visit the outer planets, by changing its velocity relative to the SUN. Energy (velocity) is transferred between the interacting bodies; in the case of a spacecraft sent past a planet on its way to someplace else in the SOLAR SYSTEM, a small amount of the planet's orbital energy is transferred to the much smaller spacecraft thus gaining much more energy, proportional to its MASS, and its speed increases dramatically. (The planet's speed also changes but by a much smaller amount because of its much larger MASS.)

gravity The attractive force exerted between two or more objects that have MASS. (*See* NEWTON'S UNIVERSAL THEORY OF GRAVITATION.)

Great Red Spot The most distinctive feature of the planet Jupiter's CLOUD BELTS, known to have persisted for at least 300 years. It is a huge weather system thousands of miles across, with a distinctly reddish hue.

greenhouse effect The gradual rising of the temperature of air in the lower ATMOSPHERE, believed to be caused by the buildup of GASES such as carbon dioxide, methane, nitrous oxide, and chlorofluorocarbons.

Herschel, William (1738–1822) German-born British astronomer best known as the first person to discover the planet Uranus. He also made telescopes—the largest and most famous being a reflecting telescope with a 40-foot (12-m) focal length, with which he discovered two of Saturn's MOONS. He also compiled an extensive catalog of NEBULAE. The European Space Agency's space telescope is named the Herschel Space Observatory in his honor.

Hertzsprung–Russell diagram A graph that plots the ABSOLUTE MAGNITUDES of STARS against their SPECTRAL CLASSIFICATION. LUMINOSITY can be substituted for absolute magnitude, and effective temperatures for spectral classification. Any star for which these values are known can be included on the diagram.

Hubble constant Named for American astronomer EDWIN HUBBLE, the Hubble constant is a ratio that expresses the rate of apparent expansion of the UNIVERSE. It is equal to the speed at which a typical GALAXY is receding from EARTH divided by its distance from Earth. The value of the Hubble constant is important in determining the age of the universe.

Hubble, Edwin (1889–1953) American astronomer who established beyond doubt the existence of GALAXIES outside the MILKY WAY. He also created a system for classifiying galaxies according to content, distance, shape, size, and brightness, and is generally credited with discovering the REDSHIFT of galaxies. The Hubble Space Telescope is named for him.

Hubble tuning fork diagram A diagram representing the various classes of GALAXY, according to the scheme drawn up by American astronomer EDWIN HUBBLE. ELLIPTICAL GALAXIES are divided into eight subclasses from E0 (a sphere) to E7 (an oblate spheroid, similar to an American football). The subclasses are placed along the handle of the tuning fork. The two prongs of the fork show three subclasses of SPIRAL GALAXIES and three of BARRED SPIRAL GALAXIES. These subclasses indicate the tightness of the SPIRAL ARMS. Where the prongs join the handle, an intermediate form of galaxy, known as a LENTICULAR GALAXY, is placed.

infrared ELECTROMAGNETIC RADIATION with a wavelength longer than that of VISIBLE LIGHT. Infrared radiation is used to study protostars and other objects deeply buried within dense interstellar clouds, because infrared light is less obscured by dust.

intercluster medium The hot tenuous GAS found between GALAXIES in a CLUSTER OF GALAXIES. It is very hot and as such emits X-RAYS.

intergalactic medium The matter found in the space between GALAXIES. It is a very tenuous GAS. Around some GALAXIES and in CLUSTERS OF GALAXIES it is denser and can be very hot (millions of degrees). (*See* INTERCLUSTER MEDIUM.)

interplanetary medium The diffuse matter found between the planets in the SOLAR SYSTEM, consisting of dust, charged particles, and neutral GAS atoms.

ion An atom that has acquired an electrical charge either by gaining or losing ELECTRONS.

IRAS (the **I**nfra**r**ed **A**stronomical **S**atellite) An INFRARED telescope launched in 1983 to observe the UNIVERSE at INFRARED wavelengths.

Kelvin A temperature scale based on a zero point at which atoms in solids stop vibrating (ABSOLUTE ZERO). Zero Kelvin corresponds to –273.16°C.

Kuiper Belt A region of the SOLAR SYSTEM, similar to the ASTEROID BELT, but beyond Neptune. The objects that occur in this belt are known as Kuiper Belt objects or trans-Neptunians. They are small icy worlds that take up to 200 years to ORBIT the SUN.

lava Molten rock or other mineral material that reaches a planet's surface, often from an erupting volcano.

lenticular galaxy An intermediate form of GALAXY between ELLIPTICAL GALAXIES and SPIRAL GALAXIES. Lenticular galaxies have flattened forms and GALACTIC DISKS but no SPIRAL ARMS.

light-year A unit of astronomical measurement equivalent to $5.87849981 \times 10^{12}$ miles (9.4605284×10^{15} m). It is the distance that is traveled by ELECTROMAGNETIC RADIATION in one year.

lithosphere The rigid outermost layer of a planet.

local group The name of the group of a few dozen GALAXIES of which our own (the MILKY WAY) is a member.

luminosity The total energy radiated into space every second by a CELESTIAL OBJECT such as a STAR.

ly (abbr.) (*See* LIGHT-YEAR.)

magnetosphere A region in space within which a planet's magnetic field is the predominant magnetic influence.

magnitude A measure of the brightness of a CELESTIAL OBJECT. APPARENT MAGNITUDE is a measure of how bright the object appears, without taking into account its distance away from us. ABSOLUTE MAGNITUDE provides a standard of comparison between objects of varying distance.

main sequence A diagonal band on the HERTZSPRUNG–RUSSELL DIAGRAM on which most STARS lie. It represents the time of a star's life when it is stable and shining steadily.

mantle The layer within a planet that lies between the thin outer CRUST and the CORE.

maser (**m**icrowave **a**mplification by **s**timulated **e**mission of **r**adiation) A high-frequency MICROWAVE amplifier or oscillator in which the signal to be amplified is used to stimulate unstable atoms into emitting energy at the same frequency. This creates a very pure color of light, and so DOPPLER shifts can be very accurately determined.

mass The amount of material in an object and a measure of its inertia: the extent to which it resists acceleration when a force is applied.

Messier, Charles (1730–1817) French astronomer who found 20 COMETS, 13 of which were original discoveries. He also devised the first comprehensive list of deep sky objects—a catalog of distant GALAXIES, STAR clusters, and stellar NEBULAE. Objects in the catalog are identified by Messier numbers. The first object—the Crab Nebula—is labelled M1, and the last—M110—is a SATELLITE galaxy located in the constellation Andromeda.

meteor A "shooting STAR"; a small METEOROID that burns up in EARTH'S ATMOSPHERE.

meteorite A rock fragment of extraterrestrial origin. Meteorites include metallic and rocky particles of a range of sizes, many of which have originated in ASTEROIDS.

meteoroid A small rocky object that ORBITS the SUN. One that enters EARTH'S ATMOSPHERE is seen as either a METEOR or a METEORITE.

microwaves ELECTROMAGNETIC RADIATION that has wavelengths in between RADIO WAVES and INFRARED on the ELECTROMAGNETIC SPECTRUM.

Milky Way The name of our GALAXY, which contains about one hundred billion STARS.

moon A naturally occurring SATELLITE in ORBIT around a planet.

Moon, the Earth's only natural SATELLITE. It is a heavily cratered body whose composition is deficient in the heavier, metallic materials.

nebula (pl. nebulae) A cloud of dust and GAS in space that is visible to observers on Earth because it either emits, reflects, or absorbs starlight.

neutron star The central remains of a STAR that has undergone a SUPERNOVA explosion.

Newton, Isaac (1642–1727) An English mathematician who established the basic laws that became the cornerstones of astronomy and physics. In about 1665 he formulated NEWTON'S UNIVERSAL THEORY OF GRAVITATION. He also made one of the earliest reflecting telescopes and proposed three laws of motion.

Newton's universal theory of gravitation Theory fomulated by ISAAC NEWTON, which states that the gravitational force between any two objects is equal to the product of their MASSES and inversely proportional to the square of the distance between them.

nova (pl. novae) A STAR that is observed to brighten suddenly by up to ten MAGNITUDES and then gradually decline back to its original magnitude over a period of months.

nuclear fusion A process by which two or more atomic NUCLEI join together to make a heavier one. This type of nuclear reaction releases energy. It occurs at the centers of STARS and causes them to shine.

nucleus (pl. nuclei) The core of an atom, consisting of neutrons and PROTONS.

Oort Cloud A spherical region, believed to surround the SOLAR SYSTEM, containing a vast number of COMETS.

optical binary A pair of STARS that lie close to one another on the CELESTIAL SPHERE because of a chance alignment. They are not physically associated with one another and probably exist at vastly different distances. (*See* BINARY STAR.)

orbit The path that an object takes under the influence of GRAVITY. Orbits follow the shape of conic sections and may therefore be circular, elliptical, parabolic, or hyperbolic.

orbital resonance (or resonant orbit) (*See* GRAVITATIONAL RESONANCE.)

organic Relating to organic chemistry. Used to describe molecules made from carbon and hydrogen with a small amount of other elements. All known life is based on organic chemistry.

parallax The apparent shift in a nearby object against a distant background when viewed from a different angle. This principle is used to measure the distance to nearby STARS.

parsec A unit used to measure distance in astronomy. It is the distance at which a STAR would show a PARALLAX shift of 1 second of arc (1/3600 of a degree) and it is equal to 3.26 LIGHT-YEARS.

perihelion The point in its ORBIT in which a member of the SOLAR SYSTEM is at its closest to the SUN.

photon An elementary particle that carries energy generated in reactions involving electromagnetism.

photosphere The surface of any STAR; a layer that emits VISIBLE LIGHT.

planetary nebula A NEBULA created when a RED GIANT sheds its outer layers of GAS. Despite the name, it has nothing to do with a planet.

planetoid A small body that resembles a planet, e.g., an ASTEROID.

plasma The fourth state of matter (after solid, liquid, and GAS) that occurs when a gas is heated until every atom and molecule becomes ionized (charged). Usually it occurs at very high temperatures. (*See* ION.)

precession The slow motion of the celestial poles on the CELESTIAL SPHERE, largely caused by EARTH "wobbling" on its axis. Because of precession the EQUINOXES change by one month every 2,000 years.

probe A spacecraft sent deep into space to visit planets, MOONS, and other CELESTIAL OBJECTS.

prominence A cloud or plume of hot, luminous GAS that is elevated into the lower solar CORONA from the top of the CHROMOSPHERE. It usually occurs over regions of solar activity, such as SUNSPOT groups.

proton An elementary particle composed of three quarks and carrying a positive electrical charge.

protoplanetary disk The disk of dust and GAS surrounding a recently formed STAR in which planets form.

pulsar A spinning NEUTRON STAR. An intense beam of radiation is emitted from its surface, which acts like the light from a lighthouse when the pulsar spins.

quasar A faraway, highly luminous ACTIVE GALACTIC NUCLEUS (abbreviation of quasi-stellar radio source). They appeared as point sources in early radio and optical surveys but did not have the correct SPECTRA, hence the term quasi-stellar—they looked like STARS, but they were not.

radio galaxy An ELLIPTICAL GALAXY that is an intense source of RADIO WAVES; a form of ACTIVE GALAXY.

radio waves ELECTROMAGNETIC RADIATION with the longest wavelengths from a few millimeters upward, bounded at the upper end of the ELECTROMAGNETIC SPECTRUM by MICROWAVES.

ram-pressure Pressure exerted on an object that is moving at supersonic speed through a fluid medium.

red dwarf A small faint STAR whose surface temperature is below 4000 K.

red giant A large, relatively cool star that is in the final stages of its life. A red giant has a SPECTRAL CLASSIFICATION of K or M. It is enlarged and appears red because of the change that it has undergone in its surface temperature.

redshift The lengthening of the wavelength of spectral lines caused either by the motion of the source away from the observer or by the motion of the observer away from the source.

reflection nebula A cloud of interstellar GAS that appears to glow because it reflects starlight.

regolith Loose rock and rock dust that lies on top of solid bedrock on a planet. On EARTH, it is mixed with biological matter and called soil.

resonance (*See* GRAVITATIONAL RESONANCE.)

retrograde rotation The motion of a planet or MOON around its rotational axis in such a way that, if it were viewed from north of the orbital plane, it would be seen to rotate in an east–west (clockwise) direction.

retrorockets Rockets used to accelerate a spacecraft against the current direction of movement, slowing it down, e.g., for ensuring a SOFT LANDING on a planet.

ring system Each of the four large outer planets (Jupiter, Saturn, Uranus, and Neptune) has a collection of mainly icy rings orbiting it. They constitute ring systems. (*See* SHEPHERD MOON.)

rocky planet (*See* TERRESTRIAL PLANET.)

satellite An object in ORBIT around a planet. Used to describe one that occurs naturally as well as human-made objects sent into space.

seismometer Apparatus used to measure and record vibrations of a rocky body's surface; used to determine the motion of the ground and hence its structure.

Seyfert, Carl (1911–60) American astronomer who first described SEYFERT GALAXIES in 1943.

Seyfert galaxy A specific class of GALAXY that has ACTIVE GALACTIC NUCLEI. They can be SPIRAL or BARRED SPIRAL GALAXIES.

shepherd moon A small rock or rock-and-ice moonlet that forms part of a planetary RING SYSTEM. Through the effect of GRAVITY and orbital resonances, it acts to sweep clean certain parts of the ring, producing more or less permanent gaps.

soft landing Occurs when a craft lands on a planetary surface without damaging or destroying itself in the process. (The terms hard landing and crash landing are interchangeable and refer to the landing of a spacecraft that is not equipped with or not using a device such as RETROROCKETS to slow it down.)

solar nebula The cloud of material from which the SOLAR SYSTEM was formed.

solar system Everything that is dominated by the SUN's gravitational field. The solar system is made up of the Sun, the planets and their MOONS, as well as minor bodies such as ASTEROIDS and COMETS.

solar wind A steady stream of PLASMA flowing away from the SUN along magnetic field lines that lead into the INTERPLANETARY MEDIUM. It is composed of charged particles such as PROTONS, ELECTRONS, and atomic NUCLEI of helium.

space shuttle NASA's reusable spacecraft used to transport objects and people into low EARTH ORBIT. Six space shuttles have been constructed: *Enterprise* (a prototype that never flew in space), *Columbia, Challenger, Discovery, Endeavour,* and *Atlantis.*

spacetime continuum A geometric framework made up of three dimensions of space and one dimension of time. Within this it is possible to locate any event in terms of space and time and determine the relationships between events.

spectral absorption lines The dark lines in a continuous SPECTRUM that are produced when ELECTROMAGNETIC RADIATION passes through a GAS cloud and certain wavelengths are absorbed by the electrons around the gas atoms.

spectral classification A method used to classify STARS based on the appearance of the SPECTRAL ABSORPTION LINES in their SPECTRA. The classes are O, B, A, F, G, K, and M.

spectrum (pl. spectra) The distribution by wavelength of all ELECTROMAGNETIC RADIATION, including visible (white light) and continues at shorter and longer wavelengths into the ULTRAVIOLET and INFRARED. White light is composed of the colors red, orange, yellow, green, blue, indigo, and violet. Measurement of spectra is possible in any regime (radio to gamma rays) using an instrument that can distinguish the luminosity at different energies (wavelengths or frequencies).

spiral arms Arms that spiral outward from the GALACTIC NUCLEUS of SPIRAL and BARRED SPIRAL GALAXIES.

spiral galaxy A GALAXY that has a spiral structure in which the SPIRAL ARMS radiate from a central bulge.

star A CELESTIAL OBJECT that shines because of the release of energy liberated in its CORE by NUCLEAR FUSION.

starburst A massive bout of STAR formation within a GALAXY. It is often triggered by giant molecular clouds colliding when galaxies merge.

stellar wind A steady stream of PLASMA flowing away from a STAR along magnetic field lines that lead into the interstellar medium.

Sun The STAR at the center of the SOLAR SYSTEM, around which EARTH and the other planets ORBIT.

sunspot A region on the PHOTOSPHERE of the SUN that is temporarily cooler than its surroundings and therefore emits less energy and is darker. It is associated with strong magnetic fields.

supergiant A STAR that has a higher LUMINOSITY and a larger radius than a giant of the same SPECTRAL CLASSIFICATION.

supermassive black hole A BLACK HOLE that has more MASS than a million suns. The largest have masses billions of times that of the SUN.

supernova (pl. supernovae) A catastrophic explosion that blows a massive STAR to pieces and marks the end of its life.

supernova remnant A NEBULA of glowing GAS. It is created either by material ejected from a SUPERNOVA radiating excess energy into space or by the shock emission caused by shock waves, produced in the explosion, traveling through the interstellar medium.

superwind The combined particle winds (streams of particles—mostly ELECTRONS and PROTONS— traveling at very high velocity) emitted by many STARS.

synchrotron radiation ELECTROMAGNETIC RADIATION that is emitted when very high-energy ELECTRONS come into contact with magnetic fields.

talus A deposit of generally large angular fragments that accumulates at the foot of a cliff or steep slope.

tectonics The deformation that affects the LITHOSPHERES of planets. The term includes folding, thrusting, shearing, and faulting.

terrestrial planet Any one of four rocky planets found within the inner SOLAR SYSTEM: Mercury, Venus, EARTH, and Mars.

tidal gravitational force A gravitational force that occurs between two or more astronomical objects (MOONS, STARS, or even GALAXIES). Such forces are responsible for ebbs and tides and can also have effects on the geological structure of an object.

torus The shape of a ring donut.

transit The passage of a smaller CELESTIAL OBJECT across another as seen from EARTH.

true-color image Any astronomical image in which the colors closely resemble what would be seen with the naked eye. Three color plates are combined to produce a true color image: red, blue, and green. (See FALSE-COLOR IMAGE.)

ultraviolet (UV) A type of ELECTROMAGNETIC RADIATION with a wavelength shorter than that of VISIBLE LIGHT.

universe The whole of space, time, and everything in it. It was created during the BIG BANG and is expanding. It is assumed by astrophysicists that the laws of physics apply equally in all parts of the universe. It is possible that other universes exist or that, if our universe is an oscillating one, there may have been prior universes and there may be subsequent ones, in which different physical laws could apply.

variable star Any STAR whose LUMINOSITY varies either because of internal processes or as a result of external phenomena, such as ECLIPSES.

visible light The range of ELECTROMAGNETIC RADIATION to which our eyes are sensitive. Visible light waves are the only electromagnetic waves we are able to see. We perceive it as colors ranging from violet through yellow to red. The range is bounded at either end by ULTRAVIOLET and INFRARED.

volcanism The geological processes associated with the rise of molten material to the surface of any planet and its emission from volcanoes.

Voyager missions The U.S. missions to the outer SOLAR SYSTEM, launched in 1979 with the uncrewed spacecraft *Voyager 1* and *Voyager 2*. They studied not only the giant planets but also a large number of their MOONS and RING SYSTEMS. *Voyager 1* reached Jupiter in March 1979 before going to Saturn, which it reached in November 1980. Meanwhile, *Voyager 2* arrived at Jupiter in 1979 and Saturn in August 1981, but went on to study Uranus (1986) and Neptune (1989). It is now passing out of the solar system.

white dwarf A highly evolved stellar remnant, the remains of the CORE of a STAR after NUCLEAR FUSION ceases. It is composed of degenerate matter and is often found in PLANETARY NEBULAE.

X-ray radiation An energetic form of ELECTROMAGNETIC RADIATION with wavelengths of around 10^{-10} m, in between ULTRAVIOLET and GAMMA RAYS.

X-rays (See X-RAY RADIATION.)

X-ray star A STAR that emits powerful X-rays, generally a BINARY STAR in which one member is a BLACK HOLE, a NEUTRON STAR, or a WHITE DWARF. Other types of stars, including the SUN, emit some X-rays from their CORONAS.Arditi, S., and M. Lachieze-Rey, *Cosmos*, Firefly Books, Richmond Hill, Ontario, Canada, 2004.

Further Resources & Acknowledgements

Carroll, B. W., and D. A. Ostlie, *An Introduction to Modern Astrophysics*, Addison-Wesley Publishing Co., Reading, MA, 1996.

Chaikin, A., *A History of Space—Exploration in Photographs*, Firefly Books, Richmond Hill, Ontario, Canada, 2004.

Chartrand, M. R., *National Audubon Society Field Guide to the Night Sky*, Alfred A. Knopf, New York, NY, 2000.

Dickinson, T., *The Universe and Beyond* (4th edn.), Firefly Books, Richmond Hill, Ontario, Canada, 2004.

Dinwiddie, R., et al., *Universe—the Definitive Visual Guide*, Dorling Kindersley, New York, NY, 2005.

Dupas, A., *Destination Mars*, Firefly Books, Richmond Hill, Ontario, Canada, 2004.

Garlick, M., *Astronomy, A Visual Guide*, Firefly Books, Richmond Hill, Ontario, Canada, 2004.

Gribbin, J., *Companion to the Cosmos*, Orion Publishing Group, London, U.K.,1997.

Hawking, S. *The Illustrated Brief History of Time* (updated and expanded edn.) Bantam Books, New York, NY, 1996.

Hawking, S., and L. Mlodinow, *A Briefer History of Time*, Random House, New York, NY, 2005.

Heudier, J-L., *The Night Sky Month-by-Month*, Firefly Books, Richmond Hill, Ontario, Canada, 2004.

Johnson, A. K., *Earth from Space*, Firefly Books, Richmond Hill, Ontario, Canada, 2004.

Moore, P., *Firefly Atlas of the Universe*, Firefly Books, Richmond Hill, Ontario, Canada, 2003.

Moore, P., *The Great Astronomical Revolution,* Albion Publishing, Chicester, U.K., 1994.

Murdin, P., and M. Penston, *The Firefly Encyclopedia of Astronomy*, Firefly Books, Richmond Hill, Ontario, Canada, 2004.

Nemiroff, R. J., and J. T. Bonnell, *The Universe 365 Days*, Thames & Hudson, London, U.K., 2003.

North, G., *Astronomy in Depth*, Springer-Verlag, London, 2002.

Sagan, C., *Cosmos*, Book Club Associates, London, 1980.

Spence, P., *Sun Observer's Guide*, Firefly Books, Richmond Hill, Ontario, Canada, 2004.

Thorne, K., *Black Holes and Time Warps: Einstein's Outrageous Legacy*, W. W. Norton & Co., New York, NY, 1994.

http://antwrp.gsfc.nasa.gov/apod/astropix.html
A very good starting point for learning about astronomical pictures, this site has a daily image with a short explanation and links to further information.

http://chandra.harvard.edu/
The Chandra X-ray Observatory is a high-resolution space telescope that has produced the most detailed images of the X-ray sky to date. The website contains superb images of some of the most energetic and exciting regions in the universe.

http://heritage.stsci.edu/
The Hubble Space Telescope has produced stunning images of the universe, the best of which have been brought together here.

http://hubblesite.org/
The Hubble Space Telescope home page contains press releases and material relating to the mission.

http://leonardo.jpl.nasa.gov/msl/home.html
This site contains information on spacecraft and space missions from all countries.

http://map.gsfc.nasa.gov/
The Wilkinson Microwave Anisotropy Probe (WMAP) home page contains images and material relating to the Cosmic Microwave Background Radiation and the beginnings of the universe.

http://science.hq.nasa.gov/missions/phase.html
A list of NASA missions past and present, with links to the relevant web pages for each mission.

http://www.ccdland.com
Nik Szymanek's astrophotography page. Images of and from several Earth-based telescopes.

http://www.eso.org/outreach/gallery/astro/
The European Southern Observatory site contains images from multiple telescopes, including the Very Large Telescope (VLT) located at the Paranal Observatory in Chile's Atacama desert.

http://www.galex.caltech.edu
The Galaxy Evolution Explorer ultraviolet space telescope can image and analyze star-forming regions in nearby galaxies. Its data will allow the study of the evolution of galaxies and reveal how they change with time.

http://www.nineplanets.org/
http://www.solarviews.com/
These two websites contain a wealth of information, images, and diagrams about objects within the solar system.

http://www.spitzer.caltech.edu/
The infrared Spitzer Space Telescope takes images of any object that emits light with wavelengths between 3 and 180 microns. The website contains images and information.

http://www.vla.nrao.edu/
The Very Large Array (VLA) is one of the world's largest telescopes—an array of 27 antennae spread over a distance of 22 miles (35 km). The website has images from the telescope and information on how radio interferometers work.

Author Acknowledgments

Jamie Wilkins
For inspiration, advice, and support, my thanks go to:

Michael and Jennet Wilkins
Robin Rees
Chris Lintott
Sir Patrick Moore
The TINDOMH Society
Duncan Thacker
Susannah Taylor
Kate Miller
The staff at Star/Messagelabs

Robert Dunn
Thanks to:
The Institute of Astronomy, Cambridge, U.K., for hosting me during this project
Andy Fabian for allowing me to work on this project as well as my research
Carolin Crawford for assistance in the final stages

Also, for answering random questions about astrophysics:

Margaret Hendry
Richard Stancliffe
Matthew Worsley
Jamie Crummy
Nina Hatch
Mary Erlund
James Graham

Index

Page numbers in *italics*, e.g., *89,* indicate that illustrations are included on those pages.

Picture Credits

abbreviations: t = top, c = center. b = bottom, r = right, l = left.

Caltech (California Institute of Technology)
CXC (Chandra X-ray Observatory Center)
ESA (European Space Agency)
ESO (European Southern Observatory)
NASA (National Aeronautics and Space Administration Flight Center)
NASA-GSFC (NASA Goddard Space Flight Center)
NASA-HQ-GRIN (NASA Headquarters Great Images In NASA)
NASA-JPL (NASA Jet Propulsion Laboratory)
NASA-JSC (NASA Johnson Space Center)
NASA-KSC (NASA Kennedy Space Center)
NASA-LaRC (NASA Langley Research Center)
NASA-MSFC (NASA Marshall Space Flight Center)
NGST (Next Generation Space Telescope)
NOAO (National Optical Astronomy Observatory)
NRAO (National Radio Astronomy Observatory)
SCETI (Schoenberg Center for Elecronoic Text & Image)
SOHO (Solar & Heliospheric Observatory). SOHO is a project of international cooperation between ESA and NASA;
STScI (Space Telescope Science Institute)
SXC (Smithsonian Astrophysical Observatory)

1, 2–3 NASA-HQ-GRIN; 7 Photos.com; 8 TopFoto /Ancient Art & Architecture Collection Ltd; 9t SCETI; 9bl Photos.com; 9br, 10tr SCETI; 10–11 Nik Szymanek; 12–13 NASA-HQ-GRIN; 14 NASA-JPL/Space Science Institute; 15t, 15b Nik Szymanek; 17t NASA-HQ-GRIN; 17b NASA-JPL; 19t, 19b SOHO (ESA & NASA); 21t NASA-JPL; 21b Nik Szymanek; 23 NASA-JPL; 25t NASA-MSFC; 25b NASA-JPL; 26t–27t NASA-HQ-GRIN; 28t–29t NASA-JPL; 29b NASA-HQ-GRIN; 31t Apollo 16/NASA; 31bl NASA-JSC; 31br NASA, ESA and J. Garvin; 32 NASA-HQ-GRIN; 33t NASA-KSC; 33b NASA-HQ-GRIN; 35t NASA, J. Bell (Cornell U.) and M. Wolff (SSI); 35bl NASA-JPL; 35br NASA/STScI/AURA; Cambridge University; 36t–37t NASA-JPL; 36b NASA-JPL/Caltech; 37b NASA-JPL; 39t NASA/ESA/ J. Parker (Southwest Research Institute), P. Thomas (Cornell University), and L. McFadden (University of Maryland, College Park; NASA-JPL/Caltech; 38 NASA-JPL; 39b NASA/GSFC Scientific Visualization Studio; 40t–41t NASA-JPL; 41b NASA/STScI; 43, 45, 46 NASA-JPL; 47t NASA, ESA, and E. Karkoschka (University of Arizona); 47b, 49t, 49b, 51t NASA-JPL; 50 NASA-JPL/Caltech; 51b, 53t NASAJPL; 52 NASA//JPL/Space Science Institute; 53b NASA, ESA, J. Clarke (Boston University), and Z. Levay (STScI); 55t NASA-HQ-GRIN; 54, 55b, 57, 59, 60, 61, 62, 63t NASA-JPL; 63b NASA-HQ-GRIN; 65tl NASA-JPL; 65tr NASA and Erich Karkoschka, University of Arizona; 65bl NASA-HQ-GRIN; 65br NASA and Erich Karkoschka, University of Arizona; 66, 67 NASA-JPL; 69t NASA-HQ-GRIN; 68, 69b NASA-JPL; 71t NASA-HQ-GRIN; 71b NASA-JPL; 73t NASA and G. Bacon (STScI); 73b NASA-JPL; 74t NASA-KSC; 74c–75c, 74b, 75b, 76tl, 76tr, 77t, 76bl, 76br, 76bc, 77b NASA-JPL; 79t, 79b NASA-HQ-GRIN; 81t NASA/JPL-Caltech; 81b CXC-NGST; 83t NASA-MSFC; 82 NASA; 83b NASA-JPL-Caltech; 84t–85t, 84b, 85bl NASA-JPL; 85br Johns Hopkins University Applied Physics Laboratory/Southwest Research Institute; 87t NASA-LaRC; 87b, 88t, 89t NASA-JPL; 88b–89b, 88c NASA-NSSDC; 90t NASA-KSC; 91t NASA-JPL; 90b, 91b SOHO (ESA & NASA); 93tl NASA-JPL; 93tr NASA-KSC; 93b, 95tl NASA-HQ-GRIN; 95tr NASA-MSFC; 95bl NASA-JPL-Arizona State University; 95br, 97tl, 97tr NASA-JPL; 97b NASA-JSC; 98 NASA-JPL; 99t NASA; 99bl NASA-JPL-Space Science Institute; 99br NASA-JPL; 101tl NASA-HQ-GRIN; 101tr, 101b NASA; 102t–103t Nik Szymanek; 104t–105t NASA-MSFC; 104b–105b NASA-JPL; 106t ESO; 107t Nik Szymanek; 107b NASA-MSFC; 109t ESO; 109b NASA/JPL-Caltech/G. Melnick (Harvard-Smithsonian CfA); 111t, 111b NASA-GSFC; 113t NASA-HQ-GRIN; 113b NASA/JPL-Caltech/N. Smith (Univ. of Colorado at Boulder); 115t NASA-HQ-GRIN; 115b NASA, ESA, and The Hubble Heritage Team (STScI/AURA); 117t NASA-HQ-GRIN; 117b NASA, ESA, Y.Nazé (University of Liege, Belgium) and Y-H. Chu (University of Illinois, Urbana); 119t NASA-JPL; 119b NASA-GSFC; 121t NASA-HQ-GRIN; 120b–121b NASA-MSFC; 123t, 123b NASA-GSFC; 125t ESO; 125b NASA, NOAO, ESA and The Hubble Heritage Team (STScI/AURA); 127t NASA and The Hubble Heritage Team (STScI/AURA); 127b ESO; 129t NASA-JPL; 128 C.R.O'Dell (Rice University)/NASA; 129b NASA, J. Bally (University of Colorado, Boulder, CO), H. Throop (Southwest Research Institute, Boulder, CO), C.R. O'Dell (Vanderbilt University, Nashville, TN); 131t Karl Stapelfeldt (JPL) and colleagues, and NASA; 131b NASA-GSFC; 132t NASA-JPL; 133t NASA, ESA, P. Kalas and J. Graham (University of California, Berkeley) and M.; Clampin (NASA/GSFC); 132b–133b NASA/JPL-Caltech/R.Hurt (SSC); 135t NASA and Greg Bacon (STSci/ AVL); 134 The Hubble Heritage Team (AURA/STScI/NASA); 135b ESO; 137t Nik Szymanek; 138t NASA/CXC/SAO; 139t NASA-JPL; 139b Wolfgang Brandner (JPL/IPAC), Eva K. Grebel (Univ. Washington), You-Hua Chu (Univ. Illinois Urbana-Champaign), and NASA; 141t Jon Morse (University of Colorado), Kris Davidson (University of Minnesota), and NASA; 140b Yves Grosdidier (University of Montreal and Observatoire de Strasbourg), Anthony Moffat (Universitie de Montreal), Gilles Joncas (Universite Laval), Agnes Acker (Observatoire de Strasbourg), and NASA; 141b NRAO/AUI/NSF; 142 Rebecca Elson and Richard Sword, Cambridge UK, and NASA; 143t, 143b NASA and The Hubble Heritage Team (STScI/AURA); 145tl NASA, ESA and G. Bacon; 145tr NASA, H.E. Bond and E. Nelan (STScI, Baltimore, Md.); M.

Barstow and M. Burleigh (University of Leicester, U.K.); and J.B. Holberg (University of Arizona); 145b D. Golimowski (Johns Hopkins University), and NASA; 145tl NASA/CXC/SAO/M.Karovska et al; 145tr NASA/CXC/SAO; 287 NASA/CXC/SAO; 147t NASA/CXC/GSFC/T. Strohmayer; 147b CXC /D. Berry; 149t NASA, John Krist (STScI); 149b NASA/CXC/M.Weiss; 150t NASA-MSFC; 151t NASA/John Trauger (JPL) and James Westphal (Caltech)); 151b NASA, ESA and AURA/Caltech; 153tl NASA; K.L. Luhman (Harvard-Smithsonian Center for Astrophysics, Cambridge, Mass.) and G. Schneider, E. Young, G. Rieke, A. Cotera, H. Chen, M. Rieke, R. Thompson (Steward Observatory, University of Arizona, Tucson, Ariz.); 153tr NASA, C.R. O'Dell and S.K. Wong (Rice University); 152b NASA-HQ-GRIN; 153b NASA and Ron Gilland(STScI); 155t NASA, ESA, and Martino Romaniello (ESO, Germany); 155b, 157t NASA and The Hubble Heritage Team (STScI/AURA); 157bl NASA-HQ-GRIN; 157br NASA, ESA, Andrew Fruchter and the ERO team (STScI); 158tl NASA-HQ-GRIN; 158tr NASA-GSFC; 159t NASA-JPL; 159b NASA-MSFC; 161t The Hubble Heritage Team (STScI/AURA/NASA); 160 NASA-HQ-GRIN; 161 NASA, ESA, C.R. O'Dell (Vanderbilt University), and M. Meixner, P. McCullough; 163t NASA/UIUC/Y. Chu et al/HST; 163b NASA-HQ-GRIN; 164 NASA/CXC/Rutgers/J.Hughes et al; 165t Nik Szymanek; 165b NASA/CXC/Rutgers/J.Warren & J.Hughes et al; 167t NASA-MSFC; 166 NRAO/AUI/NSF; 167b, 169t NASA-JPL; 169b NASA/CXC/HST/ ASU/J.Hester et al; 171t Fred Baganoff (MIT), Mark Morris (UCLA), et al, CXC, NASA; 170 NASA Jet Propulsion Laboratory; 171b NASA/CXC/UVa/C.Sarazin et al; 172t–173t ESA, NASA and P. Anders (Göttingen University Galaxy Evolution Group, Germany; 172b–173b NASA, ESA and The Hubble Heritage Team (STScI/AURA); 175b NASA-HQ-GRIN; 177t NASA, ESA and A. Nota (STScI); 176b–177b, 179t NASA-JPL; 178bl NASA, ESA and The Hubble Heritage Team (STScI/AURA); 179b NASA, ESA and The Hubble Heritage Team (AURA/ STScI)/HEIC; 178br NASA, ESA, and Martino Romaniello (ESO, Germany); 181b NASA, ESA and The Hubble Heritage Team (STScI/AURA); 183t NASA, ESA, Y. Izotov (Main Astronomical Observatory, Kyiv, UA) and T. Thuan (University of Virginia); 183b NASA and The Hubble Heritage Team (STScI/AURA); 185t NASA, ESA and The Hubble Heritage Team (STScI/AURA); 185b NASA, ESA and the GMOS Commissioning Team (Gemini Observatory); 187t Canada-France-Hawaii Telescope/J.-C. Cuillandre/Coelum; 187b NASA-HQ-GRIN; 189t, 189b NASA-JPL; 191t NASA and The Hubble Heritage Team (STScI/AURA); 191c/b UITProject, NASA/W.Waller Tufts University); 193t NASA/Hubble Space Telescope/Hubble Heritage Team; 193c NASA/JPL-Caltech/R. Kennicutt (University of Arizona) and the SINGS team; 193b, 195t NASA and The Hubble Heritage Team (STScI/AURA); 195b NASA/JPL-Caltech/M. Regan (STScI) and the SINGS team; 197t NASA and The Hubble Heritage Team (AURA/STScI); 197b NASA/JPL-Caltech/R.Kennicutt (University of Arizona)/DSS; 199t Canada-France-Hawaii Telescope/J.-C. Cuillandre/ Coelum; 199b, 200t, 201t Nik Szymanek; 201b NASA, ESA and The Hubble Heritage Team (STScI/AURA); 203t NASA and ESA; 203bl NASA and John Trauger (JPL); 203br NASA, ESA and C. Marcella Carollo (Columbia University); 205t NASA-GSFC; 204, 205b Nik Szymanek; 207t NASA-MSFC; 207bl NASA, William C. Keel (University of Alabama, Tuscaloosa); 207br NASA, Andrew S. Wilson (University of Maryland); Patrick L. Shopbell (Caltech); Chris Simpson (Subaru Telescope); Thaisa Storchi-Bergmann and F. K. B. Barbosa (UFRGS, Brazil); and Martin J. Ward (University of Leicester, U.K.); 209t NASA, ESA, R. de Grijs (Institute of Astronomy. Cambridge, UK); 209b Hubble Heritage Team (AURA/STScI/NASA); 211t G. Fritz Benedict, Andrew Howell, Inger Jorgensen, David Chapell (University of Texas), Jeffery Kenney (Yale University), and Beverly J. Smith (CASA, University of Colorado), and NASA; 211b NASA, Gerald Cecil (University of North Carolina), Sylvain Veilleux (University of Maryland), Joss Bland-Hawthorn (Anglo- Australian Observatory), and Alex Filippenko (University of California at Berkeley); 213t NASA and The Hubble Heritage Team (STScI/AURA); 212 Gary Bower, Richard Green (NOAO), the STIS Instrument Definition Team, and NASA; 213b Roeland P. van der Marel (STScI), Frank C. van den Bosch (Univ. of Washington), and NASA; 214t NASA and Jeffrey Kenney and Elizabeth Yale (Yale Universi; 215t L. Ferrarese (Johns Hopkins University) and NASA; 214bl NASA/CXC/W.Forman et al; 214br NRAO/AUI and F.N. Owen, J.A. Eilek and N.E. Kassim; 215b NASA and The Hubble Heritage Team (STScI/AURA); 217t NRAO/AUI; 217b NRAO/AUI and J.M. Uson; 219t X-ray (NASA/ CXC/M. Karovska et al.); Radioimage (NRAO/VLA/J.Condon et al.); Infrared NASA/JPL-Caltech/J.Keene (SSC/Caltech); Optical (Digitized Sky Survey U.K. Schmidt Image/STScI); 219b NASA/CXC/MPE/ S.Komossa et al; 221t NRAO/AUI; 221c NASA/UMD/A. Wilson et al; 221b NASA, H. Ford (JHU), G. Illingworth (USCS/LO), M.Clampin (STScI), G. Hartig (STScI), the ACS Science Team, and ESA; 223t NASA and The Hubble Heritage Team (STScI/AURA); 223b NASA, ESA and The Hubble Heritage Team (STScI/AURA); 225t Rodger Thompson, Marcia Rieke, Glenn Schneider (University of Arizona) and Nick Scoville (California Institute of Technology), and NASA; 225b The Hubble Heritage Team (STScI/AURA/NASA); 227tl NASA/CXC/SAO/G. Fabbiano et al; 227tr Brad Whitmore (STScI) and NASA; 227b, 229t NASA and The Hubble Heritage Team (STScI/AURA); 229b NASA, H. Ford (JHU), G. Illingworth (USCS/LO), M.Clampin (STScI), G. Hartig (STScI), the ACS Science Team, and ESA; 231t Kirk Borne (STScI), and NASA; 231b NASA-HQ-GRIN; 233t X-ray; NASA/CXC/INAF-Brera/G.Trinchieri et al; Optical; Pal.Obs. DSS; 233b NASA/CfA/J.Vrtilek et al; 235t NASA, J. English (U. Manitoba), S. Hunsberger, S. Zonak, J. Charlton, S. Gallagher (PSU), and L. Frattare (STScI); 235b The Hubble Heritage Team (STScI/AURA/NASA); 237tr NASA/CXC/UCI/A. Lewis et al); 237tl NOAO/Kitt Peak/J.Uson, D.Dale, S.Boughn, J.Kuhn; 237b NASA, ESA, Richard Ellis (Caltech) and Jean-Paul Kneib (Observatoire Midi-Pyrenees, France); 239t NASA/IoA/J.Sanders & A.Fabian; 239b NASA/ SAO/CXC/M.Markevitch et al; 241t W.N. Colley and E. Turner (Princeton University), J.A. Tyson (Bell Labs, Lucent Technologies) and NASA; 241b NASA/CXC/IoA/A.Fabian et al; 243t ESO; 243b

PICTURE CREDITS